Universitext

Universitext

Universitext is a series of textbooks that presents material from a wide variety of mathematical disciplines at master's level and beyond. The books, often well class-tested by their author, may have an informal, personal even experimental approach to their subject matter. Some of the most successful and established books in the series have evolved through several editions, always following the evolution of teaching curricula, to very polished texts.

Thus as research topics trickle down into graduate-level teaching, first textbooks written for new, cutting-edge courses may make their way into *Universitext*.

For further volumes:
www.springer.com/series/223

Leonid Koralov
Yakov G. Sinai

Theory of Probability
and Random Processes

Second Edition

 Springer

Leonid Koralov
Department of Mathematics
University of Maryland
College Park, MD, USA

Yakov G. Sinai
Department of Mathematics
Princeton University
New Jersey
Fine Hall, USA

ISBN 978-3-540-25484-3   ISBN 978-3-540-68829-7 (eBook)
DOI 10.1007/978-3-540-68829-7
Springer Heidelberg New York Dordrecht London

Library of Congress Control Number: 2012943837

Mathematics Subject Classification (2010): 60-XX

Printed on acid-free paper

Springer is part of Springer Science+Business Media (www.springer.com)

Preface

This book is primarily based on a 1-year course that has been taught for a number of years at Princeton University to advanced undergraduate and graduate students. During the last few years, a similar course has also been taught at the University of Maryland.

We would like to express our thanks to Ms. Sophie Lucas and Prof. Rafael Herrera who read the manuscript and suggested many corrections. We are particularly grateful to Prof. Boris Gurevich for making many important suggestions on both the mathematical content and style.

While writing this book, L. Koralov was supported by a National Science Foundation grant (DMS-0405152). Y. Sinai was supported by a National Science Foundation grant (DMS-0600996).

College Park, MD, USA   Leonid Koralov
Princeton, NJ, USA   Yakov Sinai

Contents

16 Strictly Stationary Random Processes 231
   16.1 Stationary Processes and Measure Preserving
      Transformations 231
   16.2 Birkhoff Ergodic Theorem 233
   16.3 Ergodicity, Mixing, and Regularity 236
   16.4 Stationary Processes with Continuous Time 241
   16.5 Problems 242

17 Generalized Random Processes 245
   17.1 Generalized Functions and Generalized Random Processes 245
   17.2 Gaussian Processes and White Noise 249

18 Brownian Motion 253
   18.1 Definition of Brownian Motion 253
   18.2 The Space $C([0, \infty))$ 255
   18.3 Existence of the Wiener Measure, Donsker Theorem 260
   18.4 Kolmogorov Theorem 264
   18.5 Some Properties of Brownian Motion 268
   18.6 Problems 271

19 Markov Processes and Markov Families 273
   19.1 Distribution of the Maximum of Brownian Motion 273
   19.2 Definition of the Markov Property 274
   19.3 Markov Property of Brownian Motion 278
   19.4 The Augmented Filtration 279
   19.5 Definition of the Strong Markov Property 281
   19.6 Strong Markov Property of Brownian Motion 283
   19.7 Problems 287

20 Stochastic Integral and the Ito Formula 289
   20.1 Quadratic Variation of Square-Integrable Martingales 289
   20.2 The Space of Integrands for the Stochastic Integral 293
   20.3 Simple Processes 295
   20.4 Definition and Basic Properties of the Stochastic Integral 296
   20.5 Further Properties of the Stochastic Integral 299
   20.6 Local Martingales 301
   20.7 Ito Formula 303
   20.8 Problems 308

21 Stochastic Differential Equations 311
   21.1 Existence of Strong Solutions to Stochastic Differential
      Equations 311
   21.2 Dirichlet Problem for the Laplace Equation 318
   21.3 Stochastic Differential Equations and PDE's 322
   21.4 Markov Property of Solutions to SDE's 331

Part I

Probability Theory

Random Variables and Their Distributions

1.1 Spaces of Elementary Outcomes, σ-Algebras, and Measures

The first object encountered in probability theory is the space of elementary outcomes. It is simply a non-empty set, usually denoted by Ω, whose elements $\omega \in \Omega$ are called elementary outcomes. Here are several simple examples.

Example. Take a finite set $X = \{x^1, \ldots, x^r\}$ and the set Ω consisting of sequences $\omega = (\omega_1, \ldots, \omega_n)$ of length $n \geq 1$, where $\omega_i \in X$ for each $1 \leq i \leq n$. In applications, ω is a result of n statistical experiments, while ω_i is the result of the i-th experiment. It is clear that $|\Omega| = r^n$, where $|\Omega|$ denotes the number of elements in the finite set Ω. If $X = \{0,1\}$, then each ω is a sequence of length n made of zeros and ones. Such a space Ω can be used to model the result of n consecutive tosses of a coin. If $X = \{1, 2, 3, 4, 5, 6\}$, then Ω can be viewed as the space of outcomes for n rolls of a die.

Example. A generalization of the previous example can be obtained as follows. Let X be a finite or countable set, and I be a finite set. Then $\Omega = X^I$ is the space of all functions from I to X.

If $X = \{0, 1\}$ and $I \subset \mathbb{Z}^d$ is a finite set, then each $\omega \in \Omega$ is a configuration of zeros and ones on a bounded subset of d-dimensional lattice. Such spaces appear in statistical physics, percolation theory, etc.

Example. Consider a lottery game where one tries to guess n distinct numbers and the order in which they will appear out of a pool of r numbers (with $n \leq r$). In order to model this game, define $X = \{1, \ldots, r\}$. Let Ω consist of sequences $\omega = (\omega_1, \ldots, \omega_n)$ of length n such that $\omega_i \in X, \omega_i \neq \omega_j$ for $i \neq j$. It is easy to show that $|\Omega| = r!/(r - n)!$.

Later in this section we shall define the notion of a probability measure, or simply probability. It is a function which ascribes real numbers between zero and one to certain (but not necessarily all!) subsets $A \subseteq \Omega$. If Ω is interpreted

L. Koralov and Y.G. Sinai, *Theory of Probability and Random Processes*, Universitext, DOI 10.1007/978-3-540-68829-7_1, © Springer-Verlag Berlin Heidelberg 2012

as the space of possible outcomes of an experiment, then the probability of A may be interpreted as the likelihood that the outcome of the experiment belongs to A. Before we introduce the notion of probability we need to discuss the classes of sets on which it will be defined.

Definition 1.1. *A collection $\mathcal{G}$ of subsets of Ω is called an algebra if it has the following three properties.*

1. $\Omega \in \mathcal{G}$.
2. $C \in \mathcal{G}$ *implies that* $\Omega \backslash C \in \mathcal{G}$.
3. $C_1, \ldots, C_n \in \mathcal{G}$ *implies that* $\bigcup_{i=1}^{n} C_i \in \mathcal{G}$.

Example. Given a set of elementary outcomes Ω, let $\underline{\mathcal{G}}$ contain two elements: the empty set and the entire set Ω, that is $\underline{\mathcal{G}} = \{\emptyset, \Omega\}$. Define $\overline{\mathcal{G}}$ as the collection of all the subsets of Ω. It is clear that both $\underline{\mathcal{G}}$ and $\overline{\mathcal{G}}$ satisfy the definition of an algebra. Let us show that if Ω is finite, then the algebra $\overline{\mathcal{G}}$ contains $2^{|\Omega|}$ elements.

Take any $C \subseteq \Omega$ and introduce the function $\chi_C(\omega)$ on Ω:

$$\chi_C(\omega) = \begin{cases} 1 & \text{if } \omega \in C, \\ 0 & \text{otherwise}, \end{cases}$$

which is called the indicator of C. It is clear that any function on Ω taking values zero and one is an indicator function of some set and determines this set uniquely. Namely, the set consists of those ω, where the function is equal to one. The number of distinct functions from Ω to the set $\{0, 1\}$ is equal to $2^{|\Omega|}$.

Lemma 1.2. *Let Ω be a space of elementary outcomes, and $\mathcal{G}$ be an algebra. Then*

1. *The empty set is an element of $\mathcal{G}$.*
2. *If $C_1, \ldots, C_n \in \mathcal{G}$, then $\bigcap_{i=1}^{n} C_i \in \mathcal{G}$.*
3. *If $C_1, C_2 \in \mathcal{G}$, then $C_1 \setminus C_2 \in \mathcal{G}$.*

Proof. Take $C = \Omega \in \mathcal{G}$ and apply the second property of Definition 1.1 to obtain that $\emptyset \in \mathcal{G}$. To prove the second statement, we note that

$$\Omega \backslash \bigcap_{i=1}^{n} C_i = \bigcup_{i=1}^{n} (\Omega \backslash C_i) \in \mathcal{G}.$$

Consequently, $\bigcap_{i=1}^{n} C_i \in \mathcal{G}$. For the third statement, we write

$$C_1 \setminus C_2 = \Omega \setminus ((\Omega \setminus C_1) \cup C_2) \in \mathcal{G}.$$

$\square$

Lemma 1.3. *If an algebra $\mathcal{G}$ is finite, then there exist non-empty sets $B_1, \ldots, B_m \in \mathcal{G}$ such that*

1. $B_i \cap B_j = \emptyset$ if $i \neq j$.
2. $\Omega = \bigcup_{i=1}^{m} B_i$.
3. For any set $C \in \mathcal{G}$ there is a set $I \subseteq \{1, \ldots, m\}$ such that $C = \bigcup_{i \in I} B_i$ (with the convention that $C = \emptyset$ if $I = \emptyset$).

Remark 1.4. The collection of sets B_i, $i = 1, \ldots, m$, defines a partition of Ω. Thus, finite algebras are generated by finite partitions.

Remark 1.5. Any finite algebra $\mathcal{G}$ has 2^m elements for some integer $m \in \mathbb{N}$. Indeed, by Lemma 1.3, there is a one-to-one correspondence between $\mathcal{G}$ and the collection of subsets of the set $\{1, \ldots, m\}$.

Proof of Lemma 1.3. Let us number all the elements of $\mathcal{G}$ in an arbitrary way:
$$\mathcal{G} = \{C_1, \ldots, C_s\}.$$
For any set $C \in \mathcal{G}$, let
$$C^1 = C, \quad C^{-1} = \Omega \backslash C.$$
Consider a sequence $b = (b_1, \ldots, b_s)$ such that each b_i is either $+1$ or -1 and set
$$B^b = \bigcap_{i=1}^{s} C_i^{b_i}.$$
From the definition of an algebra and Lemma 1.2 it follows that $B^b \in \mathcal{G}$. Furthermore, since
$$C_i = \bigcup_{b: b_i = 1} B^b,$$
any element C_i of $\mathcal{G}$ can be obtained as a union of some of the B^b. If $b' \neq b''$, then $B^{b'} \cap B^{b''} = \emptyset$. Indeed, $b' \neq b''$ means that $b_i{'} \neq b_i{''}$ for some i, say $b_i{'} = 1, b_i{''} = -1$. In the expression for $B^{b'}$ we find $C_i^1 = C_i$, so $B^{b'} \subseteq C_i$. In the expression for $B^{b''}$ we find $C_i^{-1} = \Omega \backslash C_i$, so $B^{b''} \subseteq \Omega \backslash C_i$. Therefore, all B^b are pair-wise disjoint. We can now take as B_i those B^b which are not empty. $\square$

Definition 1.6. *A collection $\mathcal{F}$ of subsets of Ω is called a σ-algebra if $\mathcal{F}$ is an algebra which is closed under countable unions, that is $C_i \in \mathcal{F}$, $i \geq 1$, implies that $\bigcup_{i=1}^{\infty} C_i \in \mathcal{F}$. The elements of $\mathcal{F}$ are called measurable sets, or events.*

As above, the simplest examples of a σ-algebra are the trivial σ-algebra, $\underline{\mathcal{F}} = \{\emptyset, \Omega\}$, and the σ-algebra $\overline{\mathcal{F}}$ which consists of all the subsets of Ω.

Definition 1.7. *A measurable space is a pair $(\Omega, \mathcal{F})$, where Ω is a space of elementary outcomes and $\mathcal{F}$ is a σ-algebra of subsets of Ω.*

Remark 1.8. A space of elementary outcomes is said to be discrete if it has a finite or countable number of elements. Whenever we consider a measurable space $(\Omega, \mathcal{F})$ with a discrete space Ω, we shall assume that $\mathcal{F}$ consists of all the subsets of Ω.

The following lemma can be proved in the same way as Lemma 1.2.

Lemma 1.9. *Let $(\Omega, \mathcal{F})$ be a measurable space. If $C_i \in \mathcal{F}$, $i \geq 1$, then $\bigcap_{i=1}^{\infty} C_i \in \mathcal{F}$.*

It may seem that there is little difference between the concepts of an algebra and a σ-algebra. However, such an appearance is deceptive. As we shall see, any interesting theory (such as measure theory or probability theory) requires the notion of a σ-algebra.

Definition 1.10. *Let $(\Omega, \mathcal{F})$ be a measurable space. A function $\xi : \Omega \to \mathbb{R}$ is said to be $\mathcal{F}$-measurable (or simply measurable) if $\{\omega : a \leq \xi(\omega) < b\} \in \mathcal{F}$ for each $a, b \in \mathbb{R}$.*

Below we shall see that linear combinations and products of measurable functions are again measurable functions. If Ω is discrete, then any real-valued function on Ω is a measurable, since $\mathcal{F}$ contains all the subsets of Ω.

In order to understand the concept of measurability better, consider the case where $\mathcal{F}$ is finite. Lemma 1.3 implies that $\mathcal{F}$ corresponds to a finite partition of Ω into subsets $B_1, \ldots, B_m$, and each $C \in \mathcal{F}$ is a union of some of the B_i.

Theorem 1.11. *If ξ is $\mathcal{F}$-measurable, then it takes a constant value on each element of the partition B_i, $1 \leq i \leq m$.*

Proof. Suppose that ξ takes at least two values, a and b, with $a < b$ on the set B_j for some $1 \leq j \leq m$. The set $\{\omega : a \leq \xi(w) < (a+b)/2\}$ must contain at least one point from B_j, yet it does not contain the entire set B_j. Thus it can not be represented as a union of some of the B_i, which contradicts the $\mathcal{F}$-measurability of the set. $\qquad\square$

Definition 1.12. *Let $(\Omega, \mathcal{F})$ be a measurable space. A function $\mu : \mathcal{F} \to [0, \infty)$ is called a finite non-negative measure if*

$$\mu(\bigcup_{i=1}^{\infty} C_i) = \sum_{i=1}^{\infty} \mu(C_i)$$

whenever $C_i \in \mathcal{F}$, $i \geq 1$, are such that $C_i \cap C_j = \emptyset$ for $i \neq j$.

The property expressed in Definition 1.12 is called the countable additivity (or the σ-additivity) of the measure.

Remark 1.13. Most often we shall omit the words finite and non-negative, and simply refer to μ as a measure. Thus, a measure is a σ-additive function on $\mathcal{F}$ with values in $\mathbb{R}^+$. In contrast, σ-finite and signed measures, to be introduced in Chap. 3, take values in $\mathbb{R}^+ \cup \{+\infty\}$ and $\mathbb{R}$, respectively.

Definition 1.14. *Let g be a binary function on Ω with values 1 (true) and 0 (false). It is said that g is true almost everywhere if there is an event C with $\mu(C) = \mu(\Omega)$ such that $g(\omega) = 1$ for all $\omega \in C$.*

Definition 1.15. *A measure P on a measurable space $(\Omega, \mathcal{F})$ is called a probability measure or a probability distribution if $P(\Omega) = 1$.*

Definition 1.16. *A probability space is a triplet $(\Omega, \mathcal{F}, P)$, where $(\Omega, \mathcal{F})$ is a measurable space and P is a probability measure. If $C \in \mathcal{F}$, then the number $P(C)$ is called the probability of C.*

Definition 1.17. *A measurable function defined on a probability space is called a random variable.*

Remark 1.18. When P is a probability measure, the term "almost surely" is often used instead of "almost everywhere".

Remark 1.19. Let us replace the σ-additivity condition in Definition 1.12 by the following: if $C_i \in \mathcal{F}$ for $1 \leq i \leq n$, where n is finite, and $C_i \cap C_j = \emptyset$ for $i \neq j$, then

$$\mu(\bigcup_{i=1}^{n} C_i) = \sum_{i=1}^{n} \mu(C_i).$$

This condition leads to the notion of a finitely additive function, instead of a measure. Notice that finite additivity implies superadditivity for infinite sequences of sets. Namely,

$$\mu(\bigcup_{i=1}^{\infty} C_i) \geq \sum_{i=1}^{\infty} \mu(C_i)$$

if the sets C_i are disjoint. Indeed, otherwise we could find a sufficiently large n such that

$$\mu(\bigcup_{i=1}^{\infty} C_i) < \sum_{i=1}^{n} \mu(C_i),$$

which would violate the finite additivity.

Let Ω be discrete. Then $p(\omega) = P(\{\omega\})$ is the probability of the elementary outcome ω. It follows from the definition of the probability measure that

1. $p(\omega) \geq 0$.
2. $\sum_{\omega \in \Omega} p(\omega) = 1$.

Lemma 1.20. *Every function $p(\omega)$ on a discrete space Ω, with the two properties above, generates a probability measure on the σ-algebra of all subsets of Ω by the formula*

$$P(C) = \sum_{\omega \in C} p(\omega).$$

Proof. It is clear that $P(C) \geq 0$ for all C, and $P(\Omega) = 1$. To verify that the σ-additivity condition of Definition 1.12 is satisfied, we need to show that if C_i, $i \geq 1$, are non-intersecting sets, then $P(\bigcup_{i=1}^{\infty} C_i) = \sum_{i=1}^{\infty} P(C_i)$.

Since the sum of a series with positive terms does not depend on the order of summation,

$$P(\bigcup_{i=1}^{\infty} C_i) = \sum_{\omega \in \bigcup_{i=1}^{\infty} C_i} p(\omega) = \sum_{i=1}^{\infty} \sum_{\omega \in C_i} p(\omega) = \sum_{i=1}^{\infty} P(C_i).$$

$\square$

Thus, we have demonstrated that in the case of a discrete probability space Ω there is a one-to-one correspondence between probability distributions on Ω and functions p with Properties 1 and 2 stated before Lemma 1.20.

Remark 1.21. If Ω is not discrete it is usually impossible to express the measure of a given set in terms of the measures of elementary outcomes. For example, in the case of the Lebesgue measure on $[0,1]$ (studied in Chap. 3) we have $P(\{\omega\}) = 0$ for every ω.

Here are some examples of probability distributions on a discrete set Ω.

1. *Uniform distribution.* Ω is finite and $p(\omega) = 1/|\Omega|$. In this case all ω have equal probabilities. For any event C we have $P(C) = |C|/|\Omega|$.
2. *Geometric distribution.* $\Omega = \mathbb{Z}^{+} = \{n : n \geq 0, \ n \text{ is an integer}\}$, and $p(n) = (1-q)q^n$, where $0 < q < 1$. This distribution is called the geometric distribution with parameter q.
3. *Poisson distribution.* The space Ω is the same as in the previous example, and $p(n) = e^{-\lambda}\lambda^n/n!$, where $\lambda > 0$. This distribution is called the Poisson distribution with parameter λ.

Let ξ be a random variable defined on a discrete probability space with values in a finite or countable set X, i.e., $\xi(\omega) \in X$ for all $\omega \in \Omega$. We can consider the events $C_x = \{\omega : \xi(\omega) = x\}$ for all x. Clearly, the intersection of C_x and C_y is empty for $x \neq y$, and $\bigcup_{x \in X} C_x = \Omega$. We can now define the probability distribution on X via $p_\xi(x) = P(C_x)$.

Definition 1.22. *The probability distribution on X defined by $p_\xi(x) = P(C_x)$ is called the probability distribution of the random variable ξ (or the probability distribution induced by the random variable ξ).*

1.2 Expectation and Variance of Random Variables on a Discrete Probability Space

Let ξ be a random variable on a discrete probability space $(\Omega, \mathcal{F}, P)$, where $\mathcal{F}$ is the collection of all subsets of Ω, and P is a probability measure. As before, we define $p(\omega) = P(\{\omega\})$. Let $X = \xi(\Omega) \subset \mathbb{R}$ be the set of values of ξ. Since Ω is discrete, X is finite or countable.

For a random variable ξ let

$$\xi_+(\omega) = \begin{cases} \xi(\omega) & \text{if } \xi(\omega) \geq 0, \\ 0 & \text{if } \xi(\omega) < 0, \end{cases}$$

$$\xi_-(\omega) = \begin{cases} -\xi(\omega) & \text{if } \xi(\omega) < 0, \\ 0 & \text{if } \xi(\omega) \geq 0. \end{cases}$$

Definition 1.23. *For a random variable ξ consider the following two series $\sum_{\omega \in \Omega} \xi_+(\omega)p(\omega)$ and $\sum_{\omega \in \Omega} \xi_-(\omega)p(\omega)$. If both series converge, then ξ is said to have a finite mathematical expectation. It is denoted by $E\xi$ and is equal to*

$$E\xi = \sum_{\omega \in \Omega} \xi_+(\omega)p(\omega) - \sum_{\omega \in \Omega} \xi_-(\omega)p(\omega) = \sum_{\omega \in \Omega} \xi(\omega)p(\omega).$$

If the first series diverges and the second one converges, then $E\xi = +\infty$. If the first series converges and the second one diverges, then $E\xi = -\infty$. If both series diverge, then $E\xi$ is not defined.

Clearly, $E\xi$ is finite if and only if $E|\xi|$ is finite.

Remark 1.24. The terms expectation, expected value, mean, and mean value are sometimes used instead of mathematical expectation.

Lemma 1.25 (Properties of the Mathematical Expectation).

1. *If $E\xi_1$ and $E\xi_2$ are finite, then for any constants a and b the expectation $E(a\xi_1 + b\xi_2)$ is finite and $E(a\xi_1 + b\xi_2) = aE\xi_1 + bE\xi_2$.*
2. *If $\xi \geq 0$, then $E\xi \geq 0$.*
3. *If $\xi \equiv 1$, then $E\xi = 1$.*
4. *If $A \leq \xi \leq B$, then $A \leq E\xi \leq B$.*
5. *$E\xi$ is finite if and only if $\sum_{x \in X} |x|p_\xi(x) < \infty$, where $p_\xi(x) = P(\{\omega : \xi(\omega) = x\})$. In this case $E\xi = \sum_{x \in X} xp_\xi(x)$.*
6. *If the random variable η is defined by $\eta = g(\xi)$, then*

$$E\eta = \sum_{x \in X} g(x)p_\xi(x),$$

and $E\eta$ is finite if and only if $\sum_{x \in X} |g(x)|p_\xi(x) < \infty$.
7. *If $|\xi| \leq |\eta|$ and $E|\eta|$ is finite, then $E|\xi|$ is finite.*

Proof. Since $E\xi_1$ and $E\xi_2$ are finite,

$$\sum_{\omega \in \Omega} |\xi_1(\omega)|p(\omega) < \infty, \quad \sum_{\omega \in \Omega} |\xi_2(\omega)|p(\omega) < \infty,$$

and

$$\sum_{\omega\in\Omega}|a\xi_1(\omega)+b\xi_2(\omega)|\mathrm{p}(\omega)\leq\sum_{\omega\in\Omega}(|a||\xi_1(\omega)|+|b||\xi_2(\omega)|)\mathrm{p}(\omega)$$

$$=|a|\sum_{\omega\in\Omega}|\xi_1(\omega)|\mathrm{p}(\omega)+|b|\sum_{\omega\in\Omega}|\xi_2(\omega)|\mathrm{p}(\omega)<\infty.$$

By using the properties of absolutely converging series, we find that

$$\sum_{\omega\in\Omega}(a\xi_1(\omega)+b\xi_2(\omega))\mathrm{p}(\omega)=a\sum_{\omega\in\Omega}\xi_1(\omega)\mathrm{p}(\omega)+b\sum_{\omega\in\Omega}\xi_2(\omega)\mathrm{p}(\omega).$$

The second and third properties are clear. Properties 1, 2, and 3 mean that expectation is a linear, non-negative, and normalized functional on the vector space of random variables.

The fourth property follows from $\xi - A \geq 0$, $B - \xi \geq 0$, which imply $E\xi - A = E(\xi - A) \geq 0$ and $B - E\xi = E(B - \xi) \geq 0$.

We now prove the sixth property, since the fifth one follows from it by setting $g(x) = x$. Let $\sum_{x\in X}|g(x)|\mathrm{p}_\xi(x) < \infty$. Since the sum of a series with non-negative terms does not depend on the order of the terms, the summation $\sum_\omega |g(\xi(\omega))|\mathrm{p}(\omega)$ can be carried out in the following way:

$$\sum_{\omega\in\Omega}|g(\xi(\omega))|\mathrm{p}(\omega)=\sum_{x\in X}\sum_{\omega:\xi(\omega)=x}|g(\xi(\omega))|\mathrm{p}(\omega)$$

$$=\sum_{x\in X}|g(x)|\sum_{\omega:\xi(\omega)=x}\mathrm{p}(\omega)=\sum_{x\in X}|g(x)|\mathrm{p}_\xi(x).$$

Thus the series $\sum_{\omega\in\Omega}|g(\xi(\omega))|\mathrm{p}(\omega)$ converges if and only if the series $\sum_{x\in X}|g(x)|\mathrm{p}_\xi(x)$ also does. If any of these series converges, then the series $\sum_{\omega\in\Omega}g(\xi(\omega))\mathrm{p}(\omega)$ converges absolutely, and its sum does not depend on the order of summation. Therefore,

$$\sum_{\omega\in\Omega}g(\xi(\omega))\mathrm{p}(\omega)=\sum_{x\in X}\sum_{\omega:\xi(\omega)=x}g(\xi(\omega))\mathrm{p}(\omega)$$

$$=\sum_{x\in X}g(x)\sum_{\omega:\xi(\omega)=x}\mathrm{p}(\omega)=\sum_{x\in X}g(x)\mathrm{p}_\xi(x),$$

and the last series also converges absolutely.

The seventh property immediately follows from the definition of the expectation.   □

Remark 1.26. The fifth property,

$$E\xi=\sum_{x\in X}x\mathrm{p}_\xi(x)\quad\text{if the series converges absolutely,}$$

can be used as a definition of expectation if ξ takes at most a countable number of values, but is defined on a probability space which is not necessarily discrete. We shall define expectation for general random variables in Chap. 3.

Lemma 1.27 (Chebyshev Inequality). *If $\xi \geq 0$ and $E\xi$ is finite, then for each $t > 0$ we have*
$$P(\xi \geq t) \leq \frac{E\xi}{t}.$$

Proof. Since $\xi \geq 0$,
$$P(\xi \geq t) = \sum_{\omega:\xi(\omega)\geq t} p(\omega) \leq \sum_{\omega:\xi(\omega)\geq t} \frac{\xi(\omega)}{t}p(\omega)$$
$$= \frac{1}{t}\sum_{\omega:\xi(\omega)\geq t} \xi(\omega)p(\omega) \leq \frac{1}{t}\sum_{\omega\in\Omega} \xi(\omega)p(\omega) = \frac{1}{t}E\xi.$$
□

Lemma 1.28 (Cauchy-Schwarz Inequality). *If $E\xi_1^2$ and $E\xi_2^2$ are finite, then $E(\xi_1\xi_2)$ is also finite and*
$$E|\xi_1\xi_2| \leq (E\xi_1^2 E\xi_2^2)^{1/2}.$$

Proof. Since, $|\xi_1\xi_2| \leq (\xi_1^2 + \xi_2^2)/2$, the expectation $E(\xi_1\xi_2)$ is finite. For every t we have
$$E(t|\xi_2| + |\xi_1|)^2 = t^2 E\xi_2^2 + 2tE|\xi_1\xi_2| + E\xi_1^2.$$
Since the left-hand side of this equality is non-negative for every t, so is the quadratic polynomial on the right-hand side, which implies that its discriminant is not positive, i.e., $E|\xi_1\xi_2| - (E\xi_1^2 E\xi_2^2)^{1/2} \leq 0$. □

Definition 1.29. *The variance of a random variable ξ is $Var\xi = E(\xi - E\xi)^2$.*

Sometimes the word dispersion is used instead of variance, leading to the alternative notation $D\xi$. The existence of the variance requires the existence of $E\xi$. Certainly there can be cases where $E\xi$ is finite, but $Var\xi$ is not finite.

Lemma 1.30 (Properties of the Variance).

1. *$Var\xi$ is finite if and only if $E\xi^2$ is finite. In this case $Var\xi = E\xi^2 - (E\xi)^2$.*
2. *If $Var\xi$ is finite, then $Var(a\xi + b) = a^2 Var\xi$ for any constants a and b.*
3. *If $A \leq \xi \leq B$, then $Var\xi \leq (\frac{B-A}{2})^2$.*

Proof. Assume first that $E\xi^2$ is finite. Then
$$(\xi - E\xi)^2 = \xi^2 - 2(E\xi)\xi + (E\xi)^2,$$
and by the first property of the mathematical expectation (see Lemma 1.25),
$$Var\xi = E\xi^2 - E(2(E\xi)\xi) + E(E\xi)^2 = E\xi^2 - 2(E\xi)^2 + (E\xi)^2 = E\xi^2 - (E\xi)^2.$$

If Varξ is finite, we have

$$\xi^2 = (\xi - E\xi)^2 + 2(E\xi) \cdot \xi - (E\xi)^2,$$

and by the first property of the expectation,

$$E\xi^2 = E(\xi - E\xi)^2 + 2(E\xi)^2 - (E\xi)^2 = \text{Var}\xi + (E\xi)^2 \ ,$$

which proves the first property of the variance. By the first property of the expectation,

$$E(a\xi + b) = aE\xi + b,$$

and therefore,

$$\text{Var}(a\xi + b) = E(a\xi + b - E(a\xi + b))^2 = E(a\xi - aE\xi)^2 = Ea^2(\xi - E\xi)^2 = a^2\text{Var}\xi,$$

which proves the second property. Let $A \leq \xi \leq B$. It follows from the second property of the variance that

$$\text{Var}\xi = E(\xi - E\xi)^2 = E(\xi - \frac{A+B}{2} - (E\xi - \frac{A+B}{2}))^2$$

$$= E(\xi - \frac{A+B}{2})^2 - (E(\xi - \frac{A+B}{2}))^2 \leq E(\xi - \frac{A+B}{2})^2 \leq (\frac{B-A}{2})^2 \ ,$$

which proves the third property. $\qquad\square$

Lemma 1.31 (Chebyshev Inequality for the Variance). *Let* Varξ *be finite. Then for each $t > 0$,*

$$P(|\xi - E\xi| \geq t) \leq \frac{\text{Var}\xi}{t^2}.$$

Proof. We apply Lemma 1.27 to the random variable $\eta = (\xi - E\xi)^2 \geq 0$. Then

$$P(|\xi - E\xi| \geq t) = P(\eta \geq t^2) \leq \frac{E\eta}{t^2} = \frac{\text{Var}\xi}{t^2}.$$

$\qquad\square$

Definition 1.32. *The covariance of the random variables ξ_1 and ξ_2 is the number* $\text{Cov}(\xi_1, \xi_2) = E(\xi_1 - m_1)(\xi_2 - m_2)$, *where $m_i = E\xi_i$ for $i = 1, 2$.*

By Lemma 1.28, if Varξ_1 and Varξ_2 are finite, then $\text{Cov}(\xi_1, \xi_2)$ is also finite. We note that

$$\text{Cov}(\xi_1, \xi_2) = E(\xi_1 - m_1)(\xi_2 - m_2)$$

$$= E(\xi_1\xi_2 - m_1\xi_2 - m_2\xi_1 + m_1m_2) = E(\xi_1\xi_2) - m_1m_2.$$

Also

$$\text{Cov}(a_1\xi_1 + b_1, a_2\xi_2 + b_2) = a_1a_2\text{Cov}(\xi_1\xi_2).$$

Let $\xi_1, \ldots, \xi_n$ be random variables and $\zeta_n = \xi_1 + \ldots + \xi_n$. If $m_i = \text{E}\xi_i$, then $\text{E}\zeta_n = \sum_{i=1}^{n} m_i$ and

$$\text{Var}\zeta_n = \text{E}(\sum_{i=1}^{n}\xi_i - \sum_{i=1}^{n}m_i)^2 = \text{E}(\sum_{i=1}^{n}(\xi_i - m_i))^2$$

$$= \sum_{i=1}^{n}\text{E}(\xi_i - m_i)^2 + 2\sum_{i<j}\text{E}(\xi_i - m_i)(\xi_j - m_j)$$

$$= \sum_{i=1}^{n}\text{Var}\xi_i + 2\sum_{i<j}\text{Cov}(\xi_i, \xi_j).$$

Definition 1.33. *The correlation coefficient of the random variables ξ_1 and ξ_2 with non-zero variances is the number $\rho(\xi_1, \xi_2) = \text{Cov}(\xi_1, \xi_2)/\sqrt{\text{Var}\xi_1\text{Var}\xi_2}$.*

It follows from the properties of the variance and the covariance that $\rho(a_1\xi_1 + b_1, a_2\xi_2 + b_2) = \rho(\xi_1, \xi_2)$ for any constants a_1, b_1, a_2, b_2.

Theorem 1.34. *Let ξ_1 and ξ_2 be random variables with non-zero variances. Then the absolute value of the correlation coefficient $\rho(\xi_1, \xi_2)$ is less than or equal to one. If $|\rho(\xi_1, \xi_2)| = 1$, then for some constants a and b the equality $\xi_2(\omega) = a\xi_1(\omega) + b$ holds almost surely.*

Proof. For every t we have

$$\text{E}(t(\xi_2 - m_2) + (\xi_1 - m_1))^2 = t^2\text{E}(\xi_2 - m_2)^2 + 2t\text{E}(\xi_1 - m_1)(\xi_2 - m_2) + \text{E}(\xi_1 - m_1)^2$$

$$= t^2\text{Var}\xi_2 + 2t\text{Cov}(\xi_1, \xi_2) + \text{Var}\xi_1.$$

Since the left-hand side of this equality is non-negative for every t, so is the quadratic polynomial on the right-hand side, which implies

$$(\text{Cov}(\xi_1, \xi_2))^2 \leq \text{Var}\xi_1\text{Var}\xi_2,$$

that is $|\rho(\xi_1, \xi_2)| \leq 1$. If $|\rho(\xi_1, \xi_2)| = 1$, then there exists $t_0 \neq 0$ such that $\text{E}(t_0(\xi_2 - m_2) + (\xi_1 - m_1))^2 = 0$, that is

$$t_0(\xi_2(\omega) - m_2) + (\xi_1(\omega) - m_1) = 0$$

almost surely. Thus $\xi_2 = m_2 + m_1/t_0 - \xi_1/t_0$. Setting $a = -1/t_0$ and $b = m_2 + m_1/t_0$, we obtain the second statement of the theorem. $\square$

1.3 Probability of a Union of Events

If $C_1, \ldots, C_n$ are disjoint events, then it follows from the definition of a probability measure that $P(\bigcup_{i=1}^{n} C_i) = \sum_{i=1}^{n} P(C_i)$. We shall derive a formula for the probability of a union of any n events.

Theorem 1.35. *Let $(\Omega, \mathcal{F}, P)$ be a probability space and $C_1, \ldots, C_n \in \mathcal{F}$. Then*

$$P\left(\bigcup_{i=1}^{n} C_i\right) = \sum_{i=1}^{n} P(C_i) - \sum_{i_1 < i_2} P(C_{i_1} \cap C_{i_2})$$

$$+ \sum_{i_1 < i_2 < i_3} P(C_{i_1} \cap C_{i_2} \cap C_{i_3}) - \sum_{i_1 < i_2 < i_3 < i_4} P(C_{i_1} \cap C_{i_2} \cap C_{i_3} \cap C_{i_4}) + \ldots$$

Proof. At first, let us assume that the space Ω is discrete. Consider the complement $\Omega \backslash (\bigcup_{i=1}^{n} C_i) = \bigcap_{i=1}^{n} (\Omega \backslash C_i)$. For any $C \in \mathcal{F}$, let χ_C be the indicator of C,

$$\chi_C(\omega) = \begin{cases} 1 & \text{if } \omega \in C, \\ 0 & \text{otherwise.} \end{cases}$$

It is easy to see that $\chi_{\Omega \backslash C}(\omega) = 1 - \chi_C(\omega)$ and

$$\chi_{\bigcap_{i=1}^{n}(\Omega \backslash C_i)}(\omega) = \prod_{i=1}^{n} \chi_{\Omega \backslash C_i}(\omega) = \prod_{i=1}^{n} (1 - \chi_{C_i}(\omega)).$$

Thus,

$$1 - P\left(\bigcup_{i=1}^{n} C_i\right) = P\left(\bigcap_{i=1}^{n} (\Omega \backslash C_i)\right)$$

$$= \sum_{\omega \in \Omega} \chi_{\bigcap_{i=1}^{n}(\Omega \backslash C_i)}(\omega) P(\{\omega\}) = \sum_{\omega \in \Omega} \prod_{i=1}^{n} (1 - \chi_{C_i}(\omega)) P(\{\omega\})$$

$$= 1 - \sum_{i=1}^{n} \sum_{\omega \in \Omega} \chi_{C_i}(\omega) P(\{\omega\}) + \sum_{i_1 < i_2} \sum_{\omega \in \Omega} \chi_{C_{i_1}}(\omega) \cdot \chi_{C_{i_2}}(\omega) P(\{\omega\}) - \ldots$$

$$= 1 - \sum_{i=1}^{n} P(C_i) + \sum_{i_1 < i_2} P(C_{i_1} \cap C_{i_2}) - \ldots ,$$

which completes the proof of the theorem for the case of discrete Ω.

In the general case, when Ω is not necessarily discrete, we can replace the sums $\sum_{\omega \in \Omega}$ with the integrals over the space Ω with respect to the measure P. This requires, however, the notion of the Lebesgue integral, which will be introduced in Chap. 3. The rest of the proof is analogous. $\square$

We shall now apply Theorem 1.35 to solve an interesting problem. Our arguments below will not be completely rigorous and are intended to develop the intuition of the reader.

Let x_1 and x_2 be two integers randomly and independently chosen from the set $\{1,\ldots,n\}$ according to the uniform distribution. This means that the space Ω_n consists of pairs $\omega = (x_1, x_2)$, where $1 \le x_1 \le n$, $1 \le x_2 \le n$. This space has n^2 elements (elementary outcomes) and the probability of each elementary outcome is $p^n(\omega) = 1/n^2$. We denote the corresponding probability measure by P^n. Let A^n be the event that x_1 and x_2 are coprime,

$$A^n = \{(x_1, x_2) \in \Omega_n : x_1 \text{ and } x_2 \text{ are coprime}\}.$$

We shall find the limit of $P^n(A^n)$ as n tends to infinity.

In our arguments below q will denote a prime number, $q > 1$. Denote by C_q^n the event in Ω_n that both x_1 and x_2 are divisible by q. Then

$$P^n(A^n) = 1 - P^n(\bigcup_{q \le n} C_q^n),$$

and by Theorem 1.35 we have

$$P^n(\bigcup_{q \le n} C_q^n) = \sum_{q \le n} P^n(C_q^n) - \sum_{q_1 < q_2 \le n} P^n(C_{q_1}^n \bigcap C_{q_2}^n)$$

$$+ \sum_{q_1 < q_2 < q_3 \le n} P^n(C_{q_1}^n \bigcap C_{q_2}^n \bigcap C_{q_3}^n) - \cdots$$

It is easy to see that

$$\lim_{n \to \infty} P^n(C_q^n) = 1/q^2, \quad \lim_{n \to \infty} P^n(C_{q_1}^n \bigcap C_{q_2}^n) = 1/q_1^2 q_2^2, \quad \text{etc.,}$$

which implies that

$$\lim_{n \to \infty} P^n(\bigcup_{q \le n} C_q^n) = \sum_q \frac{1}{q^2} - \sum_{q_1 < q_2} \frac{1}{q_1^2 q_2^2} + \sum_{q_1 < q_2 < q_3} \frac{1}{q_1^2 q_2^2 q_3^2} - \cdots$$

Since the number of terms on the right-hand side is infinite, this formula requires a more rigorous justification, which we do not provide here. We obtain

$$\lim_{n \to \infty} P^n(A^n) = 1 - \lim_{n \to \infty} P^n(\bigcup_{q \le n} C_q^n)$$

$$= 1 - \sum_q \frac{1}{q^2} + \sum_{q_1 < q_2} \frac{1}{q_1^2 q_2^2} - \sum_{q_1 < q_2 < q_3} \frac{1}{q_1^2 q_2^2 q_3^2} - \cdots = \prod_q (1 - \frac{1}{q^2}).$$

Therefore,

$$\frac{1}{\lim_{n \to \infty} P^n(A^n)} = \frac{1}{\prod_q (1 - \frac{1}{q^2})} = \prod_q \frac{1}{1 - \frac{1}{q^2}}$$

$$= \prod_q \sum_{m=0}^{\infty} \frac{1}{q^{2m}} = \sum \frac{1}{q_1^{2m_1}} \cdot \frac{1}{q_2^{2m_2}} \cdot \cdots \cdot \frac{1}{q_s^{2m_s}},$$

where the last sum is with respect to all $s \geq 0$, all finite words $(q_1, \ldots, q_s)$ with $q_1 < q_2 < \ldots < q_s$ being prime numbers, and all finite words $(m_1, \ldots, m_s)$ with $m_i \geq 1$. Since every positive integer can be written in the form $x = q_1^{m_1} q_2^{m_2} \ldots q_s^{m_s}$ in a unique way, the last sum is equal to $\sum_{x \geq 1} 1/x^2 = \pi^2/6$. Therefore,

$$\lim_{n \to \infty} P^n(A^n) = 6/\pi^2.$$

1.4 Equivalent Formulations of σ-Additivity, Borel σ-Algebras and Measurability

Let $(\Omega, \mathcal{F})$ be a measurable space.

Theorem 1.36. *Suppose that a function* P *on* $\mathcal{F}$ *has the properties of a probability measure, with* σ-additivity replaced by finite additivity, that is

1. $P(C) \geq 0$ *for any* $C \in \mathcal{F}$.
2. $P(\Omega) = 1$.
3. *If* $C_i \in \mathcal{F}$ *for* $1 \leq i \leq n$, *and* $C_i \cap C_j = \emptyset$ *for* $i \neq j$, *then*

$$P(\bigcup_{i=1}^{n} C_i) = \sum_{i=1}^{n} P(C_i).$$

Then the following four statements are equivalent.

1. P *is* σ-additive (and thus is a probability measure).
2. *For any sequence of events* $C_i \in \mathcal{F}, C_i \subseteq C_{i+1}$ *we have*

$$P(\bigcup_i C_i) = \lim_{i \to \infty} P(C_i).$$

3. *For any sequence of events* $C_i \in \mathcal{F}, C_i \supseteq C_{i+1}$ *we have*

$$P(\bigcap_i C_i) = \lim_{i \to \infty} P(C_i).$$

4. *For any sequence of events* $C_i \in \mathcal{F}, C_i \supseteq C_{i+1}, \bigcap_i C_i = \emptyset$ *we have*

$$\lim_{i \to \infty} P(C_i) = 0.$$

Proof. The equivalence of each pair of statements is proved in a similar way. For example, let us prove that the first one is equivalent to the fourth one. First, assume that P is σ-additive. Let $C_i \in \mathcal{F}, C_i \supseteq C_{i+1}, \bigcap_i C_i = \emptyset$. Consider the events $B_i = C_i \backslash C_{i+1}$. Then $B_i \cap B_j = \emptyset$ for $i \neq j$ and $C_n = \bigcup_{i \geq n} B_i$. From the σ-additivity of P we have $P(C_1) = \sum_{i=1}^{\infty} P(B_i)$. Therefore, the remainder of the series $\sum_{i=n}^{\infty} P(B_i) = P(C_n)$ tends to zero as $n \to \infty$, which gives the fourth property.

Conversely, let us prove that the fourth property implies the σ-additivity. Assume that we have a sequence of events C_i, $C_i \cap C_j = \emptyset$ for $i \neq j$. Consider $C = \bigcup_{i=1}^{\infty} C_i$. Then $C = (\bigcup_{i=1}^{n} C_i) \cup (\bigcup_{i=n+1}^{\infty} C_i)$ for any n, and by the finite additivity $P(C) = \sum_{i=1}^{n} P(C_i) + P(\bigcup_{i=n+1}^{\infty} C_i)$. The events $B_n = \bigcup_{i=n+1}^{\infty} C_i$ decrease and $\bigcap_n B_n = \emptyset$. Therefore, $P(B_n) \to 0$ as $n \to \infty$ and $P(C) = \sum_{i=1}^{\infty} P(C_i)$. $\square$

Now we shall consider some of the most important examples of σ-algebras encountered in probability theory. First we introduce the following general definition.

Definition 1.37. *Let $\mathcal{A}$ be an arbitrary collection of subsets of Ω. The intersection of all σ-algebras containing all elements of $\mathcal{A}$ is called the σ-algebra generated by $\mathcal{A}$, or the minimal σ-algebra containing $\mathcal{A}$. It is denoted by $\sigma(\mathcal{A})$. In other words,*

$$\sigma(\mathcal{A}) = \{C : C \in \mathcal{F} \text{ for each } \sigma-\text{algebra } \mathcal{F} \text{ such that } \mathcal{A} \subseteq \mathcal{F}\}. \qquad (1.1)$$

We need the following three remarks in order to make sense of this definition. First, there is at least one σ-algebra which contains $\mathcal{A}$, namely the σ-algebra of all subsets of Ω. Second, it is clear that the intersection of any collection of σ-algebras is again a σ-algebra. Therefore, the set $\sigma(\mathcal{A})$ in (1.1) is correctly defined and is a σ-algebra. Finally, it is clear that any σ-algebra $\mathcal{F}$ that contains $\mathcal{A}$ must also contain $\sigma(\mathcal{A})$. Otherwise, one could consider $\sigma(\mathcal{A}) \cap \mathcal{F}$, which would be strictly contained in $\sigma(\mathcal{A})$. In this sense $\sigma(\mathcal{A})$ is the smallest σ-algebra which contains all elements of $\mathcal{A}$.

Assume now that $\Omega = \mathbb{R}$. Consider the following families of subsets.

1. $\mathcal{A}_1$ is the collection of open intervals (a, b).
2. $\mathcal{A}_2$ is the collection of half-open intervals $[a, b)$.
3. $\mathcal{A}_3$ is the collection of half-open intervals $(a, b]$.
4. $\mathcal{A}_4$ is the collection of closed intervals $[a, b]$.
5. $\mathcal{A}_5$ is the collection of semi-infinite open intervals $(-\infty, a)$.
6. $\mathcal{A}_6$ is the collection of semi-infinite closed intervals $(-\infty, a]$.
7. $\mathcal{A}_7$ is the collection of semi-infinite open intervals (a, ∞).
8. $\mathcal{A}_8$ is the collection of semi-infinite closed intervals $[a, \infty)$.
9. $\mathcal{A}_9$ is the collection of open subsets of $\mathbb{R}$.
10. $\mathcal{A}_{10}$ is the collection of closed subsets of $\mathbb{R}$.

Theorem 1.38. *The σ-algebras generated by the above sets coincide:*

$$\sigma(\mathcal{A}_1) = \sigma(\mathcal{A}_2) = \ldots = \sigma(\mathcal{A}_9) = \sigma(\mathcal{A}_{10}).$$

This σ-algebra is called the Borel σ-algebra of $\mathbb{R}$ or the σ-algebra of Borel subsets of $\mathbb{R}$, and is usually denoted by $\mathcal{B}(\mathbb{R})$ or simply $\mathcal{B}$.

Proof. Let us prove that $\sigma(\mathcal{A}_1) = \sigma(\mathcal{A}_2)$. For any $a < b$ we can find a sequence $a_n \downarrow a$. Then $\bigcup_n [a_n, b) = (a, b)$, and therefore, $(a, b) \in \sigma(\mathcal{A}_2)$. This implies that $\sigma(\mathcal{A}_1) \subseteq \sigma(\mathcal{A}_2)$.

Conversely, for any $a < b$, we can find a sequence $a_n \uparrow a$. Then $\bigcap_n (a_n, b) = [a, b)$, and therefore, $[a, b) \in \sigma(\mathcal{A}_1)$. This implies that $\sigma(\mathcal{A}_2) \subseteq \sigma(\mathcal{A}_1)$.

The equality between $\sigma(\mathcal{A}_1), \sigma(\mathcal{A}_2), \ldots, \sigma(\mathcal{A}_7)$, and $\sigma(\mathcal{A}_8)$ is proved in a very similar way. The fact that $\sigma(\mathcal{A}_9) = \sigma(\mathcal{A}_{10})$ follows from the fact that every closed set is the complement of an open set. Also, $\sigma(\mathcal{A}_1) \subseteq \sigma(\mathcal{A}_9)$, since any interval of the form (a, b) is an open set. To show that $\sigma(\mathcal{A}_9) \subseteq \sigma(\mathcal{A}_1)$ it remains to remark that any open set can be represented as a countable union of open intervals. $\qquad\square$

Let us assume that $\Omega = X$ is a metric space.

Definition 1.39. *The Borel σ-algebra of X is the σ-algebra $\sigma(\mathcal{A})$, where $\mathcal{A}$ is the family of open subsets of X. It is usually denoted by $\mathcal{B}(X)$.*

In this definition we could take $\mathcal{A}$ to be a collection of all closed subsets, since any open set is the complement of a closed set. If X is separable, then any open set can be represented as a countable union of open balls. Thus, for a separable space X, we could define the Borel σ-algebra of X as $\sigma(\mathcal{A})$, with $\mathcal{A}$ being the family of all open balls.

Let us now define the product of measurable spaces $(\Omega_1, \mathcal{F}_1), \ldots, (\Omega_n, \mathcal{F}_n)$. The set $\Omega_1 \times \ldots \times \Omega_n$ consists of sequences $(\omega_1, \ldots, \omega_n)$, where $\omega_i \in \Omega_i$ for $1 \leq i \leq n$. The σ-algebra $\mathcal{F}_1 \times \ldots \times \mathcal{F}_n$ is defined as the minimal σ-algebra which contains all the sets of the form $A_1 \times \ldots \times A_n$, where $A_i \in \mathcal{F}_i$, $1 \leq i \leq n$. Define

$$(\Omega_1, \mathcal{F}_1) \times \ldots \times (\Omega_n, \mathcal{F}_n) = (\Omega_1 \times \ldots \times \Omega_n, \mathcal{F}_1 \times \ldots \times \mathcal{F}_n).$$

If $X_1, \ldots, X_n$ are metric spaces, then $X_1 \times \ldots \times X_n$ can be endowed with the product metric, and we can consider the Borel σ-algebra $\mathcal{B}(X_1 \times \ldots \times X_n)$. It is easy to see that for separable metric spaces it coincides with the product of σ-algebras $\mathcal{B}(X_1) \times \ldots \times \mathcal{B}(X_n)$ (see Problem 12).

Definition 1.40. *Given two measurable spaces $(\Omega, \mathcal{F})$ and $(\widetilde{\Omega}, \widetilde{\mathcal{F}})$, a function $f : \Omega \to \widetilde{\Omega}$ is called measurable if $f^{-1}(A) \in \mathcal{F}$ for every $A \in \widetilde{\mathcal{F}}$.*

When the second space is $\mathbb{R}$ with the σ-algebra $\mathcal{B}(\mathbb{R})$ of Borel sets, this definition coincides with our previous definition of measurability. Indeed, the collection of sets $A \in \widetilde{\mathcal{F}}$ for which $f^{-1}(A) \in \mathcal{F}$, forms a σ-algebra. If this σ-algebra contains all the intervals $[a, b)$, then it contains the entire Borel σ-algebra due to Theorem 1.38.

Let $g(x_1, \ldots, x_n)$ be a function of n real variables, which is measurable with respect to the Borel σ-algebra $\mathcal{B}(\mathbb{R}^n)$ (see Definition 1.39).

Lemma 1.41. *For any measurable functions $f_1(\omega), \ldots, f_n(\omega)$ the composition function $g(f_1(\omega), \ldots, f_n(\omega))$ is also measurable.*

Proof. Clearly it is sufficient to show that the pre-image of any Borel set of $\mathbb{R}^n$ under the mapping $f : \Omega \to \mathbb{R}^n$, $f(\omega) = (f_1(\omega), \ldots, f_n(\omega))$ is measurable, that is

$$f^{-1}(A) \in \mathcal{F} \qquad\qquad (1.2)$$

for any $A \in \mathcal{B}(\mathbb{R}^n)$.

If $A \subseteq \mathbb{R}^n$ is a set of the form $A = A_1 \times \ldots \times A_n$, where A_i, $i = 1, \ldots, n$, are Borel sets, then $f^{-1}(A) = \bigcap_{i=1}^n f_i^{-1}(A_i)$ is measurable. The collection of sets for which (1.2) holds is a σ-algebra. Therefore, (1.2) holds for all the sets in the smallest σ-algebra containing all the rectangles, which is easily seen to be the Borel σ-algebra of $\mathbb{R}^n$. $\qquad\square$

Applying Lemma 1.41 to the functions $g_1(x_1, \ldots, x_n) = a_1 x_1 + \ldots + a_n x_n$, $g_2(x_1, \ldots, x_n) = x_1 \cdot \ldots \cdot x_n$ and $g_3(x_1, x_2) = x_1/x_2$ we immediately obtain the following.

Lemma 1.42. *If $f_1, \ldots f_n$ are measurable functions, then their linear combination $g = a_1 f_1 + \ldots + a_n f_n$ and their product $h = f_1 \cdot \ldots \cdot f_n$ are also measurable. The ratio of two measurable functions, the second of which is not equal to zero for any ω, is also measurable.*

1.5 Distribution Functions and Densities

Definition 1.43. *For a random variable ξ on a probability space $(\Omega, \mathcal{F}, \mathrm{P})$, let $F_\xi(x)$ denote the probability that ξ does not exceed x, that is*

$$F_\xi(x) = \mathrm{P}(\{\omega : \xi(\omega) \leq x\}), \ x \in \mathbb{R}.$$

The function F_ξ is called the distribution function of the random variable ξ.

Theorem 1.44. *If F_ξ is the distribution function of a random variable ξ, then*

1. *F_ξ is non-decreasing, that is $F_\xi(x) \leq F_\xi(y)$ if $x \leq y$.*
2. *$\lim_{x \to -\infty} F_\xi(x) = 0$, $\lim_{x \to \infty} F_\xi(x) = 1$.*
3. *$F_\xi(x)$ is continuous from the right for every x, that is*

$$\lim_{y \downarrow x} F_\xi(y) = F_\xi(x).$$

Proof. The first property holds since $\{\omega : \xi(\omega) \leq x\} \subseteq \{\omega : \xi(\omega) \leq y\}$ if $x \leq y$.

In order to prove the second property, we note that the intersection of the nested events $\{\omega : \xi(\omega) \leq -n\}$, $n \geq 0$, is empty. Therefore, by Theorem 1.36, we have

$$\lim_{n\to\infty} F_\xi(-n) = \lim_{n\to\infty} P(\{\omega : \xi(\omega) \leq -n\}) = 0,$$

which implies that $\lim_{x\to-\infty} F_\xi(x) = 0$, due to monotonicity. Similarly, it is seen that $\lim_{x\to\infty} F_\xi(x) = 1$.

In order to prove the last property, we note that the intersection of the nested events $\{\omega : x < \xi(\omega) \leq x + 1/n\}$, $n \geq 0$, is empty, and therefore, by Theorem 1.36,

$$\lim_{n\to\infty} \left(F_\xi(x + \frac{1}{n}) - F_\xi(x)\right) = \lim_{n\to\infty} P(\{\omega : x < \xi(\omega) \leq x + \frac{1}{n}\}) = 0.$$

The conclusion that $\lim_{y\downarrow x} F_\xi(y) = F_\xi(x)$ follows from the monotonicity of $F_\xi(x)$. $\qquad\square$

We can now disregard the fact that $F_\xi(x)$ appears as the distribution function of a particular random variable ξ, and introduce the following definition.

Definition 1.45. *Any function F defined on the real line, which has Properties 1–3 listed in Theorem 1.44, is called a distribution function.*

Later we shall see that any distribution function defines a probability measure on $(\mathbb{R}, \mathcal{B}(\mathbb{R}))$ and is, in fact, the distribution function of the random variable $\xi(x) = x$ with respect to that measure.

In some cases (not always!) there exists a non-negative integrable function $p(t)$ on the real line such that

$$F(x) = \int_{-\infty}^{x} p(t)dt$$

for all x. In this case p is called the probability density of F or simply the density of F. If $F = F_\xi$ is the distribution function of a random variable ξ, then $p = p_\xi$ is called the probability density of ξ. While the right-hand side of the formula above in general should be understood as the Lebesgue integral (defined in Chap. 3), for continuous densities $p(t)$ it happens to be equal to the usual Riemann integral.

Note that any probability density satisfies $\int_{-\infty}^{\infty} p(t)dt = 1$. Conversely, any non-negative integrable function which has this property defines a distribution function via $F(x) = \int_{-\infty}^{x} p(t)dt$ and, thus, is a probability density.

If P is a probability distribution on $(\mathbb{R}, \mathcal{B}(\mathbb{R}))$ and p is a probability density such that

$$P((-\infty, x]) = \int_{-\infty}^{x} p(t)dt$$

for all x, then p is said to be the probability density of P. The relationship between distribution functions and probability distributions will be discussed in Sect. 3.2.

Examples of Probability Densities

1. $p(u) = \frac{1}{\sqrt{2\pi}} e^{-\frac{u^2}{2}}$, $-\infty < u < \infty$, is called the normal or Gaussian density with parameters $(0,1)$. The corresponding probability distribution function is $\Phi(x) = \frac{1}{\sqrt{2\pi}} \int_0^x e^{-\frac{u^2}{2}} du$.

2. $p(u) = \frac{1}{\sqrt{2\pi d}} e^{-\frac{(u-a)^2}{2d}}$, $-\infty < u < \infty$, is called the normal or Gaussian density with parameters (a,d). The distribution with such a density is denoted by $N(a,d)$.

3. The uniform density on the interval (a,b):

$$p(u) = \begin{cases} \frac{1}{b-a}, & u \in (a,b), \\ 0, & u \notin (a,b). \end{cases}$$

4. The function

$$p(u) = \begin{cases} \lambda e^{-\lambda u} & u \geq 0, \\ 0 & u < 0, \end{cases}$$

is called the exponential density with parameter λ.

5. $p(u) = \frac{1}{\pi(1+u^2)}$, $-\infty < u < \infty$, is called the Cauchy density or the density of the Cauchy distribution.

We shall say that ξ is a random vector if $\xi = (\xi_1, \ldots, \xi_n)$, where ξ_i, $1 \leq i \leq n$, are random variables defined on a common probability space.

Definition 1.46. *The distribution function of a random vector* $\xi = (\xi_1, \ldots, \xi_n)$ *on a probability space* $(\Omega, \mathcal{F}, P)$ *is the function* $F_\xi : \mathbb{R}^n \to \mathbb{R}$ *given by*

$$F_\xi(x_1, \ldots, x_n) = P(\{\omega : \xi_1(\omega) \leq x_1, \ldots, \xi_n(\omega) \leq x_n\}).$$

In order to formulate a multi-dimensional analogue of Theorem 1.44, we need the following notations. Let $x = (x_1, \ldots, x_n)$ and $y = (y_1, \ldots, y_n)$ be the vectors in $\mathbb{R}^n$. We'll say that $x \leq y$ if $x_i \leq y_i$ for $1 \leq i \leq n$. Given a vector $\sigma = (\sigma_1, \ldots, \sigma_n)$ with $\sigma_i \in \{0, 1\}$ for $i = 1, \ldots, n$, we can define a new vector $z = \sigma(x, y)$ with the coordinates

$$z_i = \begin{cases} y_i & \text{if} \quad \sigma_i = 0, \\ x_i & \text{if} \quad \sigma_i = 1. \end{cases}$$

Let $|\sigma| = \sigma_1 + \ldots + \sigma_n$ be the number of elements of the vector σ that are equal to one.

Theorem 1.47. *If* F_ξ *is a distribution function of a random vector* $\xi = (\xi_1, \ldots, \xi_n)$, *then*

1. *If* $x, y \in \mathbb{R}^n$ *are such that* $x \leq y$, *then*

$$\sum_{\sigma \in \{0,1\}^n} (-1)^{|\sigma|} F_\xi(\sigma(x, y)) \geq 0.$$

2. $\lim_{x_i \to -\infty} F_\xi(x) = 0$ for each $i \in \{1, \ldots, n\}$; $\lim_{x \to (+\infty, \ldots, +\infty)} F_\xi(x) = 1$.
3. $F_\xi(x)$ is continuous from above for every x, that is

$$\lim_{y \downarrow x} F_\xi(y) = F_\xi(x).$$

This theorem can be proved in the same way as Theorem 1.44. Note that the first property represents the fact that $P(x < \xi \le y) \ge 0$. From the Properties 1–3 it easily follows that F_ξ takes values in $[0, 1]$ and is non-decreasing, that is $F_\xi(x) \le F_\xi(x)$ if $x \le y$.

Definition 1.48. *Any function F defined on $\mathbb{R}^n$, which has Properties 1–3 listed in Theorem 1.47, is called a distribution function.*

1.6 Problems

1. A man's birthday is on March 1st. His father's birthday is on March 2nd. One of his grandfather's birthday is on March 3rd. How would you estimate the number of such people in the USA?
2. Suppose that n identical balls are distributed randomly among m boxes. Construct the corresponding space of elementary outcomes. Assuming that each ball is placed in a random box with equal probability, find the probability that the first box is empty.
3. A box contains 90 good items and 10 defective items. Find the probability that a sample of 10 items has no defective items.
4. Let ξ be a random variable such that $E|\xi|^m \le AC^m$ for some positive constants A and C, and all integers $m \ge 0$. Prove that $P(|\xi| > C) = 0$.
5. Suppose there are n letters addressed to n different people, and n envelopes with addresses. The letters are mixed and then randomly placed into the envelopes. Find the probability that at least one letter is in the correct envelope. Find the limit of this probability as $n \to \infty$.
6. For integers n and r, find the number of solutions of the equation

$$x_1 + \ldots + x_r = n$$

   where $x_i \ge 0$ are integers. Assuming the uniform distribution on the space of the solutions, find $P(x_1 = a)$ and its limit as $r \to \infty$, $n \to \infty$, $n/r \to \rho > 0$.
7. Find the mathematical expectation and the variance of a random variable with Poisson distribution with parameter λ.
8. Draw the graph of the distribution function of random variable ξ taking values $x_1, \ldots, x_n$ with probabilities $p_1, \ldots, p_n$.
9. Prove that if F is the distribution function of the random variable ξ, then $P(\xi = x) = F(x) - \lim_{\delta \downarrow 0} F(x - \delta)$.
10. A random variable ξ has density p. Find the density of $\eta = a\xi + b$ for $a, b \in \mathbb{R}$, $a \neq 0$.

11. A random variable ξ has uniform distribution on $[0, 2\pi]$. Find the density of the distribution of $\eta = \sin \xi$.

12. Let $(X_1, d_1), \ldots, (X_n, d_n)$ be separable metric spaces, and define $X = X_1 \times \ldots \times X_n$ to be the product space with the metric

$$d((x_1, \ldots, x_n), (y_1, \ldots, y_n)) = \sqrt{d_1^2(x_1, y_1) + \ldots + d_n^2(x_n, y_n)}.$$

Prove that $\mathcal{B}(X) = \mathcal{B}(X_1) \times \ldots \times \mathcal{B}(X_n)$.

13. An integer from 1 to 1,000 is chosen at random (with uniform distribution). What is the probability that it is an integer power (higher than the first) of an integer.

14. Let $C_1, C_2, \ldots$ be a sequence of events in a probability space $(\Omega, \mathcal{F}, \mathrm{P})$ such that $\lim_{n \to \infty} \mathrm{P}(C_n) = 0$ and $\sum_{n=1}^{\infty} \mathrm{P}(C_{n+1} \setminus C_n) < \infty$. Prove that

$$\mathrm{P}(\bigcap_{n=1}^{\infty} \bigcup_{k=n}^{\infty} C_k) = 0.$$

15. Let ξ be a random variable with continuous distribution function F. Find the distribution function of the random variable $F(\xi)$.

Sequences of Independent Trials

2.1 Law of Large Numbers and Applications

Consider a probability space $(X, \mathcal{G}, \mathrm{P}_X)$, where $\mathcal{G}$ is a σ-algebra of subsets of X and P_X is a probability measure on $(X, \mathcal{G})$. In this section we shall consider the spaces of sequences

$$\Omega = \{\omega = (\omega_1, \ldots, \omega_n), \omega_i \in X, i = 1, \ldots, n\}$$

and

$$\Omega = \{\omega = (\omega_1, \omega_2, \ldots), \omega_i \in X, i \geq 1\}.$$

In order to define the σ-algebra on Ω in the case of the space of infinite sequences, we need the notion of a finite-dimensional cylinder.

Definition 2.1. *Let $1 \leq n \leq \infty$ and Ω be the space of sequences of length n. A finite-dimensional elementary cylinder is a set of the form*

$$A = \{\omega : \omega_{t_1} \in A_1, \ldots, \omega_{t_k} \in A_k\},$$

where $t_1, \ldots, t_k \geq 1$, and $A_i \in \mathcal{G}$, $1 \leq i \leq k$.
  A finite-dimensional cylinder is a set of the form

$$A = \{\omega : (\omega_{t_1}, \ldots, \omega_{t_k}) \in B\},$$

where $t_1, \ldots, t_k \geq 1$, and $B \in \mathcal{G} \times \ldots \times \mathcal{G}$ (k times).

Clearly every cylinder belongs to the σ-algebra generated by elementary cylinders, and therefore the σ-algebras generated by elementary cylinders and all cylinders coincide. We shall denote this σ-algebra by $\mathcal{F}$. In the case of finite n it is clear that $\mathcal{F}$ is the product σ-algebra: $\mathcal{F} = \mathcal{G} \times \ldots \times \mathcal{G}$ (n times).

Definition 2.2. *A probability measure P on $(\Omega, \mathcal{F})$ corresponds to a homogeneous sequence of independent random trials if $\mathrm{P}(A) = \prod_{i=1}^{k} \mathrm{P}_X(A_i)$ for any elementary cylinder $A = \{\omega : \omega_{t_1} \in A_1, \ldots, \omega_{t_k} \in A_k\}$ with $t_1, \ldots, t_k$ distinct.*

L. Koralov and Y.G. Sinai, *Theory of Probability and Random Processes*,
Universitext, DOI 10.1007/978-3-540-68829-7_2,
© Springer-Verlag Berlin Heidelberg 2012

If n is finite, we shall see in Sect. 3.5, where the product measure is discussed, that such a measure exists and is unique. If n is infinite, the existence of such a measure on $(\Omega, \mathcal{F})$ follows from the Kolmogorov Consistency Theorem, which will be discussed in Sect. 12.2. If $X = \{x^1, \ldots, x^r\}$ is a finite set, and $n < \infty$, then the question about the existence of such a measure P does not pose any problems. Indeed, now Ω is discrete, and we can define P for each elementary outcome $\omega = (\omega_1, \ldots, \omega_n)$ by

$$P(\{\omega\}) = \prod_{i=1}^{n} P_X(\{\omega_i\}).$$

Later we shall give a more general definition of independence for families of random variables on any probability space. It will be seen that $\xi_i(\omega) = \xi_i(\omega_1, \ldots, \omega_n) = \omega_i$ form a sequence of independent random variables if the probability measure on the space $(\Omega, \mathcal{F})$ satisfies the definition above.

Let $(X, \mathcal{G}, P_X)$ be a probability space and $(\Omega, \mathcal{F}, P)$ the probability space corresponding to a finite or infinite sequence of independent random trials. Take $B \in \mathcal{G}$, and define

$$\chi^i(\omega) = \begin{cases} 1 & \text{if } \omega_i \in B, \\ 0 & \text{otherwise.} \end{cases}$$

Define ν^n to be equal to the number of occurrences of elementary outcomes from B in the sequence of the first n trials, that is

$$\nu^n(\omega) = \sum_{i=1}^{n} \chi^i(\omega).$$

Theorem 2.3. *Let $p = P_X(B)$. Then*

$$P(\nu^n = k) = \frac{n!}{k!(n-k)!} p^k (1-p)^{n-k}, \quad k = 0, \ldots, n.$$

Proof. Fix a subset $I = \{i_1, \ldots, i_k\} \subseteq \{1, \ldots, n\}$ and consider the event that $\omega_i \in B$ if and only if $i \in I$ (if $k = 0$, then I is assumed to be the empty set). Then

$$P(\{\omega : \omega_i \in B \text{ for } i \in I; \omega_i \notin B \text{ for } i \notin I\})$$

$$= \prod_{i \in I} P_X(B) \prod_{i \notin I} P_X(X \setminus B) = p^k (1-p)^{n-k} .$$

Since such events do not intersect for different I, and the number of all such subsets I is $n!/k!(n-k)!$, the result follows. $\qquad\square$

The distribution in Theorem 2.3 is called the binomial distribution with parameter p.

Theorem 2.4. *The expectation and the variance of ν^n (and therefore also of any random variable with binomial distribution with parameter p) are*

$$E(\nu^n) = np,$$

$$\mathrm{Var}(\nu^n) = np(1-p).$$

Proof. Since $\nu^n = \sum_{i=1}^{n} \chi^i$,

$$E\nu^n = \sum_{i=1}^{n} E\chi^i = \sum_{i=1}^{n} P(\omega_i \in B) = \sum_{i=1}^{n} P_X(B) = np \ .$$

For the variance,

$$\mathrm{Var}(\nu^n) = E(\nu^n - np)^2 = E(\sum_{i=1}^{n}(\chi^i - p))^2 =$$

$$\sum_{i=1}^{n} E(\chi^i - p)^2 + 2\sum_{i<j} E(\chi^i - p)(\chi^j - p) \ .$$

Since $(\chi^i)^2 = \chi^i$,

$$\sum_{i=1}^{n} E(\chi^i - p)^2 = \sum_{i=1}^{n}(E\chi^i - 2pE\chi^i + p^2) = n(p - p^2) = np(1-p) \ .$$

For $i \neq j$,

$$E(\chi^i - p)(\chi^j - p) = E\chi^i\chi^j - pE\chi^i - pE\chi^j + p^2 = p^2 - p^2 - p^2 + p^2 = 0 \ .$$

This completes the proof of the theorem.   □

The next theorem is referred to as the Law of Large Numbers for a Homogeneous Sequence of Independent Trials.

Theorem 2.5. *For any $\varepsilon > 0$,*

$$P(|\frac{\nu^n}{n} - p| < \varepsilon) \to 1 \ \text{ as } \ n \to \infty.$$

Proof. By the Chebyshev Inequality,

$$P(|\frac{\nu^n}{n} - p| \geq \varepsilon) = P(|\nu^n - np| \geq n\varepsilon)$$

$$\leq \frac{\mathrm{Var}(\nu^n)}{n^2\varepsilon^2} = \frac{np(1-p)}{n^2\varepsilon^2} = \frac{p(1-p)}{n\varepsilon^2} \to 0 \ \text{ as } \ n \to \infty.$$

    □

The Law of Large Numbers states that for a homogeneous sequence of independent trials, typical realizations are such that the frequency with which an event B appears in ω is close to the probability of this event. Later we shall encounter many other statements of this type.

Let us discuss several applications of the Law of Large Numbers.

The Law of Large Numbers for Independent Homogeneous Trials with Finitely Many Outcomes

Let $X = \{x^1, \ldots, x^r\}$ be a finite set with a probability measure P_X. Let $p_j = P_X(x^j)$, $1 \leq j \leq n$. Then the Law of Large Numbers states that for each $1 \leq j \leq r$,

$$P(|\frac{\nu_j^n}{n} - p_j| < \varepsilon) \to 1 \quad \text{as} \quad n \to \infty,$$

where $\nu_j^n(\omega)$ is the number of occurrences of x^j in the sequence of n trials $\omega = (\omega_1, \ldots, \omega_n)$. Therefore,

$$P(|\frac{\nu_j^n}{n} - p_j| < \varepsilon \quad \text{for all} \quad 1 \leq j \leq r) \to 1 \quad \text{as} \quad n \to \infty.$$

Entropy of a Distribution and MacMillan Theorem

Let $X = \{x^1, \ldots, x^r\}$ be a finite set with a probability measure P_X. Let $p_j = P_X(x^j)$, $1 \leq j \leq r$. The entropy of P_X is defined as $H = -\sum_{j=1}^r p_j \ln p_j$. If $p_j = 0$, then the product $p_j \ln p_j$ is considered to be equal to zero. It is clear that $H \geq 0$, and that $H = 0$ if and only if $p_j = 1$ for some j. Consider the non-trivial case $H > 0$. The role of entropy is seen from the following theorem.

Theorem 2.6 (MacMillan Theorem). *For every $\varepsilon > 0$ and all sufficiently large n one can find a subset $\Omega_n \subseteq \Omega$ such that*

1. *$e^{n(H-\varepsilon)} \leq |\Omega_n| \leq e^{n(H+\varepsilon)}$.*
2. *$\lim_{n \to \infty} P(\Omega_n) = 1$.*
3. *For each $\omega \in \Omega_n$ we have $e^{-n(H+\varepsilon)} \leq p(\omega) \leq e^{-n(H-\varepsilon)}$.*

Proof. Take

$$\Omega_n = \{\omega : |\frac{\nu_j^n(\omega)}{n} - p_j| \leq \delta, \ 1 \leq j \leq r\},$$

where $\delta = \delta(\varepsilon)$ will be chosen later. It follows from the Law of Large Numbers that $P(\Omega_n) \to 1$ as $n \to \infty$, which yields the second statement of the theorem.

Assume that all $p_j > 0$ (otherwise we do not consider the corresponding indices at all). Then

$$p(\omega) = p_X(\omega_1) \ldots p_X(\omega_n) = p_1^{\nu_1^n(\omega)} \ldots p_r^{\nu_r^n(\omega)} = \exp(\sum_{j=1}^r \nu_j^n(\omega) \ln p_j)$$

$$= \exp(n \sum_{j=1}^r \frac{\nu_j^n(\omega)}{n} \ln p_j) = \exp(n \sum_{j=1}^r p_j \ln p_j) \exp(n \sum_{j=1}^r (\frac{\nu_j^n(\omega)}{n} - p_j) \ln p_j)$$

$$= \exp(n(-H + \sum_{j=1}^r (\frac{\nu_j^n(\omega)}{n} - p_j) \ln p_j)) \ .$$

If δ is small enough and $\omega \in \Omega_n$, then $|\sum_{j=1}^r (\nu_j^n(\omega)/n - p_j) \ln p_j| \leq \varepsilon$, which yields the third statement of the theorem.

In order to prove the first statement, we write

$$1 \geq P(\Omega_n) = \sum_{\omega \in \Omega_n} p(\omega) \geq e^{-n(H+\varepsilon)} |\Omega_n| .$$

Therefore, $|\Omega_n| \leq e^{n(H+\varepsilon)}$. On the other hand, for sufficiently large n

$$\frac{1}{2} \leq P(\Omega_n) \leq e^{-n(H-\frac{\varepsilon}{2})} |\Omega_n|,$$

and therefore $|\Omega_n| \geq \frac{1}{2} e^{n(H-\frac{\varepsilon}{2})} \geq e^{n(H-\varepsilon)}$ for sufficiently large n. $\square$

Probabilistic Proof of the Weierstrass Theorem

Theorem 2.7. *Let f be a continuous function on the closed interval $[0,1]$. For every $\varepsilon > 0$ there exists a polynomial $b_n(x)$ of degree n such that*

$$\max_{0 \leq x \leq 1} |b_n(x) - f(x)| \leq \varepsilon.$$

The proof of this theorem, which we present now, is due to S. Bernstein.

Proof. Consider the function

$$b_n(x) = \sum_{k=0}^n \frac{n!}{k!(n-k)!} x^k (1-x)^{n-k} f(\frac{k}{n}),$$

which is called the Bernstein polynomial of the function f. We shall prove that for all sufficiently large n this polynomial has the desired property. Let $\delta > 0$ be a positive number which will be chosen later. We have

$$|b_n(x) - f(x)| = |\sum_{k=0}^n \frac{n!}{k!(n-k)!} x^k (1-x)^{n-k} (f(\frac{k}{n}) - f(x))|$$

$$\leq \sum_{k:|\frac{k}{n}-x|<\delta} \frac{n!}{k!(n-k)!} x^k (1-x)^{n-k} |f(\frac{k}{n}) - f(x)|$$

$$+ \sum_{k:|\frac{k}{n}-x|\geq\delta} \frac{n!}{k!(n-k)!} x^k (1-x)^{n-k} |f(\frac{k}{n}) - f(x)| = I_1 + I_2 .$$

Since any continuous function on $[0,1]$ is uniformly continuous, we can take δ so small that $|f(\frac{k}{n}) - f(x)| \leq \frac{\varepsilon}{2}$ whenever $|\frac{k}{n} - x| < \delta$. Therefore, $I_1 \leq \frac{\varepsilon}{2}$ since

$$\sum_{k:|\frac{k}{n}-x|<\delta} \frac{n!}{k!(n-k)!} x^k (1-x)^{n-k} \leq 1.$$

Since any continuous function on $[0,1]$ is bounded, we can find a positive constant M such that $|f(x)| \le M$ for all $0 \le x \le 1$. Therefore,

$$I_2 \le 2M \sum_{k:|\frac{k}{n}-x|\ge\delta} \frac{n!}{k!(n-k)!}x^k(1-x)^{n-k}.$$

Note that the sum on the right-hand side of this inequality is equal to the following probability (with respect to the binomial distribution with the parameter x)

$$\sum_{k:|\frac{k}{n}-x|\ge\delta} \frac{n!}{k!(n-k)!}x^k(1-x)^{n-k} = P_x(|\frac{\nu^n}{n} - x| \ge \delta).$$

By the Chebyshev inequality,

$$P_x(|\frac{\nu^n}{n} - x| \ge \delta) \le \frac{nx(1-x)}{n^2\delta^2} = \frac{x(1-x)}{n\delta^2} \le \frac{\varepsilon}{4M}$$

if n is large enough. This implies that $I_2 \le \varepsilon/2$, which completes the proof of the theorem. $\square$

Bernoulli Trials and One-Dimensional Random Walks

A homogeneous sequence of independent trials is called a sequence of Bernoulli trials if X consists of two elements.

Let $X = \{-1, 1\}$. Define $\zeta_k(\omega) = \sum_{i=1}^{k} \omega_i$. By using linear interpolation we can construct a continuous function $\zeta_s(\omega)$ of the continuous variable s, $0 \le s \le n$, with the prescribed values at integer points, whose graph is a broken line with segments having slopes ± 1. The function ζ_s can be considered as a trajectory of a walker who moves with speed ± 1. The distribution on the space of all possible functions ζ_s induced by the probability distribution of the Bernoulli trials is called a simple random walk, and a function $\zeta_s(\omega)$ is called a trajectory of a simple random walk. If X is an arbitrary finite subset of real numbers, then the same construction gives an arbitrary random walk. Its trajectory consists of segments with slopes x^j, $1 \le j \le r$. We have

$$\frac{\zeta_n}{n} = \sum_{j=1}^{r} \frac{\nu_j^n}{n}x^j = \sum_{j=1}^{r} p_j x^j + \sum_{j=1}^{r}(\frac{\nu_j^n}{n} - p_j)x^j .$$

By the Law of Large Numbers

$$P(|\sum_{j=1}^{r}(\frac{\nu_j^n}{n} - p_j)x^j| \ge \varepsilon) \to 0 \quad \text{as } n \to \infty.$$

Therefore, the sum $\sum_{j=1}^{r} p_j x^j$ characterizes the mean velocity or the drift of the random walk. If $X = \{-1, 1\}$ and $p_{-1} = p_1 = \frac{1}{2}$, then the random walk is called simple symmetric. Its drift is equal to zero. Other properties of random walks will be discussed later.

Empirical Distribution Functions and Their Convergence

Consider a sequence of n independent homogeneous trials with elementary outcomes $\omega = (\omega_1, \ldots, \omega_n)$, where ω_i are real numbers. Let us assume that a continuous function $F(t)$ is the distribution function for each ω_i.

Given $\omega = (\omega_1, \ldots, \omega_n)$, consider the distribution function $F_\omega^n(t)$, which is a right-continuous step function with jumps of size $\frac{1}{n}$ at each of the points ω_i, that is

$$F_\omega^n(t) = \frac{\#\{i : \omega_i \le t\}}{n}.$$

Definition 2.8. *The distribution function $F_\omega^n(t)$ is called the empirical distribution function.*

There are many problems in mathematical statistics where it is needed to estimate $F(t)$ by means of the observed empirical distribution function. Such estimates are based on the following theorem.

Theorem 2.9 (Glivenko-Cantelli Theorem). *If $F(t)$ is continuous, then for any $\varepsilon > 0$*

$$\mathrm{P}(\sup_{t \in \mathbb{R}} |F^n(t) - F(t)| < \varepsilon) \to 1 \quad \text{as } n \to \infty.$$

Proof. For each t the value $F^n(t)$ is a random variable and $F_\omega^n(t) = \frac{k}{n}$ if $\#\{i : \omega_i \le t\} = k$. Therefore,

$$\mathrm{P}(F^n(t) = \frac{k}{n}) = \frac{n!}{k!(n-k)!}(F(t))^k(1 - F(t))^{n-k}.$$

By the Law of Large Numbers, for any $\varepsilon > 0$

$$\mathrm{P}(|F^n(t) - F(t)| < \varepsilon) \to 1 \quad \text{as } n \to \infty.$$

We still need to prove that the same statement holds for the supremum over t. Given $\varepsilon > 0$, find a finite sequence

$$-\infty = t_1 < t_2 < \ldots < t_r = \infty,$$

such that $F(t_{i+1}) - F(t_i) < \frac{\varepsilon}{2}$ for $1 \le i \le r - 1$. Such a sequence can be found since F is continuous. As was shown above,

$$\mathrm{P}(\sup_{0 \le i \le r} |F^n(t_i) - F(t_i)| < \frac{\varepsilon}{2}) \to 1 \quad \text{as } n \to \infty. \tag{2.1}$$

For $t \in [t_i, t_{i+1}]$

$$F^n(t) - F(t) \le F^n(t_{i+1}) - F(t_i) =$$

$$F^n(t_{i+1}) - F(t_{i+1}) + (F(t_{i+1}) - F(t_i)) \le F^n(t_{i+1}) - F(t_{i+1}) + \frac{\varepsilon}{2}$$

and, similarly,

$$F^n(t) - F(t) \geq F^n(t_i) - F(t_i) - \frac{\varepsilon}{2} .$$

Therefore,

$$\sup_{t \in \mathbb{R}} |F^n(t) - F(t)| \leq \sup_{1 \leq i \leq r} |F^n(t_i) - F(t_i)| + \frac{\varepsilon}{2}.$$

By (2.1),

$$P(\sup_{t \in \mathbb{R}} |F^n(t) - F(t)| < \varepsilon) \leq P(\sup_{0 \leq i \leq r} |F^n(t_i) - F(t_i)| < \frac{\varepsilon}{2}) \to 1 \text{ as } n \to \infty.$$

$\square$

2.2 de Moivre-Laplace Limit Theorem and Applications

Consider a random variable ν^n with binomial distribution

$$P_n(k) = \frac{n!}{k!(n-k)!} p^k (1-p)^{n-k} ,$$

and let n be large. The Chebyshev Inequality implies that with probability close to one this random variable takes values in a neighborhood of size $O(\sqrt{n})$ around the point np. For this reason it is natural to expect that when k belongs to this neighborhood, the probability $P_n(k)$ decays as $O(1/\sqrt{n})$, that is, the inverse of the size of the neighborhood. The de Moivre-Laplace Theorem gives a precise formulation of this statement.

Theorem 2.10 (de Moivre-Laplace Theorem). *Let* $0 \leq k \leq n$ *and*

$$z = z(n,k) = (k - np)/\sqrt{np(1-p)}.$$

Then

$$P_n(k) = \frac{1}{\sqrt{2\pi np(1-p)}} (e^{-\frac{1}{2}z^2} + \delta_n(k)) ,$$

where $\delta_n(k)$ *uniformly tends to zero as* $n \to \infty$.

Observe that it makes sense to apply the theorem to $k = k(n)$ such that $k(n) - np = O(\sqrt{n})$, since in this case the term $\exp(-\frac{1}{2}z^2)$ is bounded from below by a positive constant, while $\delta_n(k(n))$ tends to zero.

The theorem could be easily proved with the help of the Stirling formula. We shall, instead, obtain it later as a particular case of the Local Limit Theorem (see Sect. 10.2).

Consider the random variable $\eta^n = (\nu^n - np)/\sqrt{np(1-p)}$. We have $E\eta^n = 0$ and $\text{Var}(\eta^n) = 1$. The transition from ν^n to η^n is called the normalization of the random variable ν^n. It is clear that the possible values of η^n

constitute an arithmetic progression with the step $\Delta_n = 1/\sqrt{np(1-p)}$. Note that the de Moivre-Laplace Theorem can be re-formulated as follows:

$$P(\eta^n = z) = \frac{\Delta_n}{\sqrt{2\pi}}(e^{-\frac{z^2}{2}} + \delta_n(k))$$

for any z, which can be represented as $z = (k-np)/\sqrt{np(1-p)}$ for some integer $0 \le k \le n$.

It follows that for any $C_1 < C_2$

$$\lim_{n\to\infty} P(C_1 \le \frac{\nu^n - E\nu^n}{\sqrt{\text{Var}(\nu^n)}} \le C_2) = \lim_{n\to\infty} P(C_1 \le \eta^n \le C_2)$$

$$= \lim_{n\to\infty} \sum_{C_1 \le z \le C_2} P(\eta^n = z) = \lim_{n\to\infty} \sum_{C_1 \le z \le C_2} \frac{\Delta_n}{\sqrt{2\pi}}(e^{-\frac{z^2}{2}} + \delta_n(k))$$

$$= \frac{1}{\sqrt{2\pi}} \int_{C_1}^{C_2} e^{-\frac{z^2}{2}} dz,$$

where the last equality is due to the definition of an integral as the limit of Riemann sums.

As mentioned above, $p(z) = \frac{1}{\sqrt{2\pi}}e^{-\frac{z^2}{2}}$, $z \in \mathbb{R}$, is the Gaussian density. It appears in many problems of probability theory and mathematical statistics.

The above argument shows that the distribution of the normalized number of successes in a sequence of independent random trials is almost Gaussian. This is a particular case of a more general Central Limit Theorem, which will be studied in Chap. 10.

Let us consider two applications of the de Moivre-Laplace Theorem.

Simple Symmetric Random Walk

Let $\omega = (\omega_1, \omega_2, \ldots)$ be an infinite sequence of independent homogeneous trials. We assume that each ω_i takes values $+1$ and -1, with probability $\frac{1}{2}$. Then the sequence of random variables

$$\zeta_n = \omega_1 + \ldots + \omega_n = 2\nu_1^n - n$$

is a simple symmetric random walk (which will be considered in more detail in subsequent chapters). For now we note that by the de Moivre-Laplace Theorem

$$\lim_{n\to\infty} P(C_1 \le \frac{\zeta_n}{\sqrt{n}} \le C_2) = \lim_{n\to\infty} P(C_1 \le \frac{2\nu_1^n - n}{\sqrt{n}} \le C_2)$$

$$= \lim_{n\to\infty} P(C_1 \le \frac{\nu_1^n - n/2}{\sqrt{n/4}} \le C_2) = \frac{1}{\sqrt{2\pi}} \int_{C_1}^{C_2} e^{-\frac{z^2}{2}} dz,$$

since $E\nu_1^n = n/2$ and $\text{Var}(\nu_1^n) = n/4$. This calculation shows that typical displacements of the symmetric random walk grow as $\sqrt{n}$ and, when normalized by $\sqrt{n}$, have limiting Gaussian distribution.

Empirical Distribution Functions and Their Convergence

In Sect. 2.1 we demonstrated that if F is continuous, then the empirical distribution functions $F_\omega^n(t) = |\{\omega_i : \omega_i \leq t\}|/n$ converge to the distribution function $F(t)$. With each sequence of outcomes $\omega = (\omega_1, \ldots, \omega_n)$ and each t, we can associate a new sequence $\omega' = (\omega_1', \ldots, \omega_n')$, $\omega_i' = \chi_{(-\infty,t]}(\omega_i)$, where $\chi_{(-\infty,t]}$ is the indicator function. Thus ω_i' takes value 1 (success) with probability $F(t)$ and 0 (failure) with probability $1 - F(t)$. Note that $nF_\omega^n(t) = \nu'^n(\omega')$, where $\nu'^n(\omega')$ is the number of successes in the sequence ω'. We can now apply the de Moivre-Laplace Theorem (in the integral form) to the sequence of trials with this distribution to obtain

$$\lim_{n \to \infty} \mathrm{P}\left(\frac{C_1\sqrt{F(t)(1 - F(t))}}{\sqrt{n}} \leq F^n(t) - F(t) \leq \frac{C_2\sqrt{F(t)(1 - F(t))}}{\sqrt{n}}\right)$$

$$= \lim_{n \to \infty} \mathrm{P}\left(C_1 \leq \frac{\nu'^n - \mathrm{E}\nu'^n}{\sqrt{\mathrm{Var}(\nu'^n)}} \leq C_2\right) = \frac{1}{\sqrt{2\pi}}\int_{C_1}^{C_2} e^{-\frac{z^2}{2}}\, dz \ .$$

This shows that the empirical distribution function approximates the real distribution function with the accuracy of order $1/\sqrt{n}$.

2.3 Poisson Limit Theorem

Consider a sequence of n independent trials with $X = \{0, 1\}$. Unlike the previous section, now we shall assume that the probability of success $\mathrm{P}_X(1)$ depends on n. It will be denoted by p_n.

Theorem 2.11 (Poisson Limit Theorem). *If $\lim_{n \to \infty} np_n = \lambda > 0$, then the probability that the number of occurrences of 1 in a sequence of n trials is equal to k has the following limit*

$$\lim_{n \to \infty} \mathrm{P}(\nu^n = k) = \frac{\lambda^k}{k!}e^{-\lambda}, \ k = 0, 1, \ldots \ .$$

Note that the distribution on the right-hand side is the Poisson distribution with parameter λ.

Proof. We have

$$\mathrm{P}(\nu^n = k) = \frac{n!}{k!(n-k)!}p_n^k(1 - p_n)^{n-k}$$

$$= \frac{n(n-1)\ldots(n-k+1)}{k!}p_n^k\exp((n-k)\ln(1 - p_n)) \ .$$

Here k is fixed but $n \to \infty$. Therefore,

$$\lim_{n \to \infty}(n-k)\ln(1 - p_n) = -\lim_{n \to \infty}(n-k)p_n = -\lim_{n \to \infty}p_n n\left(1 - \frac{k}{n}\right) = -\lambda.$$

Furthermore,

$$\lim_{n\to\infty} n(n-1)\dots(n-k+1)p_n^k = \lim_{n\to\infty} (np_n)^k = \lambda^k.$$

Thus,

$$\lim_{n\to\infty} \mathrm{P}(\nu^n = k) = \frac{\lambda^k}{k!}e^{-\lambda}.$$

$\square$

The Poisson Limit Theorem has an important application in statistical mechanics. Consider the following model of an ideal gas with density ρ. Let V_L be a cube with side of length L. Let $n(L)$ be the number of non-interacting particles in the cube. Their positions will be denoted by $\omega_1, \dots, \omega_{n(L)}$. We assume that $n(L) \sim \rho L^3$ as $L \to \infty$, and that each ω_k is uniformly distributed in V_L (meaning that the probability of finding a given particle in a smooth domain $U \subseteq V_L$ is equal to $\mathrm{Vol}(U)/\mathrm{Vol}(V_L)$). Fix a domain $U \subset V_L$ (U will not depend on L), and introduce the random variable $\nu_U(\omega)$ equal to the number of particles in U, that is the number of those k with $\omega_k \in U$. The Poisson Limit Theorem implies that

$$\lim_{L\to\infty} \mathrm{P}(\nu_U = k) = \frac{\lambda^k}{k!}e^{-\lambda},$$

where $\lambda = \rho\mathrm{Vol}(U)$. Indeed, since $n(L) \sim \rho L^3$, and the probability of finding a given particle in U is equal to $p_{n(L)} = \mathrm{Vol}(U)/L^3$,

$$\lim_{L\to\infty} n(L)p_{n(L)} = \lim_{L\to\infty} n(L)\mathrm{Vol}(U)/L^3 = \rho\mathrm{Vol}(U).$$

2.4 Problems

1. Find the probability that there are exactly three heads after five tosses of a symmetric coin.
2. Andrew and Bob are playing a game of table tennis. The game ends when the first player reaches 11 points if the other player has 9 points or less. However, if at any time the score is 10:10, then the game continues till one of the players is 2 points ahead. The probability that Andrew wins any given point is 60% (it's independent of what happened before during the game). What is the probability that Andrew will go on to win the game if he is currently ahead 9:8.
3. Will you consider a coin asymmetric if after $1,000$ coin tosses the number of heads is equal to 600?
4. Let ε_n be a numeric sequence such that $\varepsilon_n\sqrt{n} \to +\infty$ as $n \to \infty$ Show that for a sequence of Bernoulli trials we have

$$\mathrm{P}(|\frac{\nu^n}{n} - p| < \varepsilon_n) \to 1 \quad \text{as} \quad n \to \infty.$$

5. Using the de Moivre-Laplace Theorem, estimate the probability that during 12,000 tosses of a die the number 6 appeared between 1,900 and 2,100 times.

6. Let Ω be the space of sequences $\omega = (\omega_1, \ldots, \omega_n)$, where $\omega_i \in [0,1]$. Let P_n be the probability distribution corresponding to the homogeneous sequence of independent trials, each ω_i having uniform distribution on $[0,1]$. Let $\eta_n = \min_{1 \le i \le n} \omega_i$. Find $P_n(\eta_n \le t)$ and $\lim_{n \to \infty} P_n(n\eta_n \le t)$.

7. Two candidates were running for a post. The voting machines recorded 520,000 votes for the first candidate and 480,000 votes for the second one. Afterwards it became apparent that the voting machines were defective—they randomly and independently switched each vote for the opposite one with probability of 45%. The losing candidate asked for a re-vote. Is there a basis for a re-vote?

8. Suppose that during 1 day the price of a certain stock either goes up by 3% with probability 1/2 or goes down by 3% with probability 1/2, and that outcomes on different days are independent. Approximate the probability that after 250 days the price of the stock will be at least as high as the current price.

Lebesgue Integral and Mathematical Expectation

3.1 Definition of the Lebesgue Integral

In this section we revisit the familiar notion of mathematical expectation, but now we define it for general (not necessarily discrete) random variables. The notion of expectation is identical to the notion of the Lebesgue integral.

Let $(\Omega, \mathcal{F}, \mu)$ be a measurable space with a finite measure. A measurable function is said to be simple if it takes a finite or countable number of values. The sum, product and quotient (when the denominator does not take the value zero) of two simple functions is again a simple function.

Theorem 3.1. *Any non-negative measurable function f is a monotone limit from below of non-negative simple functions, that is $f(\omega) = \lim_{n\to\infty} f_n(\omega)$ for every ω, where f_n are non-negative simple functions and $f_n(\omega) \leq f_{n+1}(\omega)$ for every ω. Moreover, if a function f is a limit of measurable functions for all ω, then f is measurable.*

Proof. Let f_n be defined by the relations

$$f_n(\omega) = k2^{-n} \text{ if } k2^{-n} \leq f(\omega) < (k+1)2^{-n}, \ k = 0, 1, \dots .$$

The sequence f_n satisfies the requirements of the theorem.

We now prove the second statement. Given a function f which is the limit of measurable functions f_n, consider the subsets $A \subseteq \mathbb{R}$ for which $f^{-1}(A) \in \mathcal{F}$. It is easy to see that these subsets forms a σ-algebra which we shall denote by $\mathcal{R}_f$. Let us prove that open intervals $A_t = (-\infty, t)$ belong to $\mathcal{R}_f$. Indeed it is easy to check the following relation

$$f^{-1}(A_t) = \bigcup_k \bigcup_m \bigcap_{n \geq m} \{\omega : f_n(\omega) < t - \tfrac{1}{k}\} .$$

Since f_n are measurable, the sets $\{\omega : f_n(\omega) < t - \tfrac{1}{k}\}$ belong to $\mathcal{F}$, and therefore $f^{-1}(A_t) \in \mathcal{F}$. Since the smallest σ-algebra which contains all A_t is the

L. Koralov and Y.G. Sinai, *Theory of Probability and Random Processes*,   37
Universitext, DOI 10.1007/978-3-540-68829-7_3,
© Springer-Verlag Berlin Heidelberg 2012

Borel σ-algebra on $\mathbb{R}$, $f^{-1}(A) \in \mathcal{F}$ for any Borel set A of the real line. This completes the proof of the theorem. $\qquad\square$

We now introduce the Lebesgue integral of a measurable function. When f is measurable and the measure is a probability measure, we refer to the integral as the expectation of the random variable, and denote it by $\mathrm{E}f$.

We start with the case of a simple function. Let f be a simple function taking non-negative values, which we denote by $a_1, a_2, \ldots$. Let us define the events $C_i = \{\omega : f(\omega) = a_i\}$.

Definition 3.2. *The sum of the series $\sum_{i=1}^{\infty} a_i \mu(C_i)$, provided that the series converges, is called the Lebesgue integral of the function f. It is denoted by $\int_\Omega f d\mu$. If the series diverges, then it is said that the integral is equal to plus infinity.*

It is clear that the sum of the series does not depend on the order of summation. The following lemma is clear.

Lemma 3.3. *The integral of a simple non-negative function has the following properties.*

1. *$\int_\Omega f d\mu \geq 0$.*
2. *$\int_\Omega \chi_\Omega d\mu = \mu(\Omega)$, where χ_Ω is the function identically equal to 1 on Ω.*
3. *$\int_\Omega (af_1 + bf_2) d\mu = a \int_\Omega f_1 d\mu + b \int_\Omega f_2 d\mu$ for any $a, b > 0$.*
4. *$\int_\Omega f_1 d\mu \geq \int_\Omega f_2 d\mu$ if $f_1 \geq f_2 \geq 0$.*

Now let f be an arbitrary measurable function taking non-negative values. We consider the sequence f_n of non-negative simple functions which converge monotonically to f from below. It follows from the fourth property of the Lebesgue integral that the sequence $\int_\Omega f_n d\mu$ is non-decreasing and there exists a limit $\lim_{n\to\infty} \int_\Omega f_n d\mu$, which is possibly infinite.

Theorem 3.4. *Let f and f_n be as above. Then the value of $\lim_{n\to\infty} \int_\Omega f_n d\mu$ does not depend on the choice of the approximating sequence.*

We first establish the following lemma.

Lemma 3.5. *Let $g \geq 0$ be a simple function such that $g \leq f$. Assume that $f = \lim_{n\to\infty} f_n$, where f_n are non-negative simple functions such that $f_{n+1} \geq f_n$. Then $\int_\Omega g d\mu \leq \lim_{n\to\infty} \int_\Omega f_n d\mu$.*

Proof. Take an arbitrary $\varepsilon > 0$ and set $C_n = \{\omega : f_n(\omega) - g(\omega) > -\varepsilon\}$. It follows from the monotonicity of f_n that $C_n \subseteq C_{n+1}$. Since $f_n \uparrow f$ and $f \geq g$, we have $\bigcup_n C_n = \Omega$. Therefore, $\mu(C_n) \to \mu(\Omega)$ as $n \to \infty$. Let χ_{C_n} be the indicator function of the set C_n. Then $g_n = g\chi_{C_n}$ is a simple function and $g_n \leq f_n + \varepsilon$. Therefore, by the monotonicity of $\int_\Omega f_n d\mu$,

$$\int_\Omega g_n d\mu \leq \int_\Omega f_n d\mu + \varepsilon, \quad \int_\Omega g_n d\mu \leq \lim_{m\to\infty} \int_\Omega f_m d\mu + \varepsilon.$$

Since ε is arbitrary, we obtain $\int_\Omega g_n d\mu \leq \lim_{m\to\infty} \int_\Omega f_m d\mu$. It remains to prove that $\lim_{n\to\infty} \int_\Omega g_n d\mu = \int_\Omega g d\mu$.

We denote by $b_1, b_2, \ldots$ the values of the function g, and by B_i the set where the value b_i is taken, $i = 1, 2, \ldots$. Then

$$\int_\Omega g d\mu = \sum_i b_i \mu(B_i), \quad \int_\Omega g_n d\mu = \sum_i b_i \mu(B_i \cap C_n).$$

It is clear that for all i we have $\lim_{n\to\infty} \mu(B_i \cap C_n) = \mu(B_i)$. Since the series above consists of non-negative terms and the convergence is monotonic for each i, we have

$$\lim_{n\to\infty} \int_\Omega g_n d\mu = \lim_{n\to\infty} \sum_i b_i \mu(B_i \cap C_n)$$

$$= \sum_i b_i \lim_{n\to\infty} \mu(B_i \cap C_n) = \sum_i b_i \mu(B_i) = \int_\Omega g d\mu.$$

This completes the proof of the lemma. $\qquad\qquad\qquad\qquad\qquad\square$

It is now easy to prove the independence of $\lim_{n\to\infty} \int_\Omega f_n d\mu$ from the choice of the approximating sequence.

Proof of Theorem 3.4. Let there be two sequences $f_n^{(1)}$ and $f_n^{(2)}$ such that $f_{n+1}^{(1)} \geq f_n^{(1)}$ and $f_{n+1}^{(2)} \geq f_n^{(2)}$ for all n, and

$$\lim_{n\to\infty} f_n^{(1)}(\omega) = \lim_{n\to\infty} f_n^{(2)}(\omega) = f(\omega) \quad \text{for every } \omega.$$

It follows from Lemma 3.5 that for any k,

$$\int_\Omega f_k^{(1)} d\mu \leq \lim_{n\to\infty} \int_\Omega f_n^{(2)} d\mu,$$

and therefore,

$$\lim_{n\to\infty} \int_\Omega f_n^{(1)} d\mu \leq \lim_{n\to\infty} \int_\Omega f_n^{(2)} d\mu.$$

We obtain

$$\lim_{n\to\infty} \int_\Omega f_n^{(1)} d\mu \geq \lim_{n\to\infty} \int_\Omega f_n^{(2)} d\mu$$

by interchanging $f_n^{(1)}$ and $f_n^{(2)}$. Therefore,

$$\lim_{n\to\infty} \int_\Omega f_n^{(1)} d\mu = \lim_{n\to\infty} \int_\Omega f_n^{(2)} d\mu.$$

$\qquad\qquad\qquad\qquad\qquad\qquad\qquad\qquad\qquad\qquad\qquad\qquad\qquad\square$

Definition 3.6. *Let f be a non-negative measurable function and f_n a sequence of non-negative simple functions which converge monotonically to f from below. The limit $\lim_{n\to\infty} \int_\Omega f_n d\mu$ is called the Lebesgue integral of the function f. It is denoted by $\int_\Omega f d\mu$.*

In the case of a simple function f, this definition agrees with the definition of the integral for a simple function, since we can take $f_n = f$ for all n.

Now let f be an arbitrary (not necessarily positive) measurable function. We introduce the indicator functions:

$$\chi_+(\omega) = \begin{cases} 1 & \text{if } f(\omega) \geq 0, \\ 0 & \text{if } f(\omega) < 0, \end{cases}$$

$$\chi_-(\omega) = \begin{cases} 1 & \text{if } f(\omega) < 0, \\ 0 & \text{if } f(\omega) \geq 0. \end{cases}$$

Then $\chi_+(\omega) + \chi_-(\omega) \equiv 1$, $f = f\chi_+ + f\chi_- = f_+ - f_-$, where $f_+ = f\chi_+$ and $f_- = -f\chi_-$. Moreover, $f_+ \geq 0$, $f_- \geq 0$, so the integrals $\int_\Omega f_+ d\mu$ and $\int_\Omega f_- d\mu$ have already been defined.

Definition 3.7. *The function f is said to be integrable if $\int_\Omega f_+ d\mu < \infty$ and $\int_\Omega f_- d\mu < \infty$. In this case the integral is equal to $\int_\Omega f d\mu = \int_\Omega f_+ d\mu - \int_\Omega f_- d\mu$. If $\int_\Omega f_+ d\mu = \infty$ and $\int_\Omega f_- d\mu < \infty$ ($\int_\Omega f_+ d\mu < \infty$, $\int_\Omega f_- d\mu = \infty$), then $\int_\Omega f d\mu = \infty$ ($\int_\Omega f d\mu = -\infty$). If $\int_\Omega f_+ d\mu = \int_\Omega f_- d\mu = \infty$, then $\int_\Omega f d\mu$ is not defined.*

Since $|f| = f_+ + f_-$, we have $\int_\Omega |f| d\mu = \int_\Omega f_+ d\mu + \int_\Omega f_- d\mu$, and so $\int_\Omega f d\mu$ is finite if and only if $\int_\Omega |f| d\mu$ is finite. The integral has Properties 2–4 listed in Lemma 3.3.

Let $A \in \mathcal{F}$ be a measurable set and f a measurable function on $(\Omega, \mathcal{F}, \mu)$. We can define the integral of f over the set A (which is a subset of Ω) in two equivalent ways. One way is to define

$$\int_A f d\mu = \int_\Omega f\chi_A d\mu,$$

where χ_A is the indicator function of the set A. Another way is to consider the restriction of μ from Ω to A. Namely, we consider the new σ-algebra $\mathcal{F}_A$, which contains all the measurable subsets of A, and the new measure μ_A on $\mathcal{F}_A$, which agrees with μ on all the sets from $\mathcal{F}_A$. Then $(A, \mathcal{F}_A)$ is a measurable space with a measure μ_A, and we can define

$$\int_A f d\mu = \int_A f d\mu_A.$$

It can easily be seen that the above two definitions lead to the same notion of the integral over a measurable set.

Let us note another important property of the Lebesgue integral: it is a σ-additive function on $\mathcal{F}$. Namely, let $A = \bigcup_{i=1}^\infty A_i$, where $A_1, A_2, \ldots$ are measurable sets such that $A_i \cap A_j = \emptyset$ for $i \neq j$. Let f be a measurable function such that $\int_A f d\mu$ is finite. Then

$$\int_A f d\mu = \sum_{i=1}^{\infty} \int_{A_i} f d\mu.$$

To justify this statement we can first consider f to be a non-negative simple function. Then the σ-additivity follows from the fact that in an infinite series with non-negative terms the terms can be re-arranged. For an arbitrary non-negative measurable f we use the definition of the integral as a limit of integrals of simple functions. For f which is not necessarily non-negative, we use Definition 3.7.

If f is a non-negative function, the σ-additivity of the integral implies that the function $\eta(A) = \int_A f d\mu$ is itself a measure.

The mathematical expectation (which is the same as the Lebesgue integral over a probability space) has all the properties described in Chap. 1. In particular

1. $E\xi \geq 0$ if $\xi \geq 0$.
2. $E\chi_\Omega = 1$ where χ_Ω is the random variable identically equal to 1 on Ω.
3. $E(a\xi_1 + b\xi_2) = aE\xi_1 + bE\xi_2$ if $E\xi_1$ and $E\xi_2$ are finite.

The variance of the random variable ξ is defined as $E(\xi - E\xi)^2$, and the n-th order moment is defined as $E\xi^n$. Given two random variables ξ_1 and ξ_2, their covariance is defined as $\mathrm{Cov}(\xi_1, \xi_2) = E(\xi_1 - E\xi_1)(\xi_2 - E\xi_2)$. The correlation coefficient of two random variables ξ_1, ξ_2 is defined as $\rho(\xi_1, \xi_2) = \mathrm{Cov}(\xi_1, \xi_2)/\sqrt{\mathrm{Var}\xi_1 \mathrm{Var}\xi_2}$.

3.2 Induced Measures and Distribution Functions

Given a probability space $(\Omega, \mathcal{F}, P)$, a measurable space $(\widetilde{\Omega}, \widetilde{\mathcal{F}})$ and a measurable function $f : \Omega \to \widetilde{\Omega}$, we can define the induced probability measure $\widetilde{P}$ on the σ-algebra $\widetilde{\mathcal{F}}$ via the formula

$$\widetilde{P}(A) = P(f^{-1}(A)) \quad \text{for } A \in \widetilde{\mathcal{F}}.$$

Clearly $\widetilde{P}(A)$ satisfies the definition of a probability measure. The following theorem states that the change of variable is permitted in the Lebesgue integral.

Theorem 3.8. *Let $g : \widetilde{\Omega} \to \mathbb{R}$ be a random variable. Then*

$$\int_\Omega g(f(\omega))dP(\omega) = \int_{\widetilde{\Omega}} g(\widetilde{w})d\widetilde{P}(\widetilde{w}) .$$

The integral on the right-hand side is defined if and only if the integral on the left-hand side is defined.

Proof. Without loss of generality we can assume that g is non-negative. When g is a simple function, the theorem follows from the definition of the induced measure. For an arbitrary measurable function it suffices to note that any such function is a limit of a non-decreasing sequence of simple functions. □

Let us examine once again the relationship between the random variables and their distribution functions. Consider the collection of all intervals:

$$\mathcal{I} = \{(a, b), [a, b), (a, b], [a, b], \text{ where } -\infty \le a \le b \le \infty\}.$$

Let $m : \mathcal{I} \to \mathbb{R}$ be a σ-additive nonnegative function, that is

1. $m(I) \ge 0$ for any $I \in \mathcal{I}$.
2. If $I, I_i \in \mathcal{I}$, $i = 1, 2, \ldots$, $I_i \cap I_j = \emptyset$ for $i \ne j$, and $I = \bigcup_{i=1}^{\infty} I_i$, then

$$m(I) = \sum_{i=1}^{\infty} m(I_i) \,.$$

Although m is σ-additive, as required of a measure, it is not truly a measure since it is defined on the collection of intervals, which is not a σ-algebra.

We shall need the following theorem (a particular case of the theorem on the extension of a measure discussed in Sect. 3.4).

Theorem 3.9. *Let m be a σ-additive function satisfying conditions 1 and 2. Then there is a unique measure μ defined on the σ-algebra of Borel sets of the real line, which agrees with m on all the intervals, that is $\mu(I) = m(I)$ for each $I \in \mathcal{I}$.*

Consider the following three examples, which illustrate how a measure can be constructed given its values on the intervals.

Example. Let $F(x)$ be a distribution function. We define

$$m((a, b]) = F(b) - F(a), \quad m([a, b]) = F(b) - \lim_{t \uparrow a} F(t),$$

$$m((a, b)) = \lim_{t \uparrow b} F(t) - F(a), \quad m([a, b)) = \lim_{t \uparrow b} F(t) - \lim_{t \uparrow a} F(t).$$

Let us check that m is a σ-additive function. Let I, I_i, $i = 1, 2, \ldots$ be intervals of the real line (open, half-open, or closed) such that $I = \bigcup_{i=1}^{\infty} I_i$ and $I_i \cap I_j = \emptyset$ if $i \ne j$. We need to check that

$$m(I) = \sum_{i=1}^{\infty} m(I_i). \tag{3.1}$$

It is clear that $m(I) \ge \sum_{i=1}^{n} m(I_i)$ for each n, since the intervals I_i do not intersect. Therefore, $m(I) \ge \sum_{i=1}^{\infty} m(I_i)$.

In order to prove the opposite inequality, we assume that an arbitrary $\varepsilon > 0$ is given. Consider a collection of intervals $J, J_i, i = 1, 2, \ldots$ which are constructed as follows. The interval J is a closed interval, which is contained in I and satisfies $m(J) \geq m(I) - \varepsilon/2$. (In particular, if I is closed we can take $J = I$). Let J_i be an open interval, which contains I_i and satisfies $m(J_i) \leq m(I_i) + \varepsilon/2^{i+1}$. The fact that it is possible to select such intervals J and J_i follows from the definition of the function m and the continuity from the right of the function F. Note that $J \subseteq \bigcup_{i=1}^{\infty} J_i$, J is compact, and all J_i are open. Therefore, $J \subseteq \bigcup_{i=1}^{n} J_i$ for some n. Clearly $m(J) \leq \sum_{i=1}^{n} m(J_i)$. Therefore, $m(I) \leq \sum_{i=1}^{n} m(I_i) + \varepsilon$. Since ε is arbitrary, we obtain $m(I) \leq \sum_{i=1}^{\infty} m(I_i)$. Therefore, (3.1) holds, and m is a σ-additive function.

Thus any distribution function gives rise to a probability measure on the Borel σ-algebra of the real line. This measure will be denoted by μ_F. Sometimes, instead of writing $d\mu_F$ in the integral with respect to such a measure, we shall write dF.

Conversely, any probability measure μ on the Borel sets of the real line defines a distribution function via the formula $F(x) = \mu((-\infty, x])$. Thus there is a one-to-one correspondence between probability measures on the real line and distribution functions.

Remark 3.10. Similarly, there is a one-to-one correspondence between the distribution functions on $\mathbb{R}^n$ and the probability measures on the Borel sets of $\mathbb{R}^n$. Namely, the distribution function F corresponding to a measure μ is defined by $F(x_1, \ldots, x_n) = \mu((-\infty, x_1] \times \ldots \times (-\infty, x_n])$.

Example. Let f be a function defined on an interval $[a, b]$ of the real line. Let $\sigma = \{t_0, t_1, \ldots, t_n\}$, with $a = t_0 \leq t_1 \leq \ldots \leq t_n = b$, be a partition of the interval $[a, b]$ into n subintervals. We denote the length of the largest interval by $\delta(\sigma) = \max_{1 \leq i \leq n}(t_i - t_{i-1})$. The p-th variation (with $p > 0$) of f over the partition σ is defined as

$$V_{[a,b]}^p(f, \sigma) = \sum_{i=1}^{n} |f(t_i) - f(t_{i-1})|^p.$$

Definition 3.11. *The following limit*

$$V_{[a,b]}^p(f) = \limsup_{\delta(\sigma) \to 0} V_{[a,b]}^p(f, \sigma),$$

is referred to as the p-th total variation of f over the interval $[a, b]$.

Now let f be a continuous function with finite first ($p = 1$) total variation defined on an interval $[a, b]$ of the real line. Then it can be represented as a difference of two continuous non-decreasing functions, namely,

$$f(x) = V_{[a,x]}^1(f) - (V_{[a,x]}^1(f) - f(x)) = F_1(x) - F_2(x).$$

Now we can repeat the construction used in the previous example to define the measures μ_{F_1} and μ_{F_2} on the Borel subsets of $[a, b]$. Namely, we can define

$$m_i((x, y]) = m_i([x, y]) = m_i((x, y)) = m_i([x, y)) = F_i(y) - F_i(x), \quad i = 1, 2,$$

and then extend m_i to the measure μ_{F_i} using Theorem 3.9. The difference $\mu_f = \mu_{F_1} - \mu_{F_2}$ is then a signed measure (see Sect. 3.6). If g is a Borel-measurable function on $[a, b]$, its integral with respect to the signed measure μ_f, denoted by $\int_a^b g(x) df(x)$ or $\int_a^b g(x) d\mu_f(x)$, is defined as the difference of the integrals with respect to the measures μ_{F_1} and μ_{F_2},

$$\int_a^b g(x) df(x) = \int_a^b g(x) d\mu_{F_1}(x) - \int_a^b g(x) d\mu_{F_2}(x).$$

It is called the Lebesgue-Stieltjes integral of g with respect to f.

Example. For an interval I, let $I_n = I \cap [-n, n]$. Define $m_n(I)$ as the length of I_n. As in the first example, m_n is a σ-additive function. Thus m_n gives rise to a measure on the Borel sets of the real line, which will be denoted by λ_n and referred to as the Lebesgue measure on the segment $[-n, n]$. Now for any Borel set A of the real line we can define its Lebesgue measure $\lambda(A)$ via $\lambda(A) = \lim_{n \to \infty} \lambda_n(A)$. It is easily checked that λ is a σ-additive measure which, however, may take infinite values for unbounded sets A.

Remark 3.12. The Lebesgue measure on the real line is an example of a σ-finite measure. We now give the formal definition of a σ-finite measure, although most of the measures that we deal with in this book are finite (probability) measures. An integral with respect to a σ-finite measure can be defined in the same way as an integral with respect to a finite measure.

Definition 3.13. *Let $(\Omega, \mathcal{F})$ be a measurable space. A σ-finite measure is a function μ, defined on $\mathcal{F}$ with values in $[0, \infty]$, which satisfies the following conditions.*

1. *There is a sequence of measurable sets $\Omega_1 \subseteq \Omega_2 \subseteq \ldots \subseteq \Omega$ such that $\mu(\Omega_i) < \infty$ for all i, and $\bigcup_{i=1}^\infty \Omega_i = \Omega$.*
2. *If $C_i \in \mathcal{F}$, $i = 1, 2, \ldots$ and $C_i \cap C_j = \emptyset$ for $i \neq j$, then*

$$\mu\left(\bigcup_{i=1}^\infty C_i\right) = \sum_{i=1}^\infty \mu(C_i) \ .$$

If F_ξ is the distribution function of a random variable ξ, then the measure μ_{F_ξ} (also denoted by μ_ξ) coincides with the measure induced by the random variable ξ. Indeed, the values of the induced measure and of μ_ξ coincide on the intervals, and therefore on all the Borel sets due to the uniqueness part of Theorem 3.9.

Theorem 3.8 together with the fact that μ_ξ coincides with the induced measure imply the following.

Theorem 3.14. *Let ξ be a random variable and g be a Borel measurable function on $\mathbb{R}$. Then*

$$Eg(\xi) = \int_{-\infty}^{\infty} g(x)dF_\xi(x).$$

Applying this theorem to the functions $g(x) = x$, $g(x) = x^p$ and $g(x) = (x - E\xi)^2$, we obtain the following.

Corollary 3.15.

$$E\xi = \int_{-\infty}^{\infty} xdF_\xi(x), \;\; E\xi^p = \int_{-\infty}^{\infty} x^p dF_\xi(x), \;\; \text{Var}\xi = \int_{-\infty}^{\infty} (x - E\xi)^2 dF_\xi(x) .$$

3.3 Types of Measures and Distribution Functions

Let μ be a finite measure on the Borel σ-algebra of the real line. We distinguish three special types of measures.

(a) *Discrete measure.* Assume that there exists a finite or countable set $A = \{a_1, a_2, \ldots\}$ such that $\mu((-\infty, \infty)) = \mu(A)$, that is A is a set of full measure. In this case μ is called a measure of discrete type.

(b) *Singular continuous measure.* Assume that the measure of any single point is zero, $\mu(a) = 0$ for any $a \in \mathbb{R}$, and there is a Borel set B of Lebesgue measure zero which is of full measure for the measure μ, that is $\lambda(B) = 0$ and $\mu((-\infty, \infty)) = \mu(B)$. In this case μ is called a singular continuous measure.

(c) *Absolutely continuous measure.* Assume that for every set of Lebesgue measure zero the μ measure of that set is also zero, that is $\lambda(A) = 0$ implies $\mu(A) = 0$. In this case μ is called an absolutely continuous measure.

While any given measure does not necessarily belong to one of the three classes above, the following theorem states that it can be decomposed into three components, one of which is discrete, the second singular continuous, and the third absolutely continuous.

Theorem 3.16. *Given any finite measure μ on $\mathbb{R}$ there exist measures μ_1, μ_2 and μ_3, the first of which is discrete, the second singular continuous and the third absolutely continuous, such that for any Borel set C of the real line we have*

$$\mu(C) = \mu_1(C) + \mu_2(C) + \mu_3(C) .$$

Such measures μ_1, μ_2 and μ_3 are determined by the measure μ uniquely.

Proof. Let A_1 be the collection of points a such that $\mu(a) \geq 1$, let A_2 be the collection of points $a \in \mathbb{R}\backslash A_1$ such that $\mu(a) \geq \frac{1}{2}$, let A_3 be the collection of points $a \in \mathbb{R}\backslash(A_1 \bigcup A_2)$ such that $\mu(a) \geq \frac{1}{3}$, and so on. Since the measure is finite, each set A_n contains only finitely many elements. Therefore, $A = \bigcup_n A_n$ is countable. At the same time $\mu(b) = 0$ for any $b \notin A$. Let $\mu_1(C) = \mu(C \bigcap A)$.

We shall now construct the measure μ_2 and a set B of zero Lebesgue measure, but of full μ_2 measure. (Note that it may turn out that $\mu_2(B) = 0$, that is μ_2 is identically zero.) First we inductively construct sets B_n, $n \geq 1$, as follows. Take B_1 to be an empty set. Assuming that B_n has been constructed, we take B_{n+1} to be any set of Lebesgue measure zero, which does not intersect $\bigcup_{i=1}^{n} B_i$ and satisfies

$$\mu(B_{n+1}) - \mu_1(B_{n+1}) \geq \frac{1}{m} \qquad (3.2)$$

with the smallest possible m, where $m \geq 1$ is an integer. If no such m exists, then we take B_{n+1} to be the empty set. For each m there is at most a finite number of non-intersecting sets which satisfy (3.2), and therefore the set $\mathbb{R} \setminus \bigcup_{n=1}^{\infty} B_n$ contains no set C for which $\mu(C) - \mu_1(C) > 0$. We put $B = \bigcup_{n=1}^{\infty} B_n$, which is a set of Lebesgue measure zero, and define $\mu_2(C) = \mu(C \cap B) - \mu_1(C \cap B)$. Note that $\mu_2(B) = \mu_2((-\infty, \infty))$, and therefore μ_2 is singular continuous.

By the construction of μ_1 and μ_2, we have that $\mu_3(C) = \mu(C) - \mu_1(C) - \mu_2(C)$ is a measure which is equal to zero on each set of Lebesgue measure zero. Thus we have the desired decomposition. The uniqueness part is left as an easy exercise for the reader.   □

Since there is a one-to-one correspondence between probability measures on the real line and distribution functions, we can single out the classes of distribution functions corresponding to the discrete, singular continuous and absolutely continuous measures. In the discrete case $F(x) = \mu((-\infty, x])$ is a step function. The jumps occur at the points a_i of positive μ-measure.

If the distribution function F has a Lebesgue integrable density p, that is $F(x) = \int_{-\infty}^{x} p(t)dt$, then F corresponds to an absolutely continuous measure. Indeed, $\mu_F(A) = \int_A p(t)dt$ for any Borel set A, since the equality is true for all intervals, and therefore it is true for all Borel sets due to the uniqueness of the extension of the measure. The value of the integral $\int_A p(t)dt$ over any set of Lebesgue measure zero is equal to zero.

The converse is also true, i.e., any absolutely continuous measure has a Lebesgue integrable density function. This follows from the Radon-Nikodym theorem, which we shall state below.

If a measure μ does not contain a discrete component, then the distribution function is continuous. Yet if the singular continuous component is present, it cannot be represented as an integral of a density. The so-called Cantor Staircase is an example of such a distribution function. Set $F(t) = 0$ for $t \leq 0$, $F(t) = 1$ for $t \geq 1$. We construct $F(t)$ for $0 < t < 1$ inductively. At the n-th step ($n \geq 0$) we have disjoint intervals of length 3^{-n}, where the function $F(t)$ is not yet defined, although it is defined at the end points of such intervals. Let us divide every such interval into three equal parts, and set $F(t)$ on the middle interval (including the end-points) to be a constant equal to the half-sum of its values at the above-mentioned end-points. It is easy to see that the function $F(t)$ can be extended by continuity to the remaining t. The limit function is called the Cantor Staircase. It corresponds to a singular continuous probability measure. The theory of fractals is related to some classes of singular continuous measures.

3.4 Remarks on the Construction of the Lebesgue Measure

In this section we provide an abstract generalization of Theorem 3.9 on the extension of a σ-additive function. Theorem 3.9 applies to the construction of a measure on the real line which, in the case of the Lebesgue measure, can be viewed as an extension of the notion of length of an interval. In fact we can define the notion of measure starting from a σ-additive function defined on a certain collection of subsets of an abstract set.

Definition 3.17. *A collection $\mathcal{G}$ of subsets of Ω is called a semialgebra if it has the following three properties:*

1. $\Omega \in \mathcal{G}$.
2. *If $C_1, C_2 \in \mathcal{G}$, then $C_1 \bigcap C_2 \in \mathcal{G}$.*
3. *If $C_1, C_2 \in \mathcal{G}$ and $C_2 \subseteq C_1$, then there exists a finite collection of disjoint sets $A_1, \ldots, A_n \in \mathcal{G}$ such that $C_2 \bigcap A_i = \emptyset$ for $i = 1, \ldots, n$ and $C_2 \bigcup A_1 \bigcup \ldots \bigcup A_n = C_1$.*

Definition 3.18. *A non-negative function with values in $\mathbb{R}$ defined on a semialgebra $\mathcal{G}$ is said to be σ-additive if it satisfies the following condition:*
If $C = \bigcup_{i=1}^{\infty} C_i$ with $C \in \mathcal{G}$, $C_i \in \mathcal{G}$, $i = 1, 2, \ldots$, and $C_i \bigcap C_j = \emptyset$ for $i \neq j$, then

$$m(C) = \sum_{i=1}^{\infty} m(C_i) \, .$$

Theorem 3.19 (Caratheodory). *Let m be a σ-additive function defined on a semialgebra $(\Omega, \mathcal{G})$. Then there exists a measure μ defined on $(\Omega, \sigma(\mathcal{G}))$ such that $\mu(C) = m(C)$ for every $C \in \mathcal{G}$. The measure μ which has this property is unique.*

We shall only indicate a sequence of steps used in the proof of the theorem, without giving all the details. A more detailed exposition can be found in the textbook of Fomin and Kolmogorov "Elements of Theory of Functions and Functional Analysis".

Step 1. Extension of the σ-additive function from the semialgebra to the algebra. Let $\mathcal{A}$ be the collection of sets which can be obtained as finite unions of disjoint elements of $\mathcal{G}$, that is $A \in \mathcal{A}$ if $A = \bigcup_{i=1}^{n} C_i$ for some $C_i \in \mathcal{G}$, where $C_i \cap C_j = \emptyset$ if $i \neq j$. The collection of sets $\mathcal{A}$ is an algebra since it contains the set Ω and is closed under finite unions, intersections, differences, and symmetric differences. For $A = \bigcup_{i=1}^{n} C_i$ with $C_i \cap C_j = \emptyset$, $i \neq j$, we define $m(A) = \sum_{i=1}^{n} m(C_i)$. We can then show that m is still a σ-additive function on the algebra $\mathcal{A}$.

Step 2. Definition of exterior measure and of measurable sets. For any set $B \subseteq \Omega$ we can define its exterior measure as $\mu^*(B) = \inf \sum_i m(A_i)$, where the infimum is taken over all countable coverings of B by elements of the algebra $\mathcal{A}$. A set B is called measurable if for any $\varepsilon > 0$ there is $A \in \mathcal{A}$ such that $\mu^*(A \triangle B) \leq \varepsilon$. Recall that $A \triangle B$ is the notation for the symmetric difference of the sets A and B. If B is measurable we define its measure to be equal to the exterior measure: $\mu(B) = \mu^*(B)$. Denote the collection of all measurable sets by $\mathcal{B}$.

Step 3. The σ-algebra of measurable sets and σ-additivity of the measure. The main part of the proof consists of demonstrating that $\mathcal{B}$ is a σ-algebra, and that the function μ defined on it has the properties of a measure. We can then restrict the measure to the smallest σ-algebra containing the original semialgebra. The uniqueness of the measure follows easily from the non-negativity of m and from the fact that the measure is uniquely defined on the algebra $\mathcal{A}$. Alternatively, see Lemma 4.14 in Chap. 4, which also implies the uniqueness of the measure.

Remark 3.20. It is often convenient to consider the measure μ on the measurable space $(\Omega, \mathcal{B})$, rather than to restrict the measure to the σ-algebra $\sigma(\mathcal{G})$, which is usually smaller than $\mathcal{B}$. The difference is that $(\Omega, \mathcal{B})$ is always complete with respect to measure μ, while $(\Omega, \sigma(\mathcal{G}))$ does not need to be complete. We discuss the notion of completeness in the remainder of this section.

Definition 3.21. *Let $(\Omega, \mathcal{F})$ be a measurable space with a finite measure μ on it. A set $A \subseteq \Omega$ is said to be μ-negligible if there is an event $B \in \mathcal{F}$ such that $A \subseteq B$ and $\mu(B) = 0$. The space $(\Omega, \mathcal{F})$ is said to be complete with respect to μ if all μ-negligible sets belong to $\mathcal{F}$.*

Given an arbitrary measurable space $(\Omega, \mathcal{F})$ with a finite measure μ on it, we can consider an extended σ-algebra $\widetilde{\mathcal{F}}$. It consists of all sets $\widetilde{B} \subseteq \Omega$ which can be represented as $\widetilde{B} = A \cup B$, where A is a μ-negligible set and $B \in \mathcal{F}$. We define $\widetilde{\mu}(\widetilde{B}) = \mu(B)$. It is easy to see that $\widetilde{\mu}(\widetilde{B})$ does not depend on the particular representation of $\widetilde{B}$, $(\Omega, \widetilde{\mathcal{F}})$ is a measurable space, $\widetilde{\mu}$ is a finite measure, and $(\Omega, \widetilde{\mathcal{F}})$ is complete with respect to $\widetilde{\mu}$. We shall refer to $(\Omega, \widetilde{\mathcal{F}})$ as the completion of $(\Omega, \mathcal{F})$ with respect to the measure μ.

It is not difficult to see that $\widetilde{\mathcal{F}} = \sigma(\mathcal{F} \cup \mathcal{N}^\mu)$, where $\mathcal{N}^\mu$ is the collection of μ-negligible sets in Ω.

3.5 Convergence of Functions, Their Integrals, and the Fubini Theorem

Let $(\Omega, \mathcal{F}, \mu)$ be a measurable space with a finite measure. Let f and f_n, $n = 1, 2, \ldots$ be measurable functions.

Definition 3.22. *A sequence of functions f_n is said to converge to f uniformly if*

$$\lim_{n \to \infty} \sup_{\omega \in \Omega} |f_n(\omega) - f(\omega)| = 0.$$

Definition 3.23. *A sequence of functions f_n is said to converge to f in measure (or in probability, if μ is a probability measure) if for any $\delta > 0$ we have*

$$\lim_{n \to \infty} \mu(\omega : |f_n(\omega) - f(\omega)| > \delta) = 0.$$

Definition 3.24. *A sequence of functions f_n is said to converge to f almost everywhere (or almost surely) if there is a measurable set A with $\mu(\Omega \setminus A) = 0$ such that*

$$\lim_{n \to \infty} f_n(\omega) = f(\omega) \quad \text{for} \quad \omega \in A.$$

It is not difficult to demonstrate that convergence almost everywhere implies convergence in measure. The opposite implication is only true if we consider a certain subsequence of the original sequence f_n (see Problem 8). The following theorem relates the notions of convergence almost everywhere and uniform convergence.

Theorem 3.25 (Egorov Theorem). *If a sequence of measurable functions f_n converges to a measurable function f almost everywhere, then for any $\delta > 0$ there exists a measurable set $\Omega_\delta \subseteq \Omega$ such that $\mu(\Omega_\delta) \geq \mu(\Omega) - \delta$ and f_n converges to f uniformly on Ω_δ.*

Proof. Let $\delta > 0$ be fixed. Let

$$\Omega_n^m = \bigcap_{i \geq n} \{\omega : |f_i(\omega) - f(\omega)| < \frac{1}{m}\}$$

and

$$\Omega^m = \bigcup_{n=1}^{\infty} \Omega_n^m.$$

Due to the continuity of the measure (Theorem 1.36), for every m there is $n_0(m)$ such that $\mu(\Omega^m \setminus \Omega_{n_0(m)}^m) < \delta/2^m$. We define $\Omega_\delta = \bigcap_{m=1}^{\infty} \Omega_{n_0(m)}^m$. We claim that Ω_δ satisfies the requirements of the theorem.

The uniform convergence follows from the fact that $|f_i(\omega) - f(\omega)| < 1/m$ for all $\omega \in \Omega_\delta$ if $i > n_0(m)$. In order to estimate the measure of Ω_δ, we note that $f_n(\omega)$ does not converge to $f(\omega)$ if ω is outside of the set Ω^m for some m. Therefore, $\mu(\Omega \backslash \Omega^m) = 0$. This implies

$$\mu(\Omega \backslash \Omega^m_{n_0(m)}) = \mu(\Omega^m \backslash \Omega^m_{n_0(m)}) < \frac{\delta}{2^m}.$$

Therefore,

$$\mu(\Omega \backslash \Omega_\delta) = \mu(\bigcup_{m=1}^{\infty} (\Omega \backslash \Omega^m_{n_0(m)})) \leq \sum_{m=1}^{\infty} \mu(\Omega \backslash \Omega^m_{n_0(m)}) < \sum_{m=1}^{\infty} \frac{\delta}{2^m} = \delta,$$

which completes the proof of the theorem. $\square$

The following theorem justifies passage to the limit under the sign of the integral.

Theorem 3.26 (Lebesgue Dominated Convergence Theorem). *If a sequence of measurable functions f_n converges to a measurable function f almost everywhere and*

$$|f_n| \leq \varphi,$$

where φ is integrable on Ω, then the function f is integrable on Ω and

$$\lim_{n \to \infty} \int_\Omega f_n d\mu = \int_\Omega f d\mu.$$

Proof. Let some $\varepsilon > 0$ be fixed. It is easily seen that $|f(\omega)| \leq \varphi(\omega)$ for almost all ω. Therefore, as follows from the elementary properties of the integral, the function f is integrable. Let $\Omega_k = \{\omega : k - 1 \leq \varphi(\omega) < k\}$. Since the integral is a σ-additive function,

$$\int_\Omega \varphi d\mu = \sum_{k=1}^{\infty} \int_{\Omega_k} \varphi d\mu.$$

Let $m > 0$ be such that $\sum_{k=m}^{\infty} \int_{\Omega_k} \varphi d\mu < \varepsilon/5$. Let $A = \bigcup_{k=m}^{\infty} \Omega_k$. By the Egorov Theorem, we can select a set $B \subseteq \Omega \backslash A$ such that $\mu(B) \leq \varepsilon/5m$ and f_n converges to f uniformly on the set $C = (\Omega \backslash A) \backslash B$. Finally,

$$|\int_\Omega f_n d\mu - \int_\Omega f d\mu| \leq |\int_A f_n d\mu - \int_A f d\mu|$$

$$+ |\int_B f_n d\mu - \int_B f d\mu| + |\int_C f_n d\mu - \int_C f d\mu|.$$

The first term on the right-hand side can be estimated from above by $2\varepsilon/5$, since $\int_A |f_n| d\mu, \int_A |f| d\mu \leq \int_A \varphi d\mu < \varepsilon/5$. The second term does not exceed

$\mu(B) \sup_{\omega \in B}(|f_n(\omega)| + |f(\omega)|) \leq 2\varepsilon/5$. The last term can be made smaller than $\varepsilon/5$ for sufficiently large n due to the uniform convergence of f_n to f on the set C. Therefore, $|\int_\Omega f_n d\mu - \int_\Omega f d\mu| \leq \varepsilon$ for sufficiently large n, which completes the proof of the theorem. $\qquad\square$

From the Lebesgue Dominated Convergence Theorem it is easy to derive the following two statements, which we provide here without proof.

Theorem 3.27 (Levi Monotonic Convergence Theorem). *Let a sequence of measurable functions be non-decreasing almost surely, that is*

$$f_1(\omega) \leq f_2(\omega) \leq \ldots \leq f_n(\omega) \leq \ldots$$

almost surely. Assume that the integrals are bounded:

$$\int_\Omega f_n d\mu \leq K \quad \text{for all } n.$$

Then, almost surely, there exists a finite limit

$$f(\omega) = \lim_{n \to \infty} f_n(\omega),$$

the function f is integrable, and $\int_\Omega f d\mu = \lim_{n \to \infty} \int_\Omega f_n d\mu$.

Lemma 3.28 (Fatou Lemma). *If f_n is a sequence of non-negative measurable functions, then*

$$\int_\Omega \liminf_{n \to \infty} f_n d\mu \leq \liminf_{n \to \infty} \int_\Omega f_n d\mu \leq \infty.$$

Let us discuss products of σ-algebras and measures. Let $(\Omega_1, \mathcal{F}_1, \mu_1)$ and $(\Omega_2, \mathcal{F}_2, \mu_2)$ be two measurable spaces with finite measures. We shall define the product space with the product measure $(\Omega, \mathcal{F}, \mu)$ as follows. The set Ω is just a set of ordered pairs $\Omega = \Omega_1 \times \Omega_2 = \{(\omega_1, \omega_2), \omega_1 \in \Omega_1, \omega_2 \in \Omega_2\}$.

In order to define the product σ-algebra, we first consider the collection of rectangles $\mathcal{R} = \{A \times B, A \in \mathcal{F}_1, B \in \mathcal{F}_2\}$. Then $\mathcal{F}$ is defined as the smallest σ-algebra containing all the elements of $\mathcal{R}$.

Note that $\mathcal{R}$ is a semialgebra. The product measure μ on $\mathcal{F}$ is defined to be the extension to the σ-algebra of the function m defined on $\mathcal{R}$ via $m(A \times B) = \mu_1(A)\mu_2(B)$. In order to justify this extension, we need to prove that m is a σ-additive function on $\mathcal{R}$.

Lemma 3.29. *The function $m(A \times B) = \mu_1(A)\mu_2(B)$ is a σ-additive function on the semialgebra $\mathcal{R}$.*

Proof. Let $A_1 \times B_1, A_2 \times B_2, \ldots$ be a sequence of non-intersecting rectangles such that $A \times B = \bigcup_{n=1}^\infty A_n \times B_n$. Consider the sequence of functions

$f_n(\omega_1) = \sum_{i=1}^{n} \chi_{A_i}(\omega_1)\mu_2(B_i)$, where χ_{A_i} is the indicator function of the set A_i. Similarly, let $f(\omega_1) = \chi_A(\omega_1)\mu_2(B)$. Note that $f_n \leq \mu_2(B)$ for all n and $\lim_{n\to\infty} f_n(\omega_1) = f(\omega_1)$. Therefore, the Lebesgue Dominated Convergence Theorem applies. We have

$$\lim_{n\to\infty} \sum_{i=1}^{n} m(A_i \times B_i) = \lim_{n\to\infty} \sum_{i=1}^{n} \mu_1(A_i)\mu_2(B_i) = \lim_{n\to\infty} \int_{\Omega_1} f_n(\omega_1)d\mu_1(\omega_1)$$

$$= \int_{\Omega_1} f(\omega_1)d\mu_1(\omega_1) = \mu_1(A)\mu_2(B) = m(A \times B).$$

$\square$

We are now in a position to state the Fubini Theorem. If $(\Omega, \mathcal{F}, \mu)$ is a measurable space with a finite measure, and f is defined on a set of full measure $A \in \mathcal{F}$, then $\int_{\Omega} f d\mu$ will mean $\int_A f d\mu$.

Theorem 3.30 (Fubini Theorem). *Let $(\Omega_1, \mathcal{F}_1, \mu_1)$ and $(\Omega_2, \mathcal{F}_2, \mu_2)$ be two measurable spaces with finite measures, and let $(\Omega, \mathcal{F}, \mu)$ be the product space with the product measure. If a function $f(\omega_1, \omega_2)$ is integrable with respect to the measure μ, then*

$$\int_{\Omega} f(\omega_1, \omega_2)d\mu(\omega_1, \omega_2)$$

$$= \int_{\Omega_1} (\int_{\Omega_2} f(\omega_1, \omega_2)d\mu_2(\omega_2))d\mu_1(\omega_1) \qquad (3.3)$$

$$= \int_{\Omega_2} (\int_{\Omega_1} f(\omega_1, \omega_2)d\mu_1(\omega_1))d\mu_2(\omega_2).$$

In particular, the integrals inside the brackets are finite almost surely and are integrable functions of the exterior variable.

Sketch of the Proof . The fact that the theorem holds if f is an indicator function of a set $A \times B$, where $A \in \mathcal{F}_1, B \in \mathcal{F}_2$, follows from the construction of the Lebesgue measure on the product space. The fact that the theorem also holds if f is an indicator function of a measurable set then easily follows from Lemma 4.13 proved in the next chapter. $\square$

Concerning f which is not necessarily an indicator function, without loss of generality we may assume that f is non-negative. If f is a simple integrable function with a finite number of values, we can represent it as a finite linear combination of indicator functions, and therefore the theorem holds for such functions. If f is any integrable function, we can approximate it by a monotonically non-decreasing sequence of simple integrable functions with finite number of values. Then from the Levi Convergence Theorem it follows that the repeated integrals are finite and are equal to the integral on the left-hand side of (3.3).

3.6 Signed Measures and the Radon-Nikodym Theorem

In this section we state, without proof, the Radon-Nikodym Theorem and the Hahn Decomposition Theorem. Both proofs can be found in the textbook of S. Fomin and A. Kolmogorov, "Elements of Theory of Functions and Functional Analysis".

Definition 3.31. *Let $(\Omega, \mathcal{F})$ be a measurable space. A function $\eta : \mathcal{F} \to \mathbb{R}$ is called a signed measure if*

$$\eta(\bigcup_{i=1}^{\infty} C_i) = \sum_{i=1}^{\infty} \eta(C_i)$$

whenever $C_i \in \mathcal{F}$, $i \geq 1$, are such that $C_i \cap C_j = \emptyset$ for $i \neq j$.

If μ is a non-negative measure on $(\Omega, \mathcal{F})$, then an example of a signed measure is provided by the integral of a function with respect to μ,

$$\eta(A) = \int_A f d\mu,$$

where $f \in L^1(\Omega, \mathcal{F}, \mu)$. Later, when we talk about conditional expectations, it will be important to consider the converse problem—given a measure μ and a signed measure η, we would like to represent η as an integral of some function with respect to measure μ. In fact this is always possible, provided $\mu(A) = 0$ for a set $A \in \mathcal{F}$ implies that $\eta(A) = 0$ (which is, of course, true if $\eta(A)$ is an integral of some function over the set A).

To make our discussion more precise we introduce the following definition.

Definition 3.32. *Let $(\Omega, \mathcal{F})$ be a measurable space with a finite non-negative measure μ. A signed measure $\eta : \mathcal{F} \to \mathbb{R}$ is called absolutely continuous with respect to μ if $\mu(A) = 0$ implies that $\eta(A) = 0$ for $A \in \mathcal{F}$.*

Remark 3.33. An equivalent definition of absolute continuity is as follows. A signed measure $\eta : \mathcal{F} \to \mathbb{R}$ is called absolutely continuous with respect to μ if for any $\varepsilon > 0$ there is a $\delta > 0$ such that $\mu(A) < \delta$ implies that $|\eta(A)| < \varepsilon$. (In Problem 10 the reader is asked to prove the equivalence of the definitions when η is a non-negative measure.)

Theorem 3.34 (Radon-Nikodym Theorem). *Let $(\Omega, \mathcal{F})$ be a measurable space with a finite non-negative measure μ, and η a signed measure absolutely continuous with respect to μ. Then there is an integrable function f such that*

$$\eta(A) = \int_A f d\mu$$

for all $A \in \mathcal{F}$. Any two functions which have this property can be different on at most a set of μ-measure zero.

The function f is called the density or the Radon-Nikodym derivative of η with respect to the measure μ.

The following theorem implies that signed measures are simply differences of two non-negative measures.

Theorem 3.35 (Hahn Decomposition Theorem). *Let $(\Omega, \mathcal{F})$ be a measurable space with a signed measure $\eta : \mathcal{F} \to \mathbb{R}$. Then there exist two sets $\Omega^+ \in \mathcal{F}$ and $\Omega^- \in \mathcal{F}$ such that*

1. $\Omega^+ \cup \Omega^- = \Omega$ and $\Omega^+ \cap \Omega^- = \emptyset$.
2. $\eta(A \cap \Omega^+) \geq 0$ for any $A \in \mathcal{F}$.
3. $\eta(A \cap \Omega^-) \leq 0$ for any $A \in \mathcal{F}$.

If $\widetilde{\Omega}^+, \widetilde{\Omega}^-$ is another pair of sets with the same properties, then $\eta(A) = 0$ for any $A \in \mathcal{F}$ such that $A \in \Omega^+ \triangle \widetilde{\Omega}^+$ or $A \in \Omega^- \triangle \widetilde{\Omega}^-$.

Consider two non-negative measures η^+ and η^- defined by

$$\eta^+(A) = \eta(A \cap \Omega^+) \quad \text{and} \quad \eta^-(A) = -\eta(A \cap \Omega^-).$$

These are called the positive part and the negative part of η, respectively. The measure $|\eta| = \eta^+ + \eta^-$ is called the total variation of η. It easily follows from the Hahn Decomposition Theorem that η^+, η^-, and $|\eta|$ do not depend on the particular choice of Ω^+ and Ω^-. Given a measurable function f which is integrable with respect to $|\eta|$, we can define

$$\int_\Omega f d\eta = \int_\Omega f d\eta^+ - \int_\Omega f d\eta^-.$$

3.7 L^p Spaces

Let $(\Omega, \mathcal{F}, \mu)$ be a space with a finite measure. We shall call two complex-valued measurable functions f and g equivalent $(f \sim g)$ if $\mu(f \neq g) = 0$. Note that $\sim$ is indeed an equivalence relationship, i.e.,

1. $f \sim f$.
2. $f \sim g$ implies that $g \sim f$.
3. $f \sim g$ and $g \sim h$ imply that $f \sim h$.

It follows from general set theory that the set of measurable functions can be viewed as a union of non-intersecting subsets, the elements of the same subset being all equivalent, and the elements which belong to different subsets not being equivalent.

We next introduce the $L^p(\Omega, \mathcal{F}, \mu)$ spaces, whose elements are some of the equivalence classes of measurable functions. We shall not distinguish between a measurable function and the equivalence class it represents.

For $1 \leq p < \infty$ we define

$$\|f\|_p = \left(\int_\Omega |f|^p d\mu \right)^{\frac{1}{p}}.$$

The set of functions (or rather the set of equivalence classes) for which $\|f\|_p$ is finite is denoted by $L^p(\Omega, \mathcal{F}, \mu)$ or simply L^p. It readily follows that L^p is a normed linear space, with the norm $\| \cdot \|_p$, that is

1. $\|f\|_p \geq 0$, $\|f\|_p = 0$ if and only if $f = 0$.
2. $\|\alpha f\|_p = |\alpha| \|f\|_p$ for any complex number α.
3. $\|f + g\|_p \leq \|f\|_p + \|g\|_p$.

It is also not difficult to see, and we leave it for the reader as an exercise, that all the L^p spaces are complete. We also formulate the Hölder Inequality, which states that if $f \in L^p$ and $g \in L^q$ with $p, q > 1$ such that $1/p + 1/q = 1$, then $fg \in L^1$ and

$$\|fg\|_1 \leq \|f\|_p \|g\|_q.$$

When $p = q = 2$ this is also referred to as the Cauchy-Bunyakovskii Inequality. Its proof is available in many textbooks, and thus we omit it, leaving it as an exercise for the reader.

The norm in the L^2 space comes from the inner product, $\|f\|_2 = (f, f)^{1/2}$, where

$$(f, g) = \int_\Omega f \bar{g} d\mu.$$

The set L^2 equipped with this inner product is a Hilbert space.

3.8 Monte Carlo Method

Consider a bounded measurable set $U \subset \mathbb{R}^d$ and a bounded measurable function $f : U \to \mathbb{R}$. In this section we shall discuss a numerical method for evaluating the integral $I(f) = \int_U f(x) dx_1 \ldots dx_d$.

One way to evaluate such an integral is based on approximating it by Riemann sums. Namely, the set U is split into measurable subsets $U_1, \ldots, U_n$ with small diameters, and a point x_i is selected in each of the subsets U_i. Then the sum $\sum_{i=1}^n f(x_i) \lambda(U_i)$, where $\lambda(U_i)$ is the measure of U_i, serves as an approximation to the integral. This method is effective provided that f does not change much for a small change of the argument (for example, if its gradient is bounded), and if we can split the set U into a reasonably small number of subsets with small diameters (so that n is not too large for a computer to handle the summation).

On the other hand, consider the case when U is a unit cube in $\mathbb{R}^d$, and d is large (say, $d = 20$). If we try to divide U into cubes U_i, each with the side of length $1/10$ (these may still be rather large, depending on the desired accuracy of the approximation), there will be $n = 10^{20}$ of such sub-cubes, which shows that approximating the integral by the Riemann sums cannot be effective in high dimensions.

Now we describe the Monte Carlo method of numerical integration. Consider a homogeneous sequence of independent trials $\omega = (\omega_1, \omega_2, \ldots)$, where each $\omega_i \in U$ has uniform distribution in U, that is $P(\omega_i \in V) = \lambda(V)/\lambda(U)$ for any measurable set $V \subseteq U$. If U is a unit cube, such a sequence can be implemented in practice with the help of a random number generator. Let

$$I^n(\omega) = \sum_{i=1}^{n} f(\omega_i).$$

We claim that I^n/n converges (in probability) to $I(f)/\lambda(U)$.

Theorem 3.36. *For every bounded measurable function f and every $\varepsilon > 0$*

$$\lim_{n \to \infty} P(|\frac{I^n}{n} - \frac{I(f)}{\lambda(U)}| < \varepsilon) = 1 .$$

Proof. Let $\varepsilon > 0$ be fixed, and assume that $|f(x)| \le M$ for all $x \in U$ and some constant M. We split the interval $[-M, M]$ into k disjoint sub-intervals $\Delta_1, \ldots, \Delta_k$, each of length not greater than $\varepsilon/3$. The number of such intervals should not need to exceed $1 + 6M/\varepsilon$. We define the sets U_j as the pre-images of Δ_j, that is $U_j = f^{-1}(\Delta_j)$. Let us fix a point a_j in each Δ_j. Let $\nu_j^n(\omega)$ be the number of those ω_i with $1 \le i \le n$, for which $\omega_i \in U_j$. Let $J^n(\omega) = \sum_{j=1}^{k} a_j \nu_j^n(\omega)$.

Since $f(x)$ does not vary by more than $\varepsilon/3$ on each of the sets U_j,

$$|\frac{I^n(\omega)}{n} - \frac{J^n(\omega)}{n}| \le \frac{\varepsilon}{3} \quad \text{and} \quad \frac{|I(f) - \sum_{j=1}^{k} a_j \lambda(U_j)|}{\lambda(U)} \le \frac{\varepsilon}{3}.$$

Therefore, it is sufficient to demonstrate that

$$\lim_{n \to \infty} P(|\frac{J^n}{n} - \frac{\sum_{j=1}^{k} a_j \lambda(U_j)}{\lambda(U)}| < \frac{\varepsilon}{3}) = 1 ,$$

or, equivalently,

$$\lim_{n \to \infty} P(|\sum_{j=1}^{k} a_j(\frac{\nu_j^n}{n} - \frac{\lambda(U_j)}{\lambda(U)})| < \frac{\varepsilon}{3}) = 1 .$$

This follows from the law of large numbers, which states that ν_j^n/n converges in probability to $\lambda(U_j)/\lambda(U)$ for each j.   □

Remark 3.37. Later we shall prove the so-called strong law of large numbers, which will imply the almost sure convergence of the approximations in the Monte Carlo method (see Chap. 7). It is important that the convergence rate (however it is defined) can be estimated in terms of $\lambda(U)$ and $\sup_{x \in U} |f(x)|$, independently of the dimension of the space and the smoothness of the function f.

3.9 Problems

1. Let f_n, $n \geq 1$, and f be measurable functions on a measurable space $(\Omega, \mathcal{F})$. Prove that the set $\{\omega : \lim_{n \to \infty} f_n(\omega) = f(\omega)\}$ is $\mathcal{F}$-measurable. Prove that the set $\{\omega : \lim_{n \to \infty} f_n(\omega) \text{ exists}\}$ is $\mathcal{F}$-measurable.

2. Prove that if a random variable ξ taking non-negative values is such that

$$P(\xi \geq n) \geq 1/n \quad \text{for all } n \in \mathbb{N},$$

then $\mathrm{E}\xi = \infty$.

3. Construct a sequence of random variables ξ_n such that $\xi_n(\omega) \to 0$ for every ω, but $\mathrm{E}\xi_n \to \infty$ as $n \to \infty$.

4. A random variable ξ takes values in the interval $[A, B]$ and $\mathrm{Var}(\xi) = ((B - A)/2)^2$. Find the distribution of ξ.

5. Let $\{x_1, x_2, \ldots\}$ be a collection of rational points from the interval $[0, 1]$. A random variable ξ takes the value x_n with probability $1/2^n$. Prove that the distribution function $F_\xi(x)$ of ξ is continuous at every irrational point x.

6. Let ξ be a random variable with a continuous density p_ξ such that $p_\xi(0) > 0$. Find the density of η, where

$$\eta(\omega) = \begin{cases} 1/\xi(\omega) & \text{if } \xi(\omega) \neq 0, \\ 0 & \text{if } \xi(\omega) = 0. \end{cases}$$

Prove that η does not have a finite expectation.

7. Let $\xi_1, \xi_2, \ldots$ be a sequence of random variables on a probability space $(\Omega, \mathcal{F}, \mathrm{P})$ such that $\mathrm{E}|\xi_n| \leq 2^{-n}$. Prove that $\xi_n \to 0$ almost surely as $n \to \infty$.

8. Prove that if a sequence of measurable functions f_n converges to f almost surely as $n \to \infty$, then it also converges to f in measure. If f_n converges to f in measure, then there is a subsequence f_{n_k} which converges to f almost surely as $k \to \infty$.

9. Let $F(x)$ be a distribution function. Compute $\int_{-\infty}^{\infty}(F(x+10) - F(x))dx$.

10. Prove that a measure η is absolutely continuous with respect to a measure μ if and only if for any $\varepsilon > 0$ there is a $\delta > 0$ such that $\mu(A) < \delta$ implies that $\eta(A) < \varepsilon$.

11. Prove that the $L^p([0, 1], \mathcal{B}, \lambda)$ spaces are complete for $1 \leq p < \infty$. Here $\mathcal{B}$ is the σ-algebra of Borel sets, and λ is the Lebesgue measure.

12. Prove the Hölder Inequality.

13. Let $\xi_1, \xi_2, \ldots$ be a sequence of random variables on a probability space $(\Omega, \mathcal{F}, \mathrm{P})$ such that $\mathrm{E}\xi_n^2 \leq c$ for some constant c. Assume that $\xi_n \to \xi$ almost surely as $n \to \infty$. Prove that $\mathrm{E}\xi$ is finite and $\mathrm{E}\xi_n \to \mathrm{E}\xi$.

4

Conditional Probabilities and Independence

4.1 Conditional Probabilities

Let $(\Omega, \mathcal{F}, \mathrm{P})$ be a probability space, and let $A, B \in \mathcal{F}$ be two events. We assume that $\mathrm{P}(B) > 0$.

Definition 4.1. *The conditional probability of A given B is*

$$P(A|B) = \frac{P(A \bigcap B)}{P(B)} .$$

While the conditional probability depends on both A and B, this dependence has a very different nature for the two sets. As a function of A the conditional probability has the usual properties of a probability measure:

1. $P(A|B) \geq 0$.
2. $P(\Omega|B) = 1$.
3. For a finite or infinite sequence of disjoint events A_i with $A = \bigcup_i A_i$ we have
$$P(A|B) = \sum_i P(A_i|B) .$$

As a function of B, the conditional probability satisfies the so-called formula of total probability. Let $\{B_1, B_2, \ldots\}$ be a finite or countable partition of the space Ω, that is $B_i \bigcap B_j = \emptyset$ for $i \neq j$ and $\bigcup_i B_i = \Omega$. We also assume that $P(B_i) > 0$ for every i. Take $A \in \mathcal{F}$. Then

$$P(A) = \sum_i P(A \bigcap B_i) = \sum_i P(A|B_i)P(B_i) \qquad (4.1)$$

is called the formula of total probability. This formula is reminiscent of multiple integrals written as iterated integrals. The conditional probability plays the role of the inner integral and the summation over i is the analog of the outer integral.

L. Koralov and Y.G. Sinai, *Theory of Probability and Random Processes*,
Universitext, DOI 10.1007/978-3-540-68829-7_4,
© Springer-Verlag Berlin Heidelberg 2012

In mathematical statistics the events B_i are sometimes called hypotheses, and probabilities $P(B_i)$ are called prior probabilities (i.e., given pre-experiment). We assume that as a result of the trial an event A occurred. We wish, on the basis of this, to draw conclusions regarding which of the hypotheses B_i is most likely. The estimation is done by calculating the probabilities $P(B_k|A)$ which are sometimes called posterior (post-experiment) probabilities. Thus

$$P(B_k|A) = \frac{P(B_k \bigcap A)}{P(A)} = \frac{P(A|B_k)P(B_k)}{\sum_i P(B_i)P(A|B_i)} \ .$$

This relation is called Bayes' formula.

4.2 Independence of Events, σ-Algebras, and Random Variables

Definition 4.2. *Two events A_1 and A_2 are called independent if*

$$P(A_1 \bigcap A_2) = P(A_1)P(A_2) \ .$$

The events $\emptyset$ and Ω are independent of any event.

Lemma 4.3. *If (A_1, A_2) is a pair of independent events, then $(\overline{A}_1, A_2)$, $(A_1, \overline{A}_2)$, and $(\overline{A}_1, \overline{A}_2)$, where $\overline{A}_j = \Omega \backslash A_j$, $j = 1, 2$, are also pairs of independent events.*

Proof. If A_1 and A_2 are independent, then

$$P(\overline{A}_1 \bigcap A_2) = P((\Omega \backslash A_1) \bigcap A_2)) =$$

$$P(A_2) - P(A_1 \bigcap A_2) = P(A_2) - P(A_1)P(A_2) = \qquad (4.2)$$

$$(1 - P(A_1))P(A_2) = P(\overline{A}_1)P(A_2).$$

Therefore, $\overline{A}_1$ and A_2 are independent. By interchanging A_1 and A_2 in the above argument, we obtain that A_1 and $\overline{A}_2$ are independent. Finally, $\overline{A}_1$ and $\overline{A}_2$ are independent since we can replace A_2 by $\overline{A}_2$ in (4.2). □

The notion of pair-wise independence introduced above is easily generalized to the notion of independence of any finite number of events.

Definition 4.4. *The events $A_1, \ldots, A_n$ are called independent if for any $1 \leq k \leq n$ and any $1 \leq i_1 < \ldots < i_k \leq n$*

$$P(A_{i_1} \bigcap \ldots \bigcap A_{i_k}) = P(A_{i_1}) \ldots P(A_{i_k}) \ .$$

For $n \geq 3$ the pair-wise independence of events A_i and A_j for all $1 \leq i < j \leq n$ does not imply that the events $A_1, \ldots, A_n$ are independent (see Problem 5).

Consider now a collection of σ-algebras $\mathcal{F}_1, \ldots, \mathcal{F}_n$, each of which is a σ-subalgebra of $\mathcal{F}$.

Definition 4.5. *The σ-algebras $\mathcal{F}_1, \ldots, \mathcal{F}_n$ are called independent if for any $A_1 \in \mathcal{F}_1, \ldots, A_n \in \mathcal{F}_n$ the events $A_1, \ldots, A_n$ are independent.*

Take a sequence of random variables $\xi_1, \ldots, \xi_n$. Each random variable ξ_i generates the σ-algebra $\mathcal{F}_i$, where the elements of $\mathcal{F}_i$ have the form $C = \{\omega : \xi_i(\omega) \in A\}$ for some Borel set $A \subseteq \mathbb{R}$. It is easy to check that the collection of such sets is indeed a σ-algebra, since the collection of Borel subsets of $\mathbb{R}$ is a σ-algebra.

Definition 4.6. *Random variables $\xi_1, \ldots, \xi_n$ are called independent if the σ-algebras $\mathcal{F}_1, \ldots, \mathcal{F}_n$ they generate are independent.*

Finally, we can generalize the notion of independence to arbitrary families of events, σ-algebras, and random variables.

Definition 4.7. *A family of events, σ-algebras, or random variables is called independent if any finite sub-family is independent.*

We shall now prove that the expectation of a product of independent random variables is equal to the product of expectations. The converse is, in general, not true (see Problem 6).

Theorem 4.8. *If ξ and η are independent random variables with finite expectations, then the expectation of the product is also finite and $E(\xi\eta) = E\xi E\eta$.*

Proof. Let ξ_1 and ξ_2 be the positive and negative parts, respectively, of the random variable ξ, as defined above. Similarly, let η_1 and η_2 be the positive and negative parts of η. It is sufficient to prove that $E(\xi_i \eta_j) = E\xi_i E\eta_j$, $i, j = 1, 2$. We shall prove that $E(\xi_1 \eta_1) = E\xi_1 E\eta_1$, the other cases being completely similar. Define $f_n(\omega)$ and $g_n(\omega)$ by the relations

$$f_n(\omega) = k2^{-n} \text{ if } k2^{-n} \leq \xi_1(\omega) < (k+1)2^{-n},$$

$$g_n(\omega) = k2^{-n} \text{ if } k2^{-n} \leq \eta_1(\omega) < (k+1)2^{-n}.$$

Thus f_n and g_n are two sequences of simple random variables which monotonically approximate from below the variables ξ_1 and η_1 respectively. Also, the sequence of simple random variables $f_n g_n$ monotonically approximates the random variable $\xi_1 \eta_1$ from below. Therefore,

$$E\xi_1 = \lim_{n \to \infty} Ef_n, \quad E\eta_1 = \lim_{n \to \infty} Eg_n, \quad E\xi_1\eta_1 = \lim_{n \to \infty} Ef_n g_n.$$

Since the limit of a product is the product of the limits, it remains to show that $Ef_n g_n = Ef_n Eg_n$ for any n, . Let A_k^n be the event $\{k2^{-n} \leq \xi_1 < (k+1)2^{-n}\}$ and B_k^n be the event $\{k2^{-n} \leq \eta_1 < (k+1)2^{-n}\}$. Note that for any k_1 and

k_2 the events $A_{k_1}^n$ and $B_{k_2}^n$ are independent due to the independence of the random variables ξ and η. We write

$$\mathrm{E}f_n g_n = \sum_{k_1,k_2} k_1 k_2 2^{-2n} \mathrm{P}(A_{k_1}^n \bigcap B_{k_2}^n) =$$

$$\sum_{k_1} k_1 2^{-n} \mathrm{P}(A_{k_1}^n) \sum_{k_2} k_2 2^{-n} \mathrm{P}(B_{k_2}^n) = \mathrm{E}f_n \mathrm{E}g_n ,$$

which completes the proof of the theorem.   □

Consider the space Ω corresponding to the homogeneous sequence of n independent trials, $\omega = (\omega_1, \ldots, \omega_n)$, and let $\xi_i(\omega) = \omega_i$.

Lemma 4.9. *The sequence $\xi_1, \ldots, \xi_n$ is a sequence of identically distributed independent random variables.*

Proof. Each random variable ξ_i takes values in a space X with a σ-algebra $\mathcal{G}$, and the probabilities of the events $\{\omega : \xi_i(\omega) \in A\}$, $A \in \mathcal{G}$, are equal to the probability of A in the space X. Thus they are the same for different i if A is fixed, which means that ξ_i are identically distributed. Their independence follows from the definition of the sequence of independent trials.   □

4.3 π-Systems and Independence

The following notions of a π-system and of a Dynkin system are very useful when proving independence of functions and σ-algebras.

Definition 4.10. *A collection $\mathcal{K}$ of subsets of Ω is said to be a π-system if it contains the empty set and is closed under the operation of taking the intersection of two sets, that is*

1. *$\emptyset \in \mathcal{K}$.*
2. *$A, B \in \mathcal{K}$ implies that $A \bigcap B \in \mathcal{K}$.*

Definition 4.11. *A collection $\mathcal{G}$ of subsets of Ω is called a Dynkin system if it contains Ω and is closed under the operations of taking complements and finite and countable non-intersecting unions, that is*

1. *$\Omega \in \mathcal{G}$.*
2. *$A \in \mathcal{G}$ implies that $\Omega \backslash A \in \mathcal{G}$.*
3. *$A_1, A_2, \ldots \in \mathcal{G}$ and $A_n \bigcap A_m = \emptyset$ for $n \neq m$ imply that $\bigcup_n A_n \in \mathcal{G}$.*

Note that an intersection of Dynkin systems is again a Dynkin system. Therefore, it makes sense to talk about the smallest Dynkin system containing a given collection of sets $\mathcal{K}$—namely, it is the intersection of all the Dynkin systems that contain all the elements of $\mathcal{K}$.

Lemma 4.12. *Let $\mathcal{K}$ be a π-system and let $\mathcal{G}$ be the smallest Dynkin system such that $\mathcal{K} \subseteq \mathcal{G}$. Then $\mathcal{G} = \sigma(\mathcal{K})$.*

Proof. Since $\sigma(\mathcal{K})$ is a Dynkin system, we obtain $\mathcal{G} \subseteq \sigma(\mathcal{K})$. In order to prove the opposite inclusion, we first note that if a π-system is a Dynkin system, then it is also a σ-algebra. Therefore, it is sufficient to show that $\mathcal{G}$ is a π-system. Let $A \in \mathcal{G}$ and define

$$\mathcal{G}_A = \{B \in \mathcal{G} : A \bigcap B \in \mathcal{G}\}.$$

The collection of sets $\mathcal{G}_A$ obviously satisfies the first and the third conditions of Definition 4.11. It also satisfies the second condition since if $A, B \in \mathcal{G}$ and $A \bigcap B \in \mathcal{G}$, then $A \bigcap (\Omega \backslash B) = \Omega \backslash [(A \bigcap B) \bigcup (\Omega \backslash A)] \in \mathcal{G}$. Moreover, if $A \in \mathcal{K}$, then $\mathcal{K} \subseteq \mathcal{G}_A$. Thus, for $A \in \mathcal{K}$ we have $\mathcal{G}_A = \mathcal{G}$, which implies that if $A \in \mathcal{K}$, $B \in \mathcal{G}$, then $A \bigcap B \in \mathcal{G}$. This implies that $\mathcal{K} \subseteq \mathcal{G}_B$ and therefore $\mathcal{G}_B = \mathcal{G}$ for any $B \in \mathcal{G}$. Thus $\mathcal{G}$ is a π-system. □

Lemma 4.12 can be re-formulated as follows.

Lemma 4.13. *If a Dynkin system $\mathcal{G}$ contains a π-system $\mathcal{K}$, then it also contains the σ-algebra generated by $\mathcal{K}$, that is $\sigma(\mathcal{K}) \subseteq \mathcal{G}$.*

Let us consider two useful applications of this lemma.

Lemma 4.14. *If P_1 and P_2 are two probability measures which coincide on all elements of a π-system $\mathcal{K}$, then they coincide on the minimal σ-algebra which contains $\mathcal{K}$.*

Proof. Let $\mathcal{G}$ be the collection of sets A such that $P_1(A) = P_2(A)$. Then $\mathcal{G}$ is a Dynkin system, which contains $\mathcal{K}$. Consequently, $\sigma(\mathcal{K}) \subseteq \mathcal{G}$. □

In order to discuss sequences of independent random variables and the laws of large numbers, we shall need the following statement.

Lemma 4.15. *Let $\xi_1, \ldots, \xi_n$ be independent random variables, $m_1 + \ldots + m_k = n$ and $f_1, \ldots, f_k$ be measurable functions of $m_1, \ldots, m_k$ variables respectively. Then the random variables $\eta_1 = f_1(\xi_1, \ldots, \xi_{m_1})$, $\eta_2 = f_2(\xi_{m_1+1}, \ldots, \xi_{m_1+m_2}), \ldots, \eta_k = f(\xi_{m_1+\ldots+m_{k-1}+1}, \ldots, \xi_n)$ are independent.*

Proof. We shall prove the lemma in the case $k = 2$ since the general case requires only trivial modifications. Consider the sets $A = A_1 \times \ldots \times A_{m_1}$ and $B = B_1 \times \ldots \times B_{m_2}$, where $A_1, \ldots, A_{m_1}, B_1, \ldots, B_{m_2}$ are Borel subsets of $\mathbb{R}$. We shall refer to such sets as rectangles. The collections of all rectangles in $\mathbb{R}^{m_1}$ and in $\mathbb{R}^{m_2}$ are π-systems. Note that by the assumptions of the lemma,

$$P((\xi_1, \ldots, \xi_{m_1}) \in A) P((\xi_{m_1+1}, \ldots, \xi_{m_1+m_2}) \in B) =$$

$$P((\xi_1, \ldots, \xi_{m_1}) \in A, (\xi_{m_1+1}, \ldots, \xi_{m_1+m_2}) \in B).$$

(4.3)

Fix a set $B = B_1 \times \ldots \times B_{m_2}$ and notice that the collection of all the measurable sets A that satisfy (4.3) is a Dynkin system containing all the rectangles in $\mathbb{R}^{m_1}$. Therefore, the relation (4.3) is valid for all sets A in the smallest σ-algebra containing all the rectangles, which is the Borel σ-algebra on $\mathbb{R}^{m_1}$. Now we can fix a Borel set A and, using the same arguments, demonstrate that (4.3) holds for any Borel set B.

It remains to apply (4.3) to $A = f_1^{-1}(\overline{A})$ and $B = f_2^{-1}(\overline{B})$, where $\overline{A}$ and $\overline{B}$ are arbitrary Borel subsets of $\mathbb{R}$. $\qquad\square$

4.4 Problems

1. Let P be the probability distribution of the sequence of n Bernoulli trials, $\omega = (\omega_1, \ldots, \omega_n)$, $\omega_i = 1$ or 0 with probabilities p and $1-p$. Find $P(\omega_1 = 1 | \omega_1 + \ldots + \omega_n = m)$.
2. Find the distribution function of a random variable ξ which takes positive values and satisfies $P(\xi > x + y | \xi > x) = P(\xi > y)$ for all $x, y > 0$.
3. Two coins are in a bag. One is symmetric, while the other is not—if tossed it lands heads up with probability equal to 0.6. One coin is randomly pulled out of the bag and tossed. It lands heads up. What is the probability that the same coin will land heads up if tossed again?
4. Suppose that each of the random variables ξ and η takes at most two values, a and b. Prove that ξ and η are independent if $E(\xi\eta) = E\xi E\eta$.
5. Give an example of three events A_1, A_2, and A_3 which are not independent, yet pair-wise independent.
6. Give an example of two random variables ξ and η which are not independent, yet $E(\xi\eta) = E\xi E\eta$.
7. A random variable ξ has Gaussian distribution with mean zero and variance one, while a random variable η has the distribution with the density

$$p_\eta(t) = \begin{cases} te^{-\frac{t^2}{2}} & \text{if } t \geq 0 \\ 0 & \text{otherwise.} \end{cases}$$

   Find the distribution of $\zeta = \xi \cdot \eta$ assuming that ξ and η are independent.
8. Let ξ_1 and ξ_2 be two independent random variables with Gaussian distribution with mean zero and variance one. Prove that $\eta_1 = \xi_1^2 + \xi_2^2$ and $\eta_2 = \xi_1/\xi_2$ are independent.
9. Two editors were independently proof-reading the same manuscript. One found a misprints, the other found b misprints. Of those, c misprints were found by both of them. How would you estimate the total number of misprints in the manuscript?
10. Let ξ, η be independent Poisson distributed random variables with expectations λ_1 and λ_2 respectively. Find the distribution of $\zeta = \xi + \eta$.

11. Let ξ, η be independent random variables. Assume that ξ has the uniform distribution on $[0, 1]$, and η has the Poisson distribution with parameter λ. Find the distribution of $\zeta = \xi + \eta$.

12. Let $\xi_1, \xi_2, \ldots$ be independent identically distributed Gaussian random variables with mean zero and variance one. Let $\eta_1, \eta_2, \ldots$ be independent identically distributed exponential random variables with mean one. Prove that there is $n > 0$ such that

$$P(\max(\eta_1, \ldots, \eta_n) \geq \max(\xi_1, \ldots, \xi_n)) > 0.99.$$

13. Suppose that $\mathcal{A}_1$ and $\mathcal{A}_2$ are independent algebras, that is any two sets $A_1 \in \mathcal{A}_1$ and $A_2 \in \mathcal{A}_2$ are independent. Prove that the σ-algebras $\sigma(\mathcal{A}_1)$ and $\sigma(\mathcal{A}_2)$ are also independent. (Hint: use Lemma 4.12.)

14. Let $\xi_1, \xi_2, \ldots$ be independent identically distributed random variables and N be an $\mathbb{N}$-valued random variable independent of ξ_i's. Show that if ξ_1 and N have finite expectation, then

$$E \sum_{i=1}^{N} \xi_i = E(N)E(\xi_1).$$

5

Markov Chains with a Finite Number of States

5.1 Stochastic Matrices

The theory of Markov chains makes use of stochastic matrices. We therefore begin with a small digression of an algebraic nature.

Definition 5.1. *An $r \times r$ matrix $Q = (q_{ij})$ is said to be stochastic if*

1. $q_{ij} \geq 0$.
2. $\sum_{j=1}^{r} q_{ij} = 1$ for any $1 \leq i \leq r$.

A column vector $f = (f_1, \ldots, f_r)$ is said to be non-negative if $f_i \geq 0$ for $1 \leq i \leq r$. In this case we write $f \geq 0$.

Lemma 5.2. *The following statements are equivalent.*

(a) The matrix Q is stochastic.
(b1) For any $f \geq 0$ we have $Qf \geq 0$, and
(b2) If $\mathbf{1} = (1, \ldots, 1)$ is a column vector, then $Q\mathbf{1} = \mathbf{1}$, that is the vector $\mathbf{1}$ is an eigenvector of the matrix Q corresponding to the eigenvalue 1.
(c) If $\mu = (\mu_1, \ldots, \mu_r)$ is a probability distribution, that is $\mu_i \geq 0$ and $\sum_{i=1}^{r} \mu_i = 1$, then μQ is also a probability distribution.

Proof. If Q is a stochastic matrix, then (b1) and (b2) hold, and therefore (a) implies (b). We now show that (b) implies (a). Consider the column vector δ_j all of whose entries are equal to zero, except the j-th entry which is equal to one. Then $(Q\delta_j)_i = q_{ij} \geq 0$. Furthermore, $(Q\mathbf{1})_i = \sum_{j=1}^{r} q_{ij}$, and it follows from the equality $Q\mathbf{1} = \mathbf{1}$ that $\sum_{j=1}^{r} q_{ij} = 1$ for all i, and therefore (b) implies (a).

We now show that (a) implies (c). If $\mu' = \mu Q$, then $\mu'_j = \sum_{i=1}^{r} \mu_i q_{ij}$. Since Q is stochastic, we have $\mu'_j \geq 0$ and

$$\sum_{j=1}^{r} \mu'_j = \sum_{j=1}^{r} \sum_{i=1}^{r} \mu_i q_{ij} = \sum_{i=1}^{r} \sum_{j=1}^{r} \mu_i q_{ij} = \sum_{i=1}^{r} \mu_i = 1.$$

L. Koralov and Y.G. Sinai, *Theory of Probability and Random Processes*,
Universitext, DOI 10.1007/978-3-540-68829-7_5,
© Springer-Verlag Berlin Heidelberg 2012

Therefore, μ' is also a probability distribution.

Now assume that (c) holds. Consider the row vector δ_i all of whose entries are equal to zero, except the i-th entry which is equal to one. It corresponds to the probability distribution on the set $\{1, \ldots, r\}$ which is concentrated at the point i. Then $\delta_i Q$ is also a probability distribution. If follows that $q_{ij} \geq 0$ and $\sum_{j=1}^{r} q_{ij} = 1$, that is (c) implies (a). $\qquad\square$

Lemma 5.3. *Let $Q' = (q'_{ij})$ and $Q'' = (q''_{ij})$ be stochastic matrices and $Q = Q'Q'' = (q_{ij})$. Then Q is also a stochastic matrix. If $q''_{ij} > 0$ for all i, j, then $q_{ij} > 0$ for all i, j.*

Proof. We have

$$q_{ij} = \sum_{k=1}^{r} q'_{ik} q''_{kj} \ .$$

Therefore, $q_{ij} \geq 0$. If all $q''_{kj} > 0$, then $q_{ij} > 0$ since $q'_{ik} \geq 0$ and $\sum_{k=1}^{r} q'_{ik} = 1$. Furthermore,

$$\sum_{j=1}^{r} q_{ij} = \sum_{j=1}^{r} \sum_{k=1}^{r} q'_{ik} q''_{kj} = \sum_{k=1}^{r} q'_{ik} \sum_{j=1}^{r} q''_{kj} = \sum_{k=1}^{r} q'_{ik} = 1 \ .$$

$\qquad\square$

Remark 5.4. We can also consider infinite matrices $Q = (q_{ij})$, $1 \leq i, j < \infty$. An infinite matrix is said to be stochastic if

1. $q_{ij} \geq 0$, and
2. $\sum_{j=1}^{\infty} q_{ij} = 1$ for any $1 \leq i < \infty$.

It is not difficult to show that Lemmas 5.2 and 5.3 remain valid for infinite matrices.

5.2 Markov Chains

We now return to the concepts of probability theory. Let Ω be the space of sequences $(\omega_0, \ldots, \omega_n)$, where $\omega_k \in X = \{x^1, \ldots, x^r\}$, $0 \leq k \leq n$. Without loss of generality we may identify X with the set of the first r integers, $X = \{1, \ldots, r\}$.

Let P be a probability measure on Ω. Sometimes we shall denote by ω_k the random variable which assigns the value of the k-th element to the sequence $\omega = (\omega_0, \ldots, \omega_n)$. It is usually clear from the context whether ω_k stands for such a random variable or simply the k-th element of a particular sequence. We shall denote the probability of the sequence $(\omega_0, \ldots, \omega_n)$ by $\mathrm{p}(\omega_0, \ldots, \omega_n)$. Thus,

$$\mathrm{p}(i_0, \ldots, i_n) = \mathrm{P}(\omega_0 = i_0, \ldots, \omega_n = i_n).$$

Assume that we are given a probability distribution $\mu = (\mu_1, \ldots, \mu_r)$ on X and n stochastic matrices $P(1), \ldots, P(n)$ with $P(k) = (p_{ij}(k))$.

Definition 5.5. *The Markov chain with the state space X generated by the initial distribution μ on X and the stochastic matrices $P(1), \ldots, P(n)$ is the probability measure* P *on Ω such that*

$$\mathrm{P}(\omega_0 = i_0, \ldots, \omega_n = i_n) = \mu_{i_0} \cdot p_{i_0 i_1}(1) \cdot \ldots \cdot p_{i_{n-1} i_n}(n) \qquad (5.1)$$

for each $i_0, \ldots, i_n \in X$.

The elements of X are called the states of the Markov chain. Let us check that (5.1) defines a probability measure on Ω. The inequality $\mathrm{P}(\omega_0 = i_0, \ldots, \omega_n = i_n) \geq 0$ is clear. It remains to show that

$$\sum_{i_0=1}^{r} \cdots \sum_{i_n=1}^{r} \mathrm{P}(\omega_0 = i_0, \ldots, \omega_n = i_n) = 1.$$

We have

$$\sum_{i_0=1}^{r} \cdots \sum_{i_n=1}^{r} \mathrm{P}(\omega_0 = i_0, \ldots, \omega_n = i_n)$$

$$= \sum_{i_0=1}^{r} \cdots \sum_{i_n=1}^{r} \mu_{i_0} \cdot p_{i_0, i_1}(1) \cdot \ldots \cdot p_{i_{n-1} i_n}(n) \ .$$

We now perform the summation over all the values of i_n. Note that i_n is only present in the last factor in each term of the sum, and the sum $\sum_{i_n=1}^{r} p_{i_{n-1} i_n}(n)$ is equal to one, since the matrix $P(n)$ is stochastic. We then fix $i_0, \ldots, i_{n-2}$, and sum over all the values of i_{n-1}, and so on. In the end we obtain $\sum_{i_0=1}^{r} \mu_{i_0}$, which is equal to one, since μ is a probability distribution.

In the same way one can prove the following statement:

$$\mathrm{P}(\omega_0 = i_0, \ldots, \omega_k = i_k) = \mu_{i_0} \cdot p_{i_0 i_1}(1) \cdot \ldots \cdot p_{i_{k-1} i_k}(k)$$

for any $1 \leq i_0, \ldots, i_k \leq r$, $k \leq n$. This equality shows that the induced probability distribution on the space of sequences of the form $(\omega_0, \ldots, \omega_k)$ is also a Markov chain generated by the initial distribution μ and the stochastic matrices $P(1), \ldots, P(k)$.

The matrices $P(k)$ are called the transition probability matrices, and the matrix entry $p_{ij}(k)$ is called the transition probability from the state i to the state j at time k. The use of these terms is justified by the following calculation.

Assume that $\mathrm{P}(\omega_0 = i_0, \ldots, \omega_{k-2} = i_{k-2}, \omega_{k-1} = i) > 0$. We consider the conditional probability $\mathrm{P}(\omega_k = j | \omega_0 = i_0, \ldots, \omega_{k-2} = i_{k-2}, \omega_{k-1} = i)$. By the definition of the measure P,

$$P(\omega_k = j | \omega_0 = i_0, \ldots, \omega_{k-2} = i_{k-2}, \omega_{k-1} = i)$$

$$= \frac{P(\omega_0 = i_0, \ldots, \omega_{k-2} = i_{k-2}, \omega_{k-1} = i, \omega_k = j)}{P(\omega_0 = i_0, \ldots, \omega_{k-2} = i_{k-2}, \omega_{k-1} = i)}$$

$$= \frac{\mu_{i_0} \cdot p_{i_0 i_1}(1) \cdot \ldots \cdot p_{i_{k-2}i}(k-1) \cdot p_{ij}(k)}{\mu_{i_0} \cdot p_{i_0 i_1}(1) \cdot \ldots \cdot p_{i_{k-2}i}(k-1)} = p_{ij}(k).$$

The right-hand side here does not depend on $i_0, \ldots, i_{k-2}$. This property is sometimes used as a definition of a Markov chain. It is also easy to see that $P(\omega_k = j | \omega_{k-1} = i) = p_{ij}(k)$. (This is proved below for the case of a homogeneous Markov chain.)

Definition 5.6. *A Markov chain is said to be homogeneous if $P(k) = P$ for a matrix P which does not depend on k, $1 \le k \le n$.*

The notion of a homogeneous Markov chain can be understood as a generalization of the notion of a sequence of independent identical trials. Indeed, if all the rows of the stochastic matrix $P = (p_{ij})$ are equal to $(p_1, \ldots, p_r)$, where $(p_1, \ldots, p_r)$ is a probability distribution on X, then the Markov chain with such a matrix P and the initial distribution $(p_1, \ldots, p_r)$ is a sequence of independent identical trials.

In what follows we consider only homogeneous Markov chains. Such chains can be represented with the help of graphs. The vertices of the graph are the elements of X. The vertices i and j are connected by an oriented edge if $p_{ij} > 0$. A sequence of states $(i_0, i_1, \ldots, i_n)$ which has a positive probability can be represented as a path of length n on the graph starting at the point i_0, then going to the point i_1, and so on. Therefore, a homogeneous Markov chain can be represented as a probability distribution on the space of paths of length n on the graph.

Let us consider the conditional probabilities $P(\omega_{s+l} = j | \omega_l = i)$. It is assumed here that $P(\omega_l = i) > 0$. We claim that

$$P(\omega_{s+l} = j | \omega_l = i) = p_{ij}^{(s)},$$

where $p_{ij}^{(s)}$ are elements of the matrix P^s. Indeed,

$$P(\omega_{s+l} = j | \omega_l = i) = \frac{P(\omega_{s+l} = j, \omega_l = i)}{P(\omega_l = i)}$$

$$= \frac{\sum_{i_0=1}^{r} \cdots \sum_{i_{l-1}=1}^{r} \sum_{i_{l+1}=1}^{r} \cdots \sum_{i_{s+l-1}=1}^{r} P(\omega_0 = i_0, \ldots, \omega_l = i, \ldots, \omega_{s+l} = j)}{\sum_{i_0=1}^{r} \cdots \sum_{i_{l-1}=1}^{r} P(\omega_0 = i_0, \ldots, \omega_l = i)}$$

$$= \frac{\sum_{i_0=1}^{r} \cdots \sum_{i_{l-1}=1}^{r} \sum_{i_{l+1}=1}^{r} \cdots \sum_{i_{s+l-1}=1}^{r} \mu_{i_0} p_{i_0 i_1} \cdots p_{i_{l-1}i} p_{i i_{l+1}} \cdots p_{i_{s+l-1}j}}{\sum_{i_0=1}^{r} \cdots \sum_{i_{l-1}=1}^{r} \mu_{i_0} p_{i_0 i_1} \cdots p_{i_{l-1}i}}$$

$$= \frac{\sum_{i_0=1}^{r} \cdots \sum_{i_{l-1}=1}^{r} \mu_{i_0} p_{i_0 i_1} \cdots p_{i_{l-1} i} \sum_{i_{l+1}=1}^{r} \cdots \sum_{i_{s+l-1}=1}^{r} p_{i i_{l+1}} \cdots p_{i_{s+l-1} j}}{\sum_{i_0=1}^{r} \cdots \sum_{i_{l-1}=1}^{r} \mu_{i_0} p_{i_0 i_1} \cdots p_{i_{l-1} i}}$$

$$= \sum_{i_{l+1}=1}^{r} \cdots \sum_{i_{s+l-1}=1}^{r} p_{i i_{l+1}} \cdots p_{i_{s+l-1} j} = p_{ij}^{(s)}.$$

Thus the conditional probabilities $p_{ij}^{(s)} = \mathrm{P}(\omega_{s+l} = j | \omega_l = i)$ do not depend on l. They are called s-step transition probabilities. A similar calculation shows that for a homogeneous Markov chain with initial distribution μ,

$$\mathrm{P}(\omega_s = j) = (\mu P^s)_j = \sum_{i=1}^{r} \mu_i p_{ij}^{(s)}. \tag{5.2}$$

Note that by considering infinite stochastic matrices, Definition 5.5 and the argument leading to (5.2) can be generalized to the case of Markov chains with a countable number of states.

5.3 Ergodic and Non-ergodic Markov Chains

Definition 5.7. *A stochastic matrix P is said to be ergodic if there exists s such that the s-step transition probabilities $p_{ij}^{(s)}$ are positive for all i and j. A homogeneous Markov chain is said to be ergodic if it can be generated by some initial distribution and an ergodic stochastic matrix.*

By (5.2), ergodicity implies that in s steps one can, with positive probability, proceed from any initial state i to any final state j.

It is easy to provide examples of non-ergodic Markov Chains. One could consider a collection of non-intersecting sets $X_1, \ldots, X_n$, and take $X = \bigcup_{k=1}^{n} X_k$. Suppose the transition probabilities p_{ij} are such that $p_{ij} = 0$, unless i and j belong to consecutive sets, that is $i \in X_k, j \in X_{k+1}$ or $i \in X_n, j \in X_1$. Then the matrix P is block diagonal, and any power of P will contain zeros, thus P will not be ergodic.

Another example of a non-ergodic Markov chain arises when a state j cannot be reached from any other state, that is $p_{ij} = 0$ for all $i \neq j$. Then the same will be true for the s-step transition probabilities.

Finally, there may be non-intersecting sets $X_1, \ldots, X_n$ such that $X = \bigcup_{k=1}^{n} X_k$, and the transition probabilities p_{ij} are such that $p_{ij} = 0$, unless i and j belong to the same set X_k. Then the matrix is not ergodic.

The general classification of Markov chains will be discussed in Sect. 5.6.

Definition 5.8. *A probability distribution π on X is said to be stationary (or invariant) for a matrix of transition probabilities P if $\pi P = \pi$.*

Formula (5.2) means that if the initial distribution π is a stationary distribution, then the probability distribution of any ω_k is given by the same vector π and does not depend on k. Hence the term "stationary".

Theorem 5.9 (Ergodic Theorem for Markov chains). *Given a Markov chain with an ergodic matrix of transition probabilities P, there exists a unique stationary probability distribution $\pi = (\pi_1, \ldots, \pi_r)$. The n-step transition probabilities converge to the distribution π, that is*

$$\lim_{n \to \infty} p_{ij}^{(n)} = \pi_j.$$

The stationary distribution satisfies $\pi_j > 0$ for $1 \leq j \leq r$.

Proof. Let $\mu' = (\mu'_1, \ldots, \mu'_r), \mu'' = (\mu''_1, \ldots, \mu''_r)$ be two probability distributions on the space X. We set $d(\mu', \mu'') = \frac{1}{2} \sum_{i=1}^{r} |\mu'_i - \mu''_i|$. Then d can be viewed as a distance on the space of probability distributions on X, and the space of distributions with this distance is a complete metric space. We note that

$$0 = \sum_{i=1}^{r} \mu'_i - \sum_{i=1}^{r} \mu''_i = \sum_{i=1}^{r}(\mu'_i - \mu''_i) = \sideset{}{^+}\sum (\mu'_i - \mu''_i) - \sideset{}{^+}\sum (\mu''_i - \mu'_i),$$

where $\sum^+$ denotes the summation with respect to those indices i for which the terms are positive. Therefore,

$$d(\mu', \mu'') = \frac{1}{2} \sum_{i=1}^{r} |\mu'_i - \mu''_i| = \frac{1}{2}\sideset{}{^+}\sum (\mu'_i - \mu''_i) + \frac{1}{2}\sideset{}{^+}\sum (\mu''_i - \mu'_i) = \sideset{}{^+}\sum (\mu'_i - \mu''_i).$$

It is also clear that $d(\mu', \mu'') \leq 1$.

Let μ' and μ'' be two probability distributions on X and $Q = (q_{ij})$ a stochastic matrix. By Lemma 5.2, $\mu'Q$ and $\mu''Q$ are also probability distributions. Let us demonstrate that

$$d(\mu'Q, \mu''Q) \leq d(\mu', \mu''), \tag{5.3}$$

and if all $q_{ij} \geq \alpha$, then

$$d(\mu'Q, \mu''Q) \leq (1 - \alpha)d(\mu', \mu''). \tag{5.4}$$

Let J be the set of indices j for which $(\mu'Q)_j - (\mu''Q)_j > 0$. Then

$$d(\mu'Q, \mu''Q) = \sum_{j \in J}(\mu'Q - \mu''Q)_j = \sum_{j \in J}\sum_{i=1}^{r}(\mu'_i - \mu''_i)q_{ij}$$

$$\leq \sideset{}{^+}\sum_i (\mu'_i - \mu''_i)\sum_{j \in J} q_{ij} \leq \sideset{}{^+}\sum_i (\mu'_i - \mu''_i) = d(\mu', \mu''),$$

which proves (5.3). We now note that J can not contain all the indices j since both $\mu'Q$ and $\mu''Q$ are probability distributions. Therefore, at least one index

j is missing in the sum $\sum_{j \in J} q_{ij}$. Thus, if all $q_{ij} > \alpha$, then $\sum_{j \in J} q_{ij} < 1 - \alpha$ for all i, and

$$d(\mu'Q, \mu''Q) \leq (1 - \alpha) \sum_i^+ (\mu'_i - \mu''_i) = (1 - \alpha)d(\mu', \mu''),$$

which implies (5.4).

Let μ_0 be an arbitrary probability distribution on X and $\mu_n = \mu_0 P^n$. We shall show that the sequence of probability distributions μ_n is a Cauchy sequence, that is for any $\varepsilon > 0$ there exists $n_0(\varepsilon)$ such that for any $k \geq 0$ we have $d(\mu_n, \mu_{n+k}) < \varepsilon$ for $n \geq n_0(\varepsilon)$. By (5.4),

$$d(\mu_n, \mu_{n+k}) = d(\mu_0 P^n, \mu_0 P^{n+k}) \leq (1 - \alpha)d(\mu_0 P^{n-s}, \mu_0 P^{n+k-s}) \leq \ldots$$

$$\leq (1 - \alpha)^m d(\mu_0 P^{n-ms}, \mu_0 P^{n+k-ms}) \leq (1 - \alpha)^m,$$

where m is such that $0 \leq n - ms < s$. For sufficiently large n we have $(1 - \alpha)^m < \varepsilon$, which implies that μ_n is a Cauchy sequence.

Let $\pi = \lim_{n \to \infty} \mu_n$. Then

$$\pi P = \lim_{n \to \infty} \mu_n P = \lim_{n \to \infty} (\mu_0 P^n)P = \lim_{n \to \infty} (\mu_0 P^{n+1}) = \pi.$$

Let us show that the distribution π, such that $\pi P = \pi$, is unique. Let π_1 and π_2 be two distributions with $\pi_1 = \pi_1 P$ and $\pi_2 = \pi_2 P$. Then $\pi_1 = \pi_1 P^s$ and $\pi_2 = \pi_2 P^s$. Therefore, $d(\pi_1, \pi_2) = d(\pi_1 P^s, \pi_2 P^s) \leq (1 - \alpha)d(\pi_1, \pi_2)$ by (5.4). It follows that $d(\pi_1, \pi_2) = 0$, that is $\pi_1 = \pi_2$.

We have proved that for any initial distribution μ_0 the limit

$$\lim_{n \to \infty} \mu_0 P^n = \pi$$

exists and does not depend on the choice of μ_0. Let us take μ_0 to be the probability distribution which is concentrated at the point i. Then, for i fixed, $\mu_0 P^n$ is the probability distribution $(p_{ij}^{(n)})$. Therefore, $\lim_{n \to \infty} p_{ij}^{(n)} = \pi_j$.

The proof of the fact that $\pi_j > 0$ for $1 \leq j \leq r$ is left as an easy exercise for the reader. $\qquad\square$

Remark 5.10. Let μ_0 be concentrated at the point i. Then

$$d(\mu_0 P^n, \pi) = d(\mu_0 P^n, \pi P^n) \leq \ldots \leq (1-\alpha)^m d(\mu_0 P^{n-ms}, \pi P^{n-ms}) \leq (1-\alpha)^m,$$

where m is such that $0 \leq n - ms < s$. Therefore,

$$d(\mu_0 P^n, \pi) \leq (1 - \alpha)^{\frac{n}{s}-1} \leq (1 - \alpha)^{-1}\beta^n,$$

where $\beta = (1 - \alpha)^{\frac{1}{s}} < 1$. In other words, the rate of convergence of $p_{ij}^{(n)}$ to the limit π_j is exponential.

Remark 5.11. The term ergodicity comes from statistical mechanics. In our case the ergodicity of a Markov chain implies that a certain loss of memory regarding initial conditions occurs, as the probability distribution at time n becomes nearly independent of the initial distribution as $n \to \infty$. We shall discuss further the meaning of this notion in Chap. 16.

5.4 Law of Large Numbers and the Entropy of a Markov Chain

As in the case of a homogeneous sequence of independent trials, we introduce the random variable $\nu_i^n(\omega)$ equal to the number of occurrences of the state i in the sequence $\omega = (\omega_0, \ldots, \omega_n)$, that is the number of those $0 \leq k \leq n$ for which $\omega_k = i$. We also introduce the random variables $\nu_{ij}^n(\omega)$ equal to the number of those $1 \leq k \leq n$ for which $\omega_{k-1} = i, \omega_k = j$.

Theorem 5.12. *Let π be the stationary distribution of an ergodic Markov chain. Then for any $\varepsilon > 0$*

$$\lim_{n \to \infty} P(|\frac{\nu_i^n}{n} - \pi_i| \geq \varepsilon) = 0, \quad \text{for } 1 \leq i \leq r,$$

$$\lim_{n \to \infty} P(|\frac{\nu_{ij}^n}{n} - \pi_i p_{ij}| \geq \varepsilon) = 0, \quad \text{for } 1 \leq i, j \leq r.$$

Proof. Let

$$\chi_i^k(\omega) = \begin{cases} 1 & \text{if } \omega_k = i, \\ 0 & \text{if } \omega_k \neq i, \end{cases}$$

$$\chi_{ij}^k(\omega) = \begin{cases} 1 & \text{if } \omega_{k-1} = i, \, \omega_k = j, \\ 0 & \text{otherwise}, \end{cases}$$

so that

$$\nu_i^n = \sum_{k=0}^n \chi_i^k, \quad \nu_{ij}^n = \sum_{k=1}^n \chi_{ij}^k.$$

For an initial distribution μ

$$E\chi_i^k = \sum_{m=1}^r \mu_m p_{mi}^{(k)}, \quad E\chi_{ij}^k = \sum_{m=1}^r \mu_m p_{mi}^{(k)} p_{ij}.$$

As $k \to \infty$ we have $p_{mi}^{(k)} \to \pi_i$ exponentially fast. Therefore, as $k \to \infty$,

$$E\chi_i^k \to \pi_i, \quad E\chi_{ij}^k \to \pi_i p_{ij}$$

exponentially fast. Consequently

$$E\frac{\nu_i^n}{n} = E\frac{\sum_{k=0}^n \chi_i^k}{n} \to \pi_i, \quad E\frac{\nu_{ij}^n}{n} = E\frac{\sum_{k=1}^n \chi_{ij}^k}{n} \to \pi_i p_{ij}.$$

For sufficiently large n

$$\{\omega : |\frac{\nu_i^n(\omega)}{n} - \pi_i| \geq \varepsilon\} \subseteq \{\omega : |\frac{\nu_i^n(\omega)}{n} - \frac{1}{n}E\nu_i^n| \geq \frac{\varepsilon}{2}\},$$

$$\{\omega : |\frac{\nu_{ij}^n(\omega)}{n} - \pi_i p_{ij}| \geq \varepsilon\} \subseteq \{\omega : |\frac{\nu_{ij}^n(\omega)}{n} - \frac{1}{n}E\nu_{ij}^n| \geq \frac{\varepsilon}{2}\}.$$

The probabilities of the events on the right-hand side can be estimated using the Chebyshev Inequality:

$$P(|\frac{\nu_i^n}{n} - \frac{1}{n}E\nu_i^n| \geq \frac{\varepsilon}{2}) = P(|\nu_i^n - E\nu_i^n| \geq \frac{\varepsilon n}{2}) \leq \frac{4\text{Var}(\nu_i^n)}{\varepsilon^2 n^2},$$

$$P(|\frac{\nu_{ij}^n}{n} - \frac{1}{n}E\nu_{ij}^n| \geq \frac{\varepsilon}{2}) = P(|\nu_{ij}^n - E\nu_{ij}^n| \geq \frac{\varepsilon n}{2}) \leq \frac{4\text{Var}(\nu_{ij}^n)}{\varepsilon^2 n^2}.$$

Thus the matter is reduced to estimating $\text{Var}(\nu_i^n)$ and $\text{Var}(\nu_{ij}^n)$. If we set $m_i^k = E\chi_i^k = \sum_{s=1}^{r} \mu_s p_{si}^{(k)}$, then

$$\text{Var}(\nu_i^n) = E(\sum_{k=0}^{n}(\chi_i^k - m_i^k))^2 =$$

$$E\sum_{k=0}^{n}(\chi_i^k - m_i^k)^2 + 2\sum_{k_1 < k_2} E(\chi_i^{k_1} - m_i^{k_1})(\chi_i^{k_2} - m_i^{k_2}).$$

Since $0 \leq \chi_i^k \leq 1$, we have $-1 \leq \chi_i^k - m_i^k \leq 1$, $(\chi_i^k - m_i^k)^2 \leq 1$ and $\sum_{k=0}^{n} E(\chi_i^k - m_i^k)^2 \leq n + 1$. Furthermore,

$$E(\chi_i^{k_1} - m_i^{k_1})(\chi_i^{k_2} - m_i^{k_2}) = E\chi_i^{k_1}\chi_i^{k_2} - m_i^{k_1}m_i^{k_2} =$$

$$\sum_{s=1}^{r} \mu_s p_{si}^{(k_1)} p_{ii}^{(k_2-k_1)} - m_i^{k_1}m_i^{k_2} = R_{k_1,k_2}.$$

By the Ergodic Theorem (see Remark 5.10),

$$m_i^k = \pi_i + d_i^k, \quad |d_i^k| \leq c\lambda^k,$$

$$p_{si}^{(k)} = \pi_i + \beta_{s,i}^k, \quad |\beta_{s,i}^k| \leq c\lambda^k,$$

for some constants $c < \infty$ and $\lambda < 1$. This gives

$$|R_{k_1,k_2}| = |\sum_{s=1}^{r} \mu_s(\pi_i + \beta_{s,i}^{k_1})(\pi_i + \beta_{i,i}^{k_2-k_1}) - (\pi_i + d_i^{k_1})(\pi_i + d_i^{k_2})| \leq$$

$$c_1(\lambda^{k_1} + \lambda^{k_2} + \lambda^{k_2-k_1})$$

for some constant $c_1 < \infty$. Therefore, $\sum_{k_1 < k_2} R_{k_1,k_2} \leq c_2 n$, and consequently $\text{Var}(\nu_i^n) \leq c_3 n$ for some constants c_2 and c_3. The variance $\text{Var}(\nu_{ij}^n)$ can be estimated in the same way. □

We now draw a conclusion from this theorem about the entropy of a Markov chain. In the case of a homogeneous sequence of independent trials, for large n the entropy is approximately equal to $-\frac{1}{n}\ln p(\omega)$ for typical ω, that is for ω which constitute a set whose probability is arbitrarily close to one.

In order to use this property to derive a general definition of entropy, we need to study the behavior of $\ln \mathrm{p}(\omega)$ for typical ω in the case of a Markov chain. For $\omega = (\omega_0, \ldots, \omega_n)$ we have

$$\mathrm{p}(\omega) = \mu_{\omega_0} \prod_{i,j} p_{ij}^{\nu_{ij}^n(\omega)} = \exp\left(\ln \mu_{\omega_0} + \sum_{i,j} \nu_{ij}^n(\omega) \ln p_{ij}\right),$$

$$\ln \mathrm{p}(\omega) = \ln \mu_{\omega_0} + \sum_{i,j} \nu_{ij}^n(\omega) \ln p_{ij}.$$

From the Law of Large Numbers, for typical ω

$$\frac{\nu_{ij}^n(\omega)}{n} \sim \pi_i p_{ij}.$$

Therefore, for such ω

$$-\frac{1}{n}\ln \mathrm{p}(\omega) = -\frac{1}{n}\ln \mu_{\omega_0} - \sum_{i,j} \nu_{ij}^n(\omega) \ln p_{ij} \sim -\sum_{i,j} \pi_i p_{ij} \ln p_{ij}.$$

Thus it is natural to define the entropy of a Markov chain to be

$$h = -\sum_i \pi_i \sum_j p_{ij} \ln p_{ij}.$$

It is not difficult to show that with such a definition of h, the MacMillan Theorem remains true.

5.5 Products of Positive Matrices

Let $A = (a_{ij})$ be a matrix with positive entries, $1 \le i, j \le r$. Let $A^* = (a_{ij}^*)$ be the transposed matrix, that is $a_{ij}^* = a_{ji}$. Let us denote the entries of A^n by $a_{ij}^{(n)}$. We shall use the Ergodic Theorem for Markov chains in order to study the asymptotic behavior of $a_{ij}^{(n)}$ as $n \to \infty$. First, we prove the following:

Theorem 5.13 (Perron-Frobenius Theorem). *There exist a positive number λ (eigenvalue) and vectors $e = (e_1, \ldots, e_r)$ and $f = (f_1, \ldots, f_r)$ (right and left eigenvectors) such that*

1. *$e_j > 0, f_j > 0, \ 1 \le j \le r$.*
2. *$Ae = \lambda e$ and $A^* f = \lambda f$.*

If $Ae' = \lambda' e'$ and $e_j' > 0$ for $1 \le j \le r$, then $\lambda' = \lambda$ and $e' = c_1 e$ for some positive constant c_1. If $A^* f' = \lambda' f'$ and $f_j' > 0$ for $1 \le j \le r$, then $\lambda' = \lambda$ and $f' = c_2 f$ for some positive constant c_2.

Proof. Let us show that there exist $\lambda > 0$ and a positive vector e such that $Ae = \lambda e$, that is

$$\sum_{j=1}^{r} a_{ij} e_j = \lambda e_j, \ 1 \leq i \leq r.$$

Consider the convex set $\mathcal{H}$ of vectors $h = (h_1, \ldots, h_r)$ such that $h_i \geq 0, 1 \leq i \leq r$, and $\sum_{i=1}^{r} h_i = 1$. The matrix A determines a continuous transformation $\mathcal{A}$ of $\mathcal{H}$ into itself through the formula

$$(\mathcal{A}h)_i = \frac{\sum_{j=1}^{r} a_{ij} h_j}{\sum_{i=1}^{r} \sum_{j=1}^{r} a_{ij} h_j}.$$

The Brouwer Theorem states that any continuous mapping of a convex closed set in $\mathbb{R}^n$ to itself has a fixed point. Thus we can find $e \in \mathcal{H}$ such that $\mathcal{A}e = e$, that is,

$$e_i = \frac{\sum_{j=1}^{r} a_{ij} e_j}{\sum_{i=1}^{r} \sum_{j=1}^{r} a_{ij} e_j}.$$

Note that $e_i > 0$ for all $1 \leq i \leq r$. By setting $\lambda = \sum_{i=1}^{r} \sum_{j=1}^{r} a_{ij} e_j$, we obtain $\sum_{j=1}^{r} a_{ij} e_j = \lambda e_i, \ 1 \leq i \leq r$.

In the same way we can show that there is $\overline{\lambda} > 0$ and a vector f with positive entries such that $A^* f = \overline{\lambda} f$. The equalities

$$\lambda(e, f) = (Ae, f) = (e, A^* f) = (e, \overline{\lambda} f) = \overline{\lambda}(e, f)$$

show that $\lambda = \overline{\lambda}$.

We leave the uniqueness part as an exercise for the reader.   □

Let e and f be positive right and left eigenvectors, respectively, which satisfy

$$\sum_{i=1}^{r} e_i = 1 \text{ and } \sum_{i=1}^{r} e_i f_i = 1.$$

Note that these conditions determine e and f uniquely. Let $\lambda > 0$ be the corresponding eigenvalue. Set

$$p_{ij} = \frac{a_{ij} e_j}{\lambda e_i}.$$

It is easy to see that the matrix $P = (p_{ij})$ is a stochastic matrix with strictly positive entries. The stationary distribution of this matrix is $\pi_i = e_i f_i$. Indeed,

$$\sum_{i=1}^{r} \pi_i p_{ij} = \sum_{i=1}^{r} e_i f_i \frac{a_{ij} e_j}{\lambda e_i} = \frac{1}{\lambda} e_j \sum_{i=1}^{r} f_i a_{ij} = e_j f_j = \pi_j.$$

We can rewrite $a_{ij}^{(n)}$ as follows:

$$a_{ij}^{(n)} = \sum_{1 \le i_1, \dots, i_{n-1} \le r} a_{ii_1} \cdot a_{i_1 i_2} \cdot \dots \cdot a_{i_{n-2} i_{n-1}} \cdot a_{i_{n-1} j}$$

$$= \lambda^n \sum_{1 \le i_1, \dots, i_{n-1} \le r} p_{ii_1} \cdot p_{i_1 i_2} \cdot \dots \cdot p_{i_{n-2} i_{n-1}} \cdot p_{i_{n-1} j} \cdot e_i \cdot e_j^{-1} = \lambda^n e_i p_{ij}^{(n)} e_j^{-1}.$$

The Ergodic Theorem for Markov chains gives $p_{ij}^{(n)} \to \pi_j = e_j f_j$ as $n \to \infty$. Therefore,

$$\frac{a_{ij}^{(n)}}{\lambda^n} \to e_i \pi_j e_j^{-1} = e_i f_j$$

and the convergence is exponentially fast. Thus

$$a_{ij}^{(n)} \sim \lambda^n e_i f_j \quad \text{as } n \to \infty.$$

Remark 5.14. One can easily extend these arguments to the case where the matrix A^s has positive matrix elements for some integer $s > 0$.

5.6 General Markov Chains and the Doeblin Condition

Markov chains often appear as random perturbations of deterministic dynamics. Let $(X, \mathcal{G})$ be a measurable space and $f : X \to X$ a measurable mapping of X into itself. We may wish to consider the trajectory of a point $x \in X$ under the iterations of f, that is the sequence $x, f(x), f^2(x), \dots$. However, if random noise is present, then x is mapped not to $f(x)$ but to a nearby random point. This means that for each $C \in \mathcal{G}$ we must consider the transition probability from the point x to the set C. Let us give the corresponding definition.

Definition 5.15. *Let $(X, \mathcal{G})$ be a measurable space. A function $P(x, C)$, $x \in X, C \in \mathcal{G}$, is called a Markov transition function if for each fixed $x \in X$ the function $P(x, C)$, as a function of $C \in \mathcal{G}$, is a probability measure defined on $\mathcal{G}$, and for each fixed $C \in \mathcal{G}$ the function $P(x, C)$ is measurable as a function of $x \in X$.*

For x and C fixed, $P(x, C)$ is called the transition probability from the initial point x to the set C. Given a Markov transition function $P(x, C)$ and an integer $n \in \mathbb{N}$, we can define the n-step transition function

$$P^n(x, C) = \int_X \dots \int_X \int_X P(x, dy_1) \dots P(y_{n-2}, dy_{n-1}) P(y_{n-1}, C).$$

It is easy to see that P^n satisfies the definition of a Markov transition function.

A Markov transition function $P(x, C)$ defines two operators:

1. The operator P which acts on bounded measurable functions

$$(Pf)(x) = \int_X f(y)P(x, dy); \qquad (5.5)$$

2. The operator P^* which acts on the probability measures

$$(P^*\mu)(C) = \int_X P(x, C)d\mu(x). \qquad (5.6)$$

It is easy to show (see Problem 15) that the image of a bounded measurable function under the action of P is again a bounded measurable function, while the image of a probability measure μ under P^* is again a probability measure.

Remark 5.16. Note that we use the same letter P for the Markov transition function and the corresponding operator. This is partially justified by the fact that the n-th power of the operator corresponds to the n-step transition function, that is

$$(P^n f)(x) = \int_X f(y)P^n(x, dy).$$

Definition 5.17. *A probability measure π is called a stationary (or invariant) measure for the Markov transition function P if $\pi = P^*\pi$, that is*

$$\pi(C) = \int_X P(x, C)d\pi(x)$$

for all $C \in \mathcal{G}$.

Given a Markov transition function P and a probability measure μ_0 on $(X, \mathcal{G})$, we can define the corresponding homogeneous Markov chain, that is the measure on the space of sequences $\omega = (\omega_0, \ldots, \omega_n)$, $\omega_k \in X$, $k = 0, \ldots, n$. Namely, denote by $\mathcal{F}$ the σ-algebra generated by the elementary cylinders, that is by the sets of the form $A = \{\omega : \omega_0 \in A_0, \omega_1 \in A_1, \ldots, \omega_n \in A_n\}$ where $A_k \in \mathcal{G}$, $k = 0, \ldots, n$. By Theorem 3.19, if we define

$$P(A) = \int_{A_0 \times \ldots \times A_{n-1}} d\mu_0(x_0)P(x_0, dx_1) \ldots P(x_{n-2}, dx_{n-1})P(x_{n-1}, A_n),$$

there exists a measure on $\mathcal{F}$ which coincides with $P(A)$ on the elementary cylinders. Moreover, such a measure on $\mathcal{F}$ is unique.

Remark 5.18. We could also consider a measure on the space of infinite sequences $\omega = (\omega_0, \omega_1, \ldots)$ with $\mathcal{F}$ still being the σ-algebra generated by the elementary cylinders. In this case, there is still a unique measure on $\mathcal{F}$ which coincides with $P(A)$ on the elementary cylinder sets. Its existence is guaranteed by the Kolmogorov Consistency Theorem which is discussed in Chap. 12.

We have already seen that in the case of Markov chains with a finite state space the stationary measure determines the statistics of typical ω (the Law of Large Numbers). This is also true in the more general setting which we are considering now. Therefore it is important to find sufficient conditions which guarantee the existence and uniqueness of the stationary measure.

Definition 5.19. *A Markov transition function P is said to satisfy the strong Doeblin condition if there exist a probability measure ν on $(X, \mathcal{G})$ and a function $p(x, y)$ (the density of $P(x, dy)$ with respect to the measure ν) such that*

1. *$p(x, y)$ is measurable on $(X \times X, \mathcal{G} \times \mathcal{G})$.*
2. *$P(x, C) = \int_C p(x, y) d\nu(y)$ for all $x \in X$ and $C \in \mathcal{G}$.*
3. *For some constant $a > 0$ we have*

$$p(x, y) \geq a \quad \text{for all } x, y \in X.$$

Theorem 5.20. *If a Markov transition function satisfies the strong Doeblin condition, then there exists a unique stationary measure.*

Proof. By the Fubini Theorem, for any measure μ the measure $P^*\mu$ is given by the density $\int_X d\mu(x) p(x, y)$ with respect to the measure ν. Therefore, if a stationary measure exists, it is absolutely continuous with respect to ν. Let M be the space of measures which are absolutely continuous with respect to ν. For $\mu_1, \mu_2 \in M$, the distance between them is defined via $d(\mu^1, \mu^2) = \frac{1}{2} \int |m^1(y) - m^2(y)| d\nu(y)$, where m^1 and m^2 are the densities of μ^1 and μ^2 respectively. We claim that M is a complete metric space with respect to the metric d. Indeed, M is a closed subspace of $L^1(X, \mathcal{G}, \nu)$, which is a complete metric space. Let us show that the operator P^* acting on this space is a contraction.

Consider two measures μ^1 and μ^2 with the densities m^1 and m^2. Let $A^+ = \{y : m^1(y) - m^2(y) \geq 0\}$ and $A^- = X \backslash A^+$. Similarly let $B^+ = \{y : \int_X p(x, y)(m^1(x) - m^2(x)) d\nu(x) \geq 0\}$ and $B^- = X \backslash B^+$. Without loss of generality we can assume that $\nu(B^-) \geq \frac{1}{2}$ (if the contrary is true and $\nu(B^+) > \frac{1}{2}$, we can replace A^+ by A^-, B^+ by B^- and reverse the signs in some of the integrals below).

As in the discrete case, $d(\mu^1, \mu^2) = \int_{A^+} (m^1(y) - m^2(y)) d\nu(y)$. Therefore,

$$d(P^*\mu^1, P^*\mu^2) = \int_{B^+} [\int_X p(x, y)(m^1(x) - m^2(x)) d\nu(x)] d\nu(y)$$

$$\leq \int_{B^+} [\int_{A^+} p(x, y)(m^1(x) - m^2(x)) d\nu(x)] d\nu(y)$$

$$= \int_{A^+} [\int_{B^+} p(x, y) d\nu(y)] (m^1(x) - m^2(x)) d\nu(x).$$

The last expression contains the integral $\int_{B+} p(x,y)d\nu(y)$ which we estimate as follows

$$\int_{B+} p(x,y)d\nu(y) = 1 - \int_{B-} p(x,y)d\nu(y) \le 1 - a\nu(B^-) \le 1 - \frac{a}{2}.$$

This shows that

$$d(P^*\mu^1, P^*\mu^2) \le (1 - \frac{a}{2})d(\mu^1, \mu^2).$$

Therefore P^* is a contraction and has a unique fixed point, which completes the proof of the theorem. $\qquad\Box$

The strong Doeblin condition can be considerably relaxed, yet we may still be able to say something about the stationary measures. We conclude this section with a discussion of the structure of a Markov chain under the Doeblin condition. We shall restrict ourselves to formulation of results.

Definition 5.21. *We say that P satisfies the Doeblin condition if there is a finite measure μ with $\mu(X) > 0$, an integer n, and a positive ε such that for any $x \in X$*

$$P^n(x, A) \le 1 - \varepsilon \quad \text{if} \quad \mu(A) \le \varepsilon.$$

Theorem 5.22. *If a Markov transition function satisfies the Doeblin condition, then the space X can be represented as the union of non-intersecting sets:*

$$X = \bigcup_{i=1}^{k} E_i \bigcup T,$$

where the sets E_i (ergodic components) have the property $P(x, E_i) = 1$ for $x \in E_i$, and for the set T (the transient set) we have $\lim_{n\to\infty} P^n(x,T) = 0$ for all $x \in X$. The sets E_i can in turn be represented as unions of non-intersecting subsets:

$$E_i = \bigcup_{j=0}^{m_i - 1} C_i^j,$$

where C_i^j (cyclically moving subsets) have the property

$$P(x, C_i^{j+1(\mathrm{mod}\ m_i)}) = 1 \quad \text{for} \quad x \in C_i^j.$$

Note that if P is a Markov transition function on the state space X, then $P(x, A)$, $x \in E_i$, $A \subseteq E_i$ is a Markov transition function on E_i. We have the following theorem describing the stationary measures of Markov transition functions satisfying the Doeblin condition (see "Stochastic Processes" by J.L. Doob).

Theorem 5.23. *If a Markov transition function satisfies the Doeblin condition, and $X = \bigcup_{i=1}^{k} E_i \bigcup T$ is a decomposition of the state space into ergodic components and the transient set, then*

1. *The restriction of the transition function to each ergodic component has a unique stationary measure π_i.*
2. *Any stationary measure π on the space X is equal to a linear combination of the stationary measures on the ergodic components:*

$$\pi = \sum_{i=1}^{k} \alpha_i \pi_i$$

with $\alpha_i \geq 0$, $\alpha_1 + \ldots + \alpha_k = 1$.

Finally, we formulate the Strong Law of Large Numbers for Markov chains (see "Stochastic Processes" by J.L. Doob).

Theorem 5.24. *Consider a Markov transition function which satisfies the Doeblin condition and has only one ergodic component. Let π be the unique stationary measure. Consider the corresponding Markov chain (measure on the space of sequences $\omega = (\omega_0, \omega_1, \ldots)$) with some initial distribution. Then for any function $f \in L^1(X, \mathcal{G}, \pi)$ the following limit exists almost surely:*

$$\lim_{n \to \infty} \frac{\sum_{k=0}^{n} f(\omega_k)}{n+1} = \int_X f(x) d\pi(x).$$

5.7 Problems

1. Let P be a stochastic matrix. Prove that there is at least one non-negative vector π such that $\pi P = \pi$.
2. Consider a homogeneous Markov chain on a finite state space with the transition matrix P and the initial distribution μ. Prove that for any $0 < k < n$ the induced probability distribution on the space of sequences $(\omega_k, \omega_{k+1}, \ldots, \omega_n)$ is also a homogeneous Markov chain. Find its initial distribution and the matrix of transition probabilities.
3. Consider a homogeneous Markov chain on a finite state space X with transition matrix P and the initial distribution δ_x, $x \in X$, that is $P(\omega_0 = x) = 1$. Let τ be the first k such that $\omega_k \neq x$. Find the probability distribution of τ.
4. Consider the one-dimensional simple symmetric random walk (Markov chain on the state space $\mathbb{Z}$ with transition probabilities $p_{i,i+1} = p_{i,i-1} = 1/2$). Prove that it does not have a stationary distribution.
5. For a homogeneous Markov chain on a finite state space X with transition matrix P and initial distribution μ, find $P(\omega_n = x^1 | \omega_0 = x^2, \omega_{2n} = x^3)$, where $x^1, x^2, x^3 \in X$.
6. Consider a homogeneous ergodic Markov chain on the finite state space $X = \{1, \ldots, r\}$ with the transition matrix P and the stationary distribution π. Assuming that π is also the initial distribution, find the following limit

$$\lim_{n \to \infty} \frac{\ln P(\omega_i \neq 1 \text{ for } 0 \leq i \leq n)}{n}.$$

7. Consider a homogeneous ergodic Markov chain on the finite state space $X = \{1, \ldots, r\}$. Define the random variables τ_n, $n \geq 1$, as the consecutive times when the Markov chain is in the state 1, that is

$$\tau_1 = \inf(i \geq 0 : \omega_i = 1),$$

$$\tau_n = \inf(i > \tau_{n-1} : \omega_i = 1), \ n > 1.$$

Prove that $\tau_1, \tau_2 - \tau_1, \tau_3 - \tau_2, \ldots$ is a sequence of independent random variables.

8. Consider a homogeneous ergodic Markov chain on a finite state space with the transition matrix P and the stationary distribution π. Assuming that π is also the initial distribution, prove that the distribution of the inverse process $(\omega_n, \omega_{n-1}, \ldots, \omega_1, \omega_0)$ is also a homogeneous Markov chain. Find its matrix of transition probabilities and stationary distribution.

9. Find the stationary distribution of the Markov chain with the countable state space $\{0, 1, 2, \ldots, n, \ldots\}$, where each point, including 0, can either return to 0 with probability $1/2$ or move to the right $n \mapsto n + 1$ with probability $1/2$.

10. Let P be a matrix of transition probabilities of a homogeneous ergodic Markov chain on a finite state space such that $p_{ij} = p_{ji}$. Find its stationary distribution.

11. Consider a homogeneous Markov chain on the finite state space $X = \{1, \ldots, r\}$. Assume that all the elements of the transition matrix are positive. Prove that for any $k \geq 0$ and any $x^0, x^1, \ldots, x^k \in X$,

$$P(\text{there is } n \text{ such that } \omega_n = x^0, \omega_{n+1} = x^1, \ldots, \omega_{n+k} = x^k) = 1.$$

12. Consider a Markov chain on a finite state space. Let k_1, k_2, l_1 and l_2 be integers such that $0 \leq k_1 < l_1 \leq l_2 < k_2$. Consider the conditional probabilities
$$f(i_{k_1}, \ldots, i_{l_1-1}, i_{l_2+1}, \ldots, i_{k_2}) =$$
$$P(\omega_{l_1} = i_{l_1}, \ldots, \omega_{l_2} = i_{l_2} | \omega_{k_1} = i_{k_1}, \ldots, \omega_{l_1-1} = i_{l_1-1}, \omega_{l_2+1}$$
$$= i_{l_2+1}, \ldots, \omega_{k_2} = i_{k_2})$$

with $i_{l_1}, \ldots, i_{l_2}$ fixed. Prove that whenever f is defined, it depends only on i_{l_1-1} and i_{l_2+1}.

13. Consider a Markov chain whose state space is $\mathbb{R}$. Let $P(x, A)$, $x \in \mathbb{R}$, $A \in \mathcal{B}(\mathbb{R})$, be the following Markov transition function,

$$P(x, A) = \lambda([x - 1/2, x + 1/2] \cap A),$$

where λ is the Lebesgue measure. Assuming that the initial distribution is concentrated at the origin, find $P(|\omega_2| \leq 1/4)$.

14. Let p_{ij}, $i,j \in \mathbb{Z}$, be the transition probabilities of a Markov chain on the state space $\mathbb{Z}$. Suppose that

$$p_{i,i-1} = 1 - p_{i,i+1} = r(i)$$

for all $i \in \mathbb{Z}$, where $r(i) = r_- < 1/2$ if $i < 0$, $r(0) = 1/2$, and $r(i) = r_+ > 1/2$ if $i > 0$. Find the stationary distribution for this Markov chain. Does this Markov chain satisfy the Doeblin condition?

15. For a given Markov transition function, let P and P^* be the operators defined by (5.5) and (5.6), respectively. Prove that the image of a bounded measurable function under the action of P is again a bounded measurable function, while the image of a probability measure μ under P^* is again a probability measure.

16. Consider a Markov chain whose state space is the unit circle. Let the density of the transition function $P(x, dy)$ be given by

$$p(x,y) = \begin{cases} 1/(2\varepsilon) & \text{if } \text{angle } (y,x) < \varepsilon, \\ 0 & \text{otherwise,} \end{cases}$$

where $\varepsilon > 0$. Find the stationary measure for this Markov chain.

Random Walks on the Lattice $\mathbb{Z}^d$

6.1 Recurrent and Transient Random Walks

In this section we study random walks on the lattice $\mathbb{Z}^d$. The lattice $\mathbb{Z}^d$ is a collection of points $x = (x_1, \ldots, x_d)$ where x_i are integers, $1 \leq i \leq d$.

Definition 6.1. *A random walk on $\mathbb{Z}^d$ is a homogeneous Markov chain whose state space is $X = \mathbb{Z}^d$.*

Thus we have here an example of a Markov chain with a countable state space. Let $P = (p_{xy})$, $x, y \in \mathbb{Z}^d$, be the infinite stochastic matrix of transition probabilities.

Definition 6.2. *A random walk is said to be spatially homogeneous if $p_{xy} = p_{y-x}$, where $p = \{p_z, z \in \mathbb{Z}^d\}$ is a probability distribution on the lattice $\mathbb{Z}^d$.*

From now on we shall consider only spatially homogeneous random walks. We shall refer to the number of steps $1 \leq n \leq \infty$ as the length of the walk. The function $i \to \omega_i$, $0 \leq i \leq n$ ($0 \leq i < \infty$ if $n = \infty$), will be referred to as the path or the trajectory of the random walk.

Spatially homogeneous random walks are closely connected to homogeneous sequences of independent trials. Indeed, let $\omega = (\omega_0, \ldots, \omega_n)$ be a trajectory of the random walk. Then $\mathrm{p}(\omega) = \mu_{\omega_0} p_{\omega_0 \omega_1} \cdots p_{\omega_{n-1} \omega_n} = \mu_{\omega_0} p_{\omega_1'} \cdots p_{\omega_n'}$, where $\omega_1' = \omega_1 - \omega_0, \ldots, \omega_n' = \omega_n - \omega_{n-1}$ are the increments of the walk. In order to find the probability of a given sequence of increments, we need only take the sum over ω_0 in the last expression. This yields $p_{\omega_1'} \cdots p_{\omega_n'}$, which is exactly the probability of a given outcome in the sequence of independent homogeneous trials. We shall repeatedly make use of this property.

Let us take $\mu = \delta(0)$, that is consider a random walk starting at the origin. Assume that $\omega_i \neq 0$ for $1 \leq i \leq n - 1$ and $\omega_0 = \omega_n = 0$. In this case we say that the trajectory of the random walk returns to the initial point for the first time at the n-th step. The set of such ω will be denoted by A_n. We set $f_0 = 0$ and $f_n = \sum_{\omega \in A_n} \mathrm{p}(\omega)$ for $n > 0$.

L. Koralov and Y.G. Sinai, *Theory of Probability and Random Processes*, 85
Universitext, DOI 10.1007/978-3-540-68829-7_6,
© Springer-Verlag Berlin Heidelberg 2012

Definition 6.3. *A random walk is called recurrent if $\sum_{n=1}^{\infty} f_n = 1$. If this sum is less than one, the random walk is called transient.*

The definition of recurrence means that the probability of the set of those trajectories which return to the initial point is equal to 1. Here we introduce a general criterion for the recurrence of a random walk. Let B_n consist of those sequences $\omega = (\omega_0, \ldots, \omega_n)$ for which $\omega_0 = \omega_n = 0$. For elements of B_n it is possible that $\omega_i = 0$ for some $i, 1 \le i \le n-1$. Consequently $A_n \subseteq B_n$. We set $u_0 = 1$ and $u_n = \sum_{\omega \in B_n} p(\omega)$ for $n \ge 1$.

Lemma 6.4 (Criterion for Recurrence). *A random walk is recurrent if and only if $\sum_{n \ge 0} u_n = \infty$.*

Proof. We first prove an important formula which relates f_n and u_n:

$$u_n = f_n u_0 + f_{n-1} u_1 + \ldots + f_0 u_n \quad \text{for } n \ge 1. \tag{6.1}$$

We have $B_n = \bigcup_{i=1}^{n} C_i$, where

$$C_i = \{\omega : \omega \in B_n, \omega_i = 0 \text{ and } \omega_j \ne 0, 1 \le j < i\}.$$

Since the sets C_i are pair-wise disjoint,

$$u_n = \sum_{i=1}^{n} \mathrm{P}(C_i).$$

We note that

$$\mathrm{P}(C_i) = \sum_{\omega \in C_i} p_{\omega_1'} \cdots p_{\omega_n'} = \sum_{\omega \in A_i} p_{\omega_1'} \cdots p_{\omega_i'} \sum_{\omega : \omega_{i+1}' + \ldots + \omega_n' = 0} p_{\omega_{i+1}'} \cdots p_{\omega_n'} = f_i u_{n-i}.$$

Since $f_0 = 0$ and $u_0 = 1$,

$$u_n = \sum_{i=0}^{n} f_i u_{n-i} \quad \text{for } n \ge 1; \quad u_0 = 1. \tag{6.2}$$

This completes the proof of (6.1).

Now we need the notion of a generating function. Let a_n, $n \ge 0$, be an arbitrary bounded sequence. The generating function of the sequence a_n is the sum of the power series $A(z) = \sum_{n \ge 0} a_n z^n$, which is an analytic function of the complex variable z in the domain $|z| < 1$. The essential fact for us is that $A(z)$ uniquely determines the sequence a_n since

$$a_n = \frac{1}{n!} \frac{d^n}{dz^n} A(z)|_{z=0}.$$

Returning to our random walk, consider the generating functions

$$F(z) = \sum_{n\geq 0} f_n z^n, \quad U(z) = \sum_{n\geq 0} u_n z^n.$$

Let us multiply the left and right sides of (6.2) by z^n, and sum with respect to n from 0 to ∞. We get $U(z)$ on the left, and $1 + U(z)F(z)$ on the right, that is

$$U(z) = 1 + U(z)F(z),$$

which can be also written as $F(z) = 1 - 1/U(z)$. We now note that

$$\sum_{n=1}^{\infty} f_n = F(1) = \lim_{z\to 1} F(z) = 1 - \lim_{z\to 1} \frac{1}{U(z)}.$$

Here and below, z tends to one from the left on the real axis.

We first assume that $\sum_{n=0}^{\infty} u_n < \infty$. Then

$$\lim_{z\to 1} U(z) = U(1) = \sum_{n=0}^{\infty} u_n < \infty,$$

and

$$\lim_{z\to 1} \frac{1}{U(z)} = \frac{1}{\sum_{n=0}^{\infty} u_n} > 0.$$

Therefore $\sum_{n=1}^{\infty} f_n < 1$, which means the random walk is transient.

When $\sum_{n=0}^{\infty} u_n = \infty$ we show that $\lim_{z\to 1}(1/U(z)) = 0$. Let us fix $\varepsilon > 0$ and find $N = N(\varepsilon)$ such that $\sum_{n=0}^{N} u_n \geq 2/\varepsilon$. Then for z sufficiently close to 1 we have $\sum_{n=0}^{N} u_n z^n \geq 1/\varepsilon$. Consequently, for such z

$$\frac{1}{U(z)} \leq \frac{1}{\sum_{n=0}^{N} u_n z^n} \leq \varepsilon.$$

This means that $\lim_{z\to 1}(1/U(z)) = 0$, and consequently $\sum_{n=1}^{\infty} f_n = 1$. In other words, the random walk is recurrent. $\qquad\square$

We now consider an application of this criterion. Let $e_1, \ldots, e_d$ be the unit coordinate vectors and let $p_y = 1/(2d)$ if $y = \pm e_s$, $1 \leq s \leq d$, and 0 otherwise. Such a random walk is called simple symmetric.

Theorem 6.5 (Polya). *A simple symmetric random walk is recurrent for $d = 1, 2$ and is transient for $d \geq 3$.*

Sketch of the Proof. The probability u_{2n} is the probability that $\sum_{k=1}^{2n} w'_k = 0$. For $d = 1$, the de Moivre-Laplace Theorem gives $u_{2n} \sim \frac{1}{\sqrt{\pi n}}$ as $n \to \infty$. Also $u_{2n+1} = 0$, $0 \leq n < \infty$. In the multi-dimensional case u_{2n} decreases as $cn^{-\frac{d}{2}}$

(we shall demonstrate this fact when we discuss the Local Limit Theorem in Sect. 10.2). Therefore the series $\sum_{n=0}^{\infty} u_n$ diverges for $d = 1, 2$ and converges otherwise.

$\square$

In dimension $d = 3$, it easily follows from the Polya Theorem that "typical" trajectories of a random walk go off to infinity as $n \to \infty$. One can ask a variety of questions about the asymptotic properties of such trajectories. For example, for each n consider the unit vector $v_n = \omega_n / ||\omega_n||$, which is the projection of the random walk on the unit sphere. One question is: does $\lim_{n \to \infty} v_n$ exist for typical trajectories? This would imply that a typical trajectory goes off to infinity in a given direction. In fact, this is not the case, as there is no such limit. Furthermore, the vectors v_n tend to be uniformly distributed on the unit sphere. Such a phenomenon is possible because the random walk is typically located at a distance of order $O(\sqrt{n})$ away from the origin after n steps, and therefore manages to move in all directions.

Spatially homogeneous random walks on $\mathbb{Z}^d$ are special cases of homogeneous random walks on groups. Let G be a countable group and $p = \{p_g, g \in G\}$ be a probability distribution on G. We consider a Markov chain in which the state space is the group G and the transition probability is $p_{xy} = p_{y-x}$, $x, y \in G$. As with the usual lattice $\mathbb{Z}^d$ we can formulate the definition for the recurrence of a random walk and prove an analogous criterion. In the case of simple random walks, the answer to the question as to whether the walk is transient, depends substantially on the group G. For example, if G is the free group with two generators, a and b, and if the probability distribution p is concentrated on the four points a, b, a^{-1} and b^{-1}, then such a random walk will always be transient.

There are also interesting problems in connection with continuous groups. The groups $SL(m, \mathbb{R})$ of matrices of order m with real elements and determinant equal to one arise particularly often. Special methods have been devised to study random walks on such groups. We shall study one such problem in more detail in Sect. 11.2.

6.2 Random Walk on $\mathbb{Z}$ and the Reflection Principle

In this section and the next we shall make several observations about the simple symmetric random walk on $\mathbb{Z}$, some of them of combinatorial nature, which will be useful in understanding the statistics of typical trajectories of the walk. In particular we shall see that while the random walk is symmetric, the proportion of time that it spends to the right of the origin does not tend to a deterministic limit (which one could imagine to be equal to $1/2$), but has a non-trivial limiting distribution (arcsine law).

The first observation we make concerns the probability that the walk returns to the origin after $2n$ steps. In order for the walk to return to the origin after $2n$ steps, the number of steps to the right should be exactly equal to n. There are $(2n)!/(n!)^2$ ways to place n symbols $+1$ in a sequence composed of n symbols $+1$ and n symbols -1. Since there are 2^{2n} possibilities for the trajectory of the random walk of $2n$ steps, all of which are equally probable, we obtain that $u_{2n} = 2^{-2n}(2n)!/(n!)^2$. Clearly the trajectory cannot return to the origin after an odd number of steps.

Let us now derive a formula for the probability that the time $2n$ is the moment of the first return to the origin. Note that the generating function

$$U(z) = \sum_{n \geq 0} u_n z^n = \sum_{n \geq 0} \frac{(2n)!}{(n!)^2} 2^{-2n} z^{2n}$$

is equal to $1/\sqrt{1 - z^2}$. Indeed, the function $1/\sqrt{1 - z^2}$ is analytic in the unit disc, and the coefficients in its Taylor series are equal to the coefficients of the sum $U(z)$.

Since $U(z) = 1 + U(z)F(z)$,

$$F(z) = 1 - \frac{1}{U(z)} = 1 - \sqrt{1 - z^2}.$$

This function is also analytic in the unit disc, and can be written as the sum of its Taylor series

$$F(z) = \sum_{n \geq 1} \frac{(2n)!}{(2n - 1)(n!)^2} 2^{-2n} z^{2n}.$$

Therefore

$$f_{2n} = \frac{(2n)!}{(2n - 1)(n!)^2} 2^{-2n} = \frac{u_{2n}}{2n - 1}. \tag{6.3}$$

The next lemma is called the reflection principle. Let $x, y > 0$, where x is the initial point for the random walk. We say that a path of the random walk contains the origin if $\omega_k = 0$ for some $0 \leq k \leq n$.

Lemma 6.6 (Reflection Principle). *The number of paths of the random walk of length n which start at $\omega_0 = x > 0$, end at $\omega_n = y > 0$ and contain the origin is equal to the number of all paths from $-x$ to y.*

Proof. Let us exhibit a one-to-one correspondence between the two sets of paths. For each path $(\omega_0, \ldots, \omega_n)$ which starts at $\omega_0 = x$ and contains the origin, let k be the first time when the path reaches the origin, that is $k = \min\{i : \omega_i = 0\}$. The corresponding path which starts at $-x$ is $(-\omega_0, \ldots, -\omega_{k-1}, \omega_k, \omega_{k+1}, \ldots, \omega_n)$. Clearly this is a one-to-one correspondence. $\square$

As an application of the reflection principle let us consider the following problem. Let $x(n)$ and $y(n)$ be integer-valued functions of n such that $x(n) \sim a\sqrt{n}$ and $y(n) \sim b\sqrt{n}$ as $n \to \infty$ for some positive constants a and b. For a path of the random walk which starts at $x(n)$, we shall estimate the probability that it ends at $y(n)$ after n steps, subject to the condition that it always stays to the right of the origin. We require that $y(n) - x(n) - n$ is even, since otherwise the probability is zero.

Thus we are interested in the relation between the number of paths which go from x to y in n steps while staying to the right of the origin (denoted by $M(x, y, n)$), and the number of paths which start at x, stay to the right of the origin and end anywhere on the positive semi-axis (denoted by $M(x, n)$).

Let $N(x, y, n)$ denote the number of paths of length n which go from x to y and let $N(x, n)$ denote the number of paths which start at x and end on the positive semi-axis. Recall that the total number of paths of length n is 2^n.

By the de Moivre-Laplace Theorem,

$$\frac{N(x(n), y(n), n)}{2^n} \sim \sqrt{\frac{2}{\pi n}} e^{-\frac{(y(n)-x(n))^2}{2n}} \quad \text{as } n \to \infty.$$

The integral version of the de Moivre-Laplace Theorem implies

$$\frac{N(x(n), n)}{2^n} \sim \frac{1}{\sqrt{2\pi}} \int_{-\frac{x(n)}{\sqrt{n}}}^{\infty} e^{-\frac{z^2}{2}} dz \quad \text{as } n \to \infty,$$

and by the reflection principle the desired probability is equal to

$$\frac{M(x(n), y(n), n)}{M(x(n), n)} = \frac{N(x(n), y(n), n) - N(-x(n), y(n), n)}{N(x(n), n) - N(-x(n), n)}$$

$$\sim \frac{2(e^{-\frac{(b-a)^2}{2}} - e^{-\frac{(a+b)^2}{2}})}{\sqrt{n} \int_{-a}^{a} e^{-\frac{z^2}{2}} dz}.$$

6.3 Arcsine Law

In this section we shall consider the asymptotics of several quantities related to the statistics of the one-dimensional simple symmetric random walk.[1]For a random walk of length $2n$ we shall study the distribution of the last visit to the origin. We shall also examine the proportion of time spent by a path of the random walk on one side of the origin, say, on the positive semi-axis. In order to make the description symmetric, we say that the path is on the positive semi-axis at time $k > 0$ if $\omega_k > 0$ or if $\omega_k = 0$ and $\omega_{k-1} > 0$. Similarly, we say that the path is on the negative semi-axis at time $k > 0$ if $\omega_k < 0$ or if $\omega_k = 0$ and $\omega_{k-1} < 0$.

---

[1] This section can be omitted during the first reading.

Consider the random walk of length $2n$, and let $a_{2k,2n}$ be the probability that the last visit to the origin occurs at time $2k$. Let $b_{2k,2n}$ be the probability that a path is on the positive semi-axis exactly $2k$ times. Let s_{2n} be the probability that a path does not return to the origin by time $2n$, that is $s_{2n} = \mathrm{P}(\omega_k \neq 0 \text{ for } 1 \leq k \leq 2n)$.

Lemma 6.7. *The probability that a path which starts at the origin does not return to the origin by time $2n$ is equal to the probability that a path returns to the origin at time $2n$, that is*

$$s_{2n} = u_{2n}. \tag{6.4}$$

Proof. Let $n, x \geq 0$ be integers and let $N_{n,x}$ be the number of paths of length n which start at the origin and end at x. Then

$$N_{n,x} = \frac{n!}{(\frac{n+x}{2})!(\frac{n-x}{2})!} \quad \text{if } n \geq x, \ n - x \text{ is even},$$

and $N_{n,x} = 0$ otherwise.

Let us now find the number of paths of length n from the origin to the point $x > 0$ such that $\omega_i > 0$ for $1 \leq i \leq n$. It is equal to the number of paths of length $n - 1$ which start at the point 1, end at the point x, and do not contain the origin. By the reflection principle, this is equal to

$$N_{n-1,x-1} - N_{n-1,x+1}.$$

In order to calculate s_{2n}, let us consider all possible values of ω_{2n}, taking into account the symmetry with respect to the origin:

$$s_{2n} = 2\mathrm{P}(\omega_1 > 0, \ldots, \omega_{2n} > 0) = 2\sum_{x=1}^{\infty} \mathrm{P}(\omega_1 > 0, \ldots, \omega_{2n-1} > 0, \omega_{2n} = 2x)$$

$$= \frac{2\sum_{x=1}^{\infty}(N_{2n-1,2x-1} - N_{2n-1,2x+1})}{2^{2n}} = \frac{2N_{2n-1,1}}{2^{2n}} = \frac{(2n)!}{(n!)^2}2^{-2n} = u_{2n}.$$

$\square$

Lemma 6.7 implies

$$b_{0,2n} = b_{2n,2n} = u_{2n}. \tag{6.5}$$

The first equality follows from the definition of $b_{2k,2n}$. To demonstrate the second one we note that since ω_{2n} is even,

$$\mathrm{P}(\omega_1 > 0, \ldots, \omega_{2n} > 0) = \mathrm{P}(\omega_1 > 0, \ldots, \omega_{2n} > 0, \omega_{2n+1} > 0).$$

By taking the point 1 as the new origin, each path of length $2n + 1$ starting at zero for which $\omega_1 > 0, \ldots, \omega_{2n+1} > 0$ can be identified with a path of length $2n$ for which $\omega_1 \geq 0, \ldots, \omega_{2n} \geq 0$. Therefore,

$$b_{2n,2n} = \mathrm{P}(\omega_1 \geq 0, \ldots, \omega_{2n} \geq 0) = 2\mathrm{P}(\omega_1 > 0, \ldots, \omega_{2n+1} > 0)$$

$$= 2\mathrm{P}(\omega_1 > 0, \ldots, \omega_{2n} > 0) = u_{2n},$$

which implies (6.5).

We shall prove that

$$a_{2k,2n} = u_{2k}u_{2n-2k} \tag{6.6}$$

and

$$b_{2k,2n} = u_{2k}u_{2n-2k}. \tag{6.7}$$

The probability of the last visit occurring at time $2k$ equals the product of u_{2k} and the probability that a path which starts at the origin at time $2k$ does not return to the origin by the time $2n$. Therefore, due to (6.4), we have $a_{2k,2n} = u_{2k}u_{2n-2k}$.

From (6.5) it also follows that (6.7) is true for $k = 0$ and for $k = n$, and thus we need to demonstrate it for $1 \leq k \leq n-1$. We shall argue by induction. Assume that (6.7) has been demonstrated for $n < n_0$ for all k. Now let $n = n_0$ and $1 \leq k \leq n-1$.

Let r be such that the path returns to the origin for the first time at step $2r$. Consider the paths which return to the origin for the first time at step $2r$, are on the positive semi-axis till the time $2r$, and are on the positive semi-axis for a total of $2k$ times out of the first $2n$ steps. The number of such paths is equal to $\frac{1}{2}2^{2r}f_{2r}2^{2n-2r}b_{2k-2r,2n-2r}$. Similarly, the number of paths which return to the origin for the first time at step $2r$, are on the negative semi-axis till time $2r$, and are on the positive semi-axis for a total of $2k$ times out of the first $2n$ steps, is equal to $\frac{1}{2}2^{2r}f_{2r}2^{2n-2r}b_{2k,2n-2r}$. Dividing by 2^n and taking the sum in r, we obtain

$$b_{2k,2n} = \frac{1}{2}\sum_{r=1}^{k} f_{2r}b_{2k-2r,2n-2r} + \frac{1}{2}\sum_{r=1}^{n-k} f_{2r}b_{2k,2n-2r}.$$

Since $1 \leq r \leq k \leq n-1$, by the induction hypothesis the last expression is equal to

$$b_{2k,2n} = \frac{1}{2}u_{2n-2k}\sum_{r=1}^{k} f_{2r}u_{2k-2r} + \frac{1}{2}u_{2k}\sum_{r=1}^{n-k} f_{2r}u_{2n-2k-2r}.$$

Note that due to (6.1) the first sum equals u_{2k}, while the second one equals u_{2n-2k}. This proves (6.7) for $n = n_0$, which means that (6.7) is true for all n.

Let $0 \leq x \leq 1$ be fixed and let

$$F_n(x) = \sum_{k \leq xn} a_{2k,2n}.$$

Thus $F_n(x)$ is the probability that the path does not visit the origin after time $2xn$. In other words, F_n is the distribution function of the random variable which is equal to the fraction of time that the path spends before the last visit to the origin. Due to (6.6) and (6.7) we have $F_n(x) = \sum_{k \leq xn} b_{2k,2n}$. Therefore, F_n is also the distribution function for the random variable which is equal to the fraction of time that the path spends on the positive semi-axis.

Lemma 6.8 (Arcsine Law). *For each* $0 \leq x \leq 1$,

$$\lim_{n \to \infty} F_n(x) = \frac{2}{\pi} \arcsin(\sqrt{x}).$$

Proof. By the de Moivre-Laplace Theorem,

$$u_{2n} \sim \frac{1}{\sqrt{\pi n}} \quad \text{as } n \to \infty.$$

Fix two numbers x_1 and x_2 such that $0 < x_1 < x_2 < 1$. Let $k = k(n)$ satisfy $x_1 n \leq k \leq x_2 n$. Thus, by (6.6)

$$a_{2k,2n} = u_{2k} u_{2n-2k} = \frac{1}{\pi \sqrt{k(n-k)}} + o\left(\frac{1}{n}\right) \quad \text{as } n \to \infty,$$

and therefore

$$\lim_{n \to \infty} (F_n(x_2) - F_n(x_1)) = \int_{x_1}^{x_2} \frac{1}{\pi \sqrt{x(1-x)}} dx.$$

Let

$$F(y) = \int_0^y \frac{1}{\pi \sqrt{x(1-x)}} dx = \frac{2}{\pi} \arcsin(\sqrt{y}).$$

Thus

$$\lim_{n \to \infty} (F_n(x_2) - F_n(x_1)) = F(x_2) - F(x_1). \tag{6.8}$$

Note that $F(0) = 0$, $F(1) = 1$, and F is continuous on the interval $[0,1]$. Given $\varepsilon > 0$, find $\delta > 0$ such that $F(\delta) \leq \varepsilon$ and $F(1-\delta) \geq 1 - \varepsilon$. By (6.8),

$$F_n(1 - \delta) - F_n(\delta) \geq 1 - 3\varepsilon$$

for all sufficiently large n. Since F_n is a distribution function, $F_n(\delta) \leq 3\varepsilon$ for all sufficiently large n. Therefore, by (6.8) with $x_1 = \delta$,

$$|F_n(x_2) - F(x_2)| \leq 4\varepsilon$$

for all sufficiently large n. Since $\varepsilon > 0$ is arbitrary,

$$\lim_{n \to \infty} F_n(x_2) = F(x_2).$$

$\square$

6.4 Gambler's Ruin Problem

Let us consider a random walk (of infinite length) on the one-dimensional lattice with transition probabilities $p(x, x + 1) = p, p(x, x - 1) = 1 - p = q$, and $p(x, y) = 0$ if $|x - y| \neq 1$. This means that the probability of making one step to the right is p, the probability of making one step to the left is q, and all the other probabilities are zero. Let us assume that $0 < p < 1$. We shall consider the measures P_z on the space of elementary outcomes which correspond to the walk starting at a point z, that is $P_z(\omega_0 = z) = 1$.

Given a pair of integers z and A such that $z \in [0, A]$, we shall study the distribution of the number of steps needed for the random walk starting at z to reach one of the end-points of the interval $[0, A]$. We shall also be interested in finding the probability of the random walk reaching the right (or left) end-point of the interval before reaching the other end-point.

These questions can be given the following simple interpretation. Imagine a gambler, whose initial fortune is z, placing bets with the unit stake at integer moments of time. The fortune of the gambler after n time steps can be represented as the position, after n steps, of a random walk starting at z. The game stops when the gambler's fortune becomes equal to A or zero, whichever happens first. We shall be interested in the distribution of length of the game and in the probabilities of the gambler either losing the entire fortune or reaching the goal of accumulating the fortune equal to A.

Let $R(z, n)$ be the probability that a trajectory of the random walk starting at z does not reach the end-points of the interval during the first n time steps. Obviously, $R(z, 0) = 1$ for $0 < z < A$. Let us set $R(0, n) = R(A, n) = 0$ for $n \geq 0$ (which is in agreement with the fact that a game which starts with the fortune 0 or A lasts zero steps). If $0 < z < A$ and $n > 0$, then

$$R(z, n) = P_z(0 < \omega_i < A, i = 0, \dots, n)$$

$$= P_z(\omega_1 = z + 1, 0 < \omega_i < A, i = 0, \dots, n)$$

$$+ P_z(\omega_1 = z - 1, 0 < \omega_i < A, i = 0, \dots, n)$$

$$= p P_{z+1}(0 < \omega_i < A, i = 0, \dots, n - 1) + q P_{z-1}(0 < \omega_i < A, i = 0, \dots, n - 1)$$

$$= p R(z + 1, n - 1) + q R(z - 1, n - 1).$$

We have thus demonstrated that $R(z, t)$ satisfies the following partial difference equation

$$R(z, n) = p R(z + 1, n - 1) + q R(z - 1, n - 1), \quad 0 < z < A, \quad n > 0. \quad (6.9)$$

In general, one could study this equation with any initial and boundary conditions

$$R(z, 0) = \varphi(z) \text{ for } 0 < z < A, \quad (6.10)$$

$$R(0, n) = \psi_0(n), R(A, n) = \psi_A(n) \text{ for } n \geq 0. \quad (6.11)$$

In our case, $\varphi(z) \equiv 1$ and $\psi_0(n) = \psi_A(n) \equiv 0$. Let us note several properties of solutions to Eq. (6.9):

(a) Equation (6.9) (with any initial and boundary conditions) has a unique solution, since it can be solved recursively.

(b) If the boundary conditions are $\psi_0(n) = \psi_A(n) \equiv 0$, then the solution depends monotonically on the initial conditions. Namely, if R^i are the solutions with the initial conditions $R^i(z,0) = \varphi^i(z)$ for $0 < z < A$, $i = 1, 2$, and $\varphi^1(z) \le \varphi^2(z)$ for $0 < z < A$, then $R^1(z,n) \le R^2(z,n)$ for all z, n, as can be checked by induction on n.

(c) If the boundary conditions are $\psi_0(n) = \psi_A(n) \equiv 0$, then the solution depends linearly on the initial conditions. Namely, if R^i are the solutions with the initial conditions $R^i(z,0) = \varphi^i(z)$ for $0 < z < A$, $i = 1, 2$, and c_1, c_2 are any constants, then $c_1 R^1 + c_2 R^2$ is the solution with the initial condition $c_1\varphi^1(z) + c_2\varphi^2(z)$. This follows immediately from Eq. (6.9).

(d) Since $p + q = 1$ and $0 < p, q < 1$, from (6.9) it follows that if the boundary conditions are $\psi_0(n) = \psi_A(n) \equiv 0$, then

$$\max_{z \in [0,A]} R(z,n) \le \max_{z \in [0,A]} R(z, n-1), \quad n > 0. \qquad (6.12)$$

(e) Consider the initial and boundary conditions $\varphi(z) \equiv 1$ and $\psi_0(n) = \psi_A(n) \equiv 0$. We claim that $\max_{z \in [0,A]} R(z,n)$ decays exponentially in n. For each $z \in [0, A]$ the random walk starting at z reaches one of the endpoints of the segment in A steps or fewer with positive probability, since the event that the first A steps are to the right has positive probability. Therefore,

$$\max_{z \in [0,A]} R(z, A) \le r < 1. \qquad (6.13)$$

If we replace the initial condition $\varphi(z) \equiv 1$ by some other function $\tilde{\varphi}(z)$ with $0 \le \tilde{\varphi}(z) \le 1$, then (6.13) will still hold, since the solution depends monotonically on the initial conditions. Furthermore, if $0 \le \tilde{\varphi}(z) \le c$, then

$$\max_{z \in [0,A]} R(z, A) \le cr \qquad (6.14)$$

with the same r as in (6.13), since the solution depends linearly on the initial conditions. Observe that $\tilde{R}(z,n) = R(z, A+n)$ is the solution of (6.9) with the initial condition $\tilde{\varphi}(z) = R(z, A)$ and zero boundary conditions. Therefore,

$$\max_{z \in [0,A]} R(z, 2A) = \max_{z \in [0,A]} \tilde{R}(z, A) \le r \max_{z \in [0,A]} R(z, A) \le r^2.$$

Proceeding by induction, we can show that $\max_{z \in [0,A]} R(z, kA) \le r^k$ for any integer $k \ge 0$. Coupled with (6.12) this implies

$$\max_{z \in [0,A]} R(z, n) \le r^{\lfloor \frac{n}{A} \rfloor}.$$

We have thus demonstrated that the probability of the game lasting longer than n steps decays exponentially with n. In particular, the expectation of the length of the game is finite for all $z \in [0, A]$.

Let us study the asymptotics of $R(z,n)$ as $n \to \infty$ in more detail. Let M be the $(A-1) \times (A-1)$ matrix with entries above the diagonal equal to p, those below the diagonal equal to q, and the remaining entries equal to zero,

$$M_{i,i+1} = p, \quad M_{i,i-1} = q, \quad \text{and} \quad M_{i,j} = 0 \quad \text{if } |i-j| \neq 1.$$

Define the sequence of vectors v_n, $n \geq 0$, as $v_n = (R(1,n), \ldots, R(A-1,n))$. From (6.9) to (6.11) we see that $v_n = Mv_{n-1}$, and therefore $v_n = M^n v_0$. We could try to use the analysis of Sect. 5.5 to study the asymptotics of M^n. However, now for any s some of the entries of M^s will be equal to zero ($M^s_{i,j} = 0$ if $i - j - s$ is odd). While it is possible to extend the results of Sect. 5.5 to our situation, we shall instead examine the particular case $p = q = \frac{1}{2}$ directly.

If $p = q = \frac{1}{2}$, we can exhibit all the eigenvectors and eigenvalues of the matrix M. Namely, there are $A - 1$ eigenvectors $w_k(z) = \sin(\frac{kz}{A}\pi)$, $k = 1, \ldots, A-1$, where z labels the components of the eigenvectors. The corresponding eigenvalues are $\lambda_k = \cos(\frac{k}{A}\pi)$. To verify this it is enough to note that

$$\frac{1}{2}\sin(\frac{k(z+1)}{A}\pi) + \frac{1}{2}\sin(\frac{k(z-1)}{A}\pi) = \cos(\frac{k}{A}\pi)\sin(\frac{kz}{A}\pi).$$

Let $v_0 = a_1 w_1 + \ldots + a_{A-1} w_{A-1}$ be the representation of v_0 in the basis of eigenvectors. Then

$$M^n v_0 = a_1 \lambda_1^n w_1 + \ldots + a_{A-1}\lambda_{A-1}^n w_{A-1}.$$

Note that λ_1 and λ_{A-1} are the eigenvalues with the largest absolute values, $\lambda_1 = -\lambda_{A-1} = \cos(\frac{\pi}{A})$, while $|\lambda_k| < \cos(\frac{\pi}{A})$ for $1 < k < A - 1$, where we have assumed that $A \geq 3$. Therefore,

$$M^n v_0 = \lambda_1^n(a_1 w_1 + (-1)^n a_{A-1} w_{A-1}) + o(\lambda_1^n) \quad \text{as } n \to \infty.$$

The values of a_1 and a_{A-1} can easily be calculated explicitly given that the eigenvectors form an orthogonal basis. We have thus demonstrated that the main term of the asymptotics of $R(z,n) = v_n(z) = M^n v_0(z)$ decays as $\cos^n(\frac{\pi}{A})$ when $n \to \infty$.

Let $S(z)$ be the expectation of the length of the game which starts at z and $T(z)$ the probability that the gambler will win the fortune A before going broke (the random walk reaching A before reaching 0). Thus, for $0 < z < A$

$$S(z) = \sum_{n=1}^{\infty} n\mathrm{P}_z(0 < \omega_i < A, i = 0, \ldots, n-1, \omega_n \notin (0,A)),$$

$$T(z) = \mathrm{P}_z(0 < \omega_i < A, i = 0, \ldots, n-1, \omega_n = A, \quad \text{for some } n > 0).$$

We have shown that the game ends in finite time with probability one. Therefore the probability of the event that the game ends with the gambler going broke before accumulating the fortune A is equal to $1 - T(z)$. Hence we do not need to study this case separately.

In exactly the same way we obtained (6.9), we can obtain the following equations for $S(z)$ and $T(z)$:

$$S(z) = pS(z+1) + qS(z-1) + 1, \quad 0 < z < A, \tag{6.15}$$

$$T(z) = pT(z+1) + qT(z-1), \quad 0 < z < A, \tag{6.16}$$

with the boundary conditions

$$S(0) = S(A) = 0, \tag{6.17}$$

$$T(0) = 0, \quad T(A) = 1. \tag{6.18}$$

The difference equations (6.15) and (6.16) are time-independent, in contrast to (6.9). Let us demonstrate that both equations have at most one solution (with given boundary conditions). Indeed, suppose that either both $u_1(z)$ and $u_2(z)$ satisfy (6.15) or both satisfy (6.16), and that $u_1(0) = u_2(0)$, $u_1(A) = u_2(A)$. Then the difference $u(z) = u_1(z) - u_2(z)$ satisfies

$$u(z) = pu(z+1) + qu(z-1), \quad 0 < z < A,$$

with the boundary conditions $u(0) = u(A) = 0$.

If $u(z)$ is not identically zero, then there is either a point $0 < z_0 < A$ such that $u(z_0) = \max_{0<z<A} u(z) > 0$, or a point $0 < z_0 < A$ such that $u(z_0) = \min_{0<z<A} u(z) < 0$. Without loss of generality we may assume that the former is the case. Let z_1 be the smallest value of z where the maximum is achieved, so $u(z_1 - 1) < u(z_1)$. Then $u(z_1) > pu(z_1 + 1) + qu(z_1 - 1)$, since $p + q = 1$ and $q > 0$. This contradicts the fact that u is a solution of the equation and thus proves the uniqueness.

We can exhibit explicit formulas for the solutions of (6.15), (6.17) and (6.16), (6.18). Namely, if $p \neq q$, then

$$S(z) = \frac{1}{p-q}\left(\frac{A((\frac{q}{p})^z - 1)}{(\frac{q}{p})^A - 1} - z\right),$$

$$T(z) = \frac{(\frac{q}{p})^z - 1}{(\frac{q}{p})^A - 1}.$$

If $p = q = \frac{1}{2}$, then

$$S(z) = z(A - z),$$

$$T(z) = \frac{z}{A}.$$

Although, by substituting the formulas for $S(z)$ and $T(z)$ into the respective equations, it is easy to verify that these are indeed the required solutions, it is worth explaining how to arrive at the above formulas.

If $p \neq q$, then any linear combination $c_1 u_1(z) + c_2 u_2(z)$ of the functions $u_1(z) = (\frac{q}{p})^z$ and $u_2(z) = 1$ solves the equation

$$f(z) = pf(z+1) + qf(z-1).$$

The function $w(z) = \frac{-z}{p-q}$ solves the non-homogeneous equation

$$f(z) = pf(z+1) + qf(z-1) + 1.$$

We can now look for solutions to (6.15) and (6.16) in the form

$$S(z) = c_1 u_1(z) + c_2 u_2(z) + w \quad \text{and} \quad T(z) = k_1 u_1(z) + k_2 u_2(z),$$

where the constants c_1, c_2, k_1, and k_2 can be found from the respective boundary conditions. If $p = q = \frac{1}{2}$, then we need to take $u_1 = z, u_2 = 1$, and $w = -z^2$.

If the game is fair, that is $p = q = \frac{1}{2}$, then the probability that the gambler will win the fortune A before going broke is directly proportional to the gambler's initial fortune and inversely proportional to A,

$$T(z) = \frac{z}{A}.$$

This is not the case if $p \neq q$. For example, if the game is not favorable for the gambler, that is $p < q$, and the initial fortune is equal to $\frac{A}{2}$, then $T(\frac{A}{2})$ decays exponentially in A.

If $p = q = \frac{1}{2}$ and $z = \frac{A}{2}$, then the expected length of the game is $S(z) = \frac{A^2}{4}$. This is not surprising, since we have already seen that for symmetric random walks the typical displacement has order of the square root of the length of the walk.

6.5 Problems

1. For the three-dimensional simple symmetric random walk which starts at the origin, find the probability of those $\omega = (\omega_0, \omega_1, \ldots)$ for which there is a unique $k \geq 1$ such that $\omega_k = 0$.
2. Prove that the spatially homogeneous one-dimensional random walk with $p_1 = 1 - p_{-1} \neq 1/2$ is non-recurrent.
3. Prove that a spatially homogeneous random walk does not have a stationary probability measure unless $p_0 = 1$.
4. Let t_n be a sequence such that $t_n \sim n$ as $n \to \infty$. Let $(\omega_0, \ldots, \omega_{2n})$ be a trajectory of a simple symmetric random walk on $\mathbb{Z}$. Find the limit of the following conditional probabilities

$$\lim_{n \to \infty} P(a \leq \frac{\omega_{t_n}}{\sqrt{n}} \leq b | \omega_0 = \omega_{2n} = 0),$$

where a and b are fixed numbers.

5. Derive the expression (6.3) for f_{2n} using the reflection principle.
6. Suppose that in the gambler's ruin problem the stake is reduced from 1 to $1/2$. How will that affect the probability of the gambler accumulating the fortune A before going broke? Examine each of the cases $p < q$, $p = q$, and $p > q$ separately.
7. Let us modify the gambler's ruin problem to allow for a possibility of a draw. That is, the gambler wins with probability p, looses with probability q, and draws with probability r, where $p + q + r = 1$. Let $S(z)$ be the expectation of the length of the game which starts at z and $T(z)$ the probability that the gambler will win the fortune A before going broke. Find $S(z)$ and $T(z)$.
8. Consider a homogeneous Markov chain on a finite state space $X = \{x^1, \ldots, x^r\}$ with the transition matrix P and the initial distribution μ. Assume that all the elements of P are positive. Let $\tau = \min\{n : w_n = x^1\}$. Find $E\tau$.

7

Laws of Large Numbers

7.1 Definitions, the Borel-Cantelli Lemmas, and the Kolmogorov Inequality

We again turn our discussion to sequences of independent random variables. Let $\xi_1, \xi_2, \ldots$ be a sequence of random variables with finite expectations $m_n = \mathrm{E}\xi_n$, $n = 1, 2, \ldots$. Let $\zeta_n = (\xi_1 + \ldots + \xi_n)/n$ and $\overline{\zeta}_n = (m_1 + \ldots + m_n)/n$.

Definition 7.1. *The sequence of random variables ξ_n satisfies the Law of Large Numbers if $\zeta_n - \overline{\zeta}_n$ converges to zero in probability, that is $\mathrm{P}(|\zeta_n - \overline{\zeta}_n| > \varepsilon) \to 0$ as $n \to \infty$ for any $\varepsilon > 0$.*

It satisfies the Strong Law of Large Numbers if $\zeta_n - \overline{\zeta}_n$ converges to zero almost surely, that is $\lim_{n \to \infty}(\zeta_n - \overline{\zeta}_n) = 0$ for almost all ω.

If the random variables ξ_n are independent, and if $\mathrm{Var}(\xi_i) \le V < \infty$, then by the Chebyshev Inequality, the Law of Large Numbers holds:

$$\mathrm{P}(|\zeta_n - \overline{\zeta}_n| > \varepsilon) = \mathrm{P}(|\xi_1 + \ldots + \xi_n - (m_1 + \ldots + m_n)| \ge \varepsilon n)$$
$$\le \frac{\mathrm{Var}(\xi_1 + \ldots + \xi_n)}{\varepsilon^2 n^2} \le \frac{V}{\varepsilon^2 n},$$

which tends to zero as $n \to \infty$. There is a stronger statement due to Khinchin:

Theorem 7.2 (Khinchin). *A sequence ξ_n of independent identically distributed random variables with finite mathematical expectation satisfies the Law of Large Numbers.*

Historically, the Khinchin Theorem was one of the first theorems related to the Law of Large Numbers. We shall not prove it now, but obtain it later as a consequence of the Birkhoff Ergodic Theorem, which will be discussed in Chap. 16.

We shall need the following three general statements.

L. Koralov and Y.G. Sinai, *Theory of Probability and Random Processes*, Universitext, DOI 10.1007/978-3-540-68829-7_7,
© Springer-Verlag Berlin Heidelberg 2012

Lemma 7.3 (First Borel-Cantelli Lemma). *Let* $(\Omega, \mathcal{F}, P)$ *be a proba-bility space and* $\{A_n\}$ *an infinite sequence of events,* $A_n \subseteq \Omega$, *such that* $\sum_{n=1}^{\infty} P(A_n) < \infty$. *Define*

$$A = \{\omega : \text{there is an infinite sequence } n_i(\omega) \text{ such that } \omega \in A_{n_i}, i = 1, 2, \ldots\}.$$

Then $P(A) = 0$.

Proof. Clearly,

$$A = \bigcap_{k=1}^{\infty} \bigcup_{n=k}^{\infty} A_n.$$

Then $P(A) \leq P(\bigcup_{n=k}^{\infty} A_n) \leq \sum_{n=k}^{\infty} P(A_n) \to 0$ as $k \to \infty$.   □

Lemma 7.4 (Second Borel-Cantelli Lemma). *Let* A_n *be an infinite sequence of independent events with* $\sum_{n=1}^{\infty} P(A_n) = \infty$, *and let*

$$A = \{\omega : \text{there is an infinite sequence } n_i(\omega) \text{ such that } \omega \in A_{n_i}, i = 1, 2, \ldots\}.$$

Then $P(A) = 1$.

Proof. We have $\Omega \backslash A = \bigcup_{k=1}^{\infty} \bigcap_{n=k}^{\infty} (\Omega \backslash A_n)$. Then

$$P(\Omega \backslash A) \leq \sum_{k=1}^{\infty} P(\bigcap_{n=k}^{\infty} (\Omega \backslash A_n))$$

for any n. By the independence of A_n we have the independence of $\Omega \backslash A_n$, and therefore

$$P(\bigcap_{n=k}^{\infty} (\Omega \backslash A_n)) = \prod_{n=k}^{\infty} (1 - P(A_n)).$$

The fact that $\sum_{n=k}^{\infty} P(A_n) = \infty$ for any k implies that $\prod_{n=k}^{\infty} (1 - P(A_n)) = 0$ (see Problem 1).   □

Theorem 7.5 (Kolmogorov Inequality). *Let* $\xi_1, \xi_2, \ldots$ *be a sequence of independent random variables which have finite mathematical expectations and variances,* $m_i = E\xi_i$, $V_i = \text{Var}(\xi_i)$. *Then*

$$P(\max_{1 \leq k \leq n} |(\xi_1 + \ldots + \xi_k) - (m_1 + \ldots + m_k)| \geq t) \leq \frac{1}{t^2} \sum_{i=1}^{n} V_i.$$

Proof. We consider the events $C_k = \{\omega : |(\xi_1 + \ldots + \xi_i) - (m_1 + \ldots + m_i)| < t$ for $1 \leq i < k, |(\xi_1 + \ldots + \xi_k) - (m_1 + \ldots + m_k)| \geq t\}$, $C = \bigcup_{k=1}^{n} C_k$. It is clear that C is the event whose probability is estimated in the Kolmogorov Inequality, and that C_k are pair-wise disjoint. Thus

$$\sum_{i=1}^{n} V_i = \text{Var}(\xi_1 + \ldots + \xi_n) = \int_{\Omega} ((\xi_1 + \ldots + \xi_n) - (m_1 + \ldots + m_n))^2 dP \geq$$

$$\sum_{k=1}^{n} \int_{C_k} ((\xi_1 + \ldots + \xi_n) - (m_1 + \ldots + m_n))^2 dP =$$

$$\sum_{k=1}^{n} [\int_{C_k} ((\xi_1 + \ldots + \xi_k) - (m_1 + \ldots + m_k))^2 dP +$$

$$2 \int_{C_k} ((\xi_1 + \ldots + \xi_k) - (m_1 + \ldots + m_k))((\xi_{k+1} + \ldots + \xi_n) - (m_{k+1} + \ldots + m_n)) dP +$$

$$\int_{C_k} ((\xi_{k+1} + \ldots + \xi_n) - (m_{k+1} + \ldots + m_n))^2 dP].$$

The last integral on the right-hand side is non-negative. Most importantly, the middle integral is equal to zero. Indeed, by Lemma 4.15, the random variables

$$\eta_1 = ((\xi_1 + \ldots + \xi_k) - (m_1 + \ldots + m_k)) \chi_{C_k}$$

and

$$\eta_2 = (\xi_{k+1} + \ldots + \xi_n) - (m_{k+1} + \ldots + m_n)$$

are independent. By Theorem 4.8, the expectation of their product is equal to the product of the expectations. Thus, the middle integral is equal to

$$E(\eta_1 \eta_2) = E\eta_1 E\eta_2 = 0.$$

Therefore,

$$\sum_{i=1}^{n} V_i \geq \sum_{k=1}^{n} \int_{C_k} ((\xi_1 + \ldots + \xi_k) - (m_1 + \ldots + m_k))^2 dP \geq$$

$$t^2 \sum_{k=1}^{n} P(C_k) = t^2 P(C).$$

That is $P(C) \leq \frac{1}{t^2} \sum_{i=1}^{n} V_i.$ □

7.2 Kolmogorov Theorems on the Strong Law of Large Numbers

Theorem 7.6 (First Kolmogorov Theorem). *A sequence of independent random variables ξ_i, such that $\sum_{i=1}^{\infty} \text{Var}(\xi_i)/i^2 < \infty$, satisfies the Strong Law of Large Numbers.*

Proof. Without loss of generality we may assume that $m_i = \mathrm{E}\xi_i = 0$ for all i. Otherwise we could define a new sequence of random variables $\xi_i' = \xi_i - m_i$. We need to show that $\zeta_n = (\xi_1 + \ldots + \xi_n)/n \to 0$ almost surely. Let $\varepsilon > 0$, and consider the event

$$B(\varepsilon) = \{\omega : \text{there is } N = N(\omega) \text{ such that for all } n \geq N(\omega) \text{ we have } |\zeta_n| < \varepsilon\}.$$

Clearly

$$B(\varepsilon) = \bigcup_{N=1}^{\infty} \bigcap_{n=N}^{\infty} \{\omega : |\zeta_n| < \varepsilon\}.$$

Let

$$B_k(\varepsilon) = \{\omega : \max_{2^{k-1} \leq n < 2^k} |\zeta_n| \geq \varepsilon\}.$$

By the Kolmogorov Inequality,

$$\mathrm{P}(B_k(\varepsilon)) = \mathrm{P}(\max_{2^{k-1} \leq n < 2^k} \frac{1}{n}|\sum_{i=1}^{n} \xi_i| \geq \varepsilon|) \leq$$

$$\mathrm{P}(\max_{2^{k-1} \leq n < 2^k} |\sum_{i=1}^{n} \xi_i| \geq \varepsilon 2^{k-1}) \leq$$

$$\mathrm{P}(\max_{1 \leq n < 2^k} |\sum_{i=1}^{n} \xi_i| \geq \varepsilon 2^{k-1}) \leq \frac{1}{\varepsilon^2 2^{2k-2}} \sum_{i=1}^{2^k} \mathrm{Var}(\xi_i).$$

Therefore,

$$\sum_{k=1}^{\infty} \mathrm{P}(B_k(\varepsilon)) \leq \sum_{k=1}^{\infty} \frac{1}{\varepsilon^2 2^{2k-2}} \sum_{i=1}^{2^k} \mathrm{Var}(\xi_i) =$$

$$\frac{1}{\varepsilon^2} \sum_{i=1}^{\infty} \mathrm{Var}(\xi_i) \sum_{k \geq [\log_2 i]} \frac{1}{2^{2k-2}} \leq \frac{c}{\varepsilon^2} \sum_{i=1}^{\infty} \frac{\mathrm{Var}(\xi_i)}{i^2} < \infty,$$

where c is some constant. By the First Borel-Cantelli Lemma, for almost every ω there exists an integer $k_0 = k_0(\omega)$ such that $\max_{2^{k-1} \leq n \leq 2^k} |\zeta_n| < \varepsilon$ for all $k \geq k_0$. Therefore $\mathrm{P}(B(\varepsilon)) = 1$ for any $\varepsilon > 0$. In particular $\mathrm{P}(B(\frac{1}{m})) = 1$ and $\mathrm{P}(\bigcap_m B(\frac{1}{m})) = 1$. But if $\omega \in \bigcap_m B(\frac{1}{m})$, then for any m there exists $N = N(\omega, m)$ such that for all $n \geq N(\omega, m)$ we have $|\zeta_n| < \frac{1}{m}$. In other words, $\lim_{n \to \infty} \zeta_n = 0$ for such ω.   □

Theorem 7.7 (Second Kolmogorov Theorem). *A sequence ξ_i of independent identically distributed random variables with finite mathematical expectation $m = \mathrm{E}\xi_i$ satisfies the Strong Law of Large Numbers.*

This theorem follows from the Birkhoff Ergodic Theorem, which is discussed in Chap. 16. For this reason we do not provide its proof now.

The Law of Large Numbers, as well as the Strong Law of Large Numbers, is related to theorems known as Ergodic Theorems. These theorems give general conditions under which the averages of random variables have a limit.

Both Laws of Large Numbers state that for a sequence of random variables ξ_n, the average $\frac{1}{n}\sum_{i=1}^n \xi_i$ is close to its mathematical expectation, and therefore does not depend asymptotically on ω, i.e., it is not random. In other words, deterministic regularity appears with high probability in long series of random variables.

Let c be a constant and define

$$\xi^c(\omega) = \begin{cases} \xi(\omega) & \text{if } |\xi(\omega)| \leq c, \\ 0 & \text{if } |\xi(\omega)| > c. \end{cases}$$

Theorem 7.8 (Three Series Theorem). *Let ξ_i be a sequence of independent random variables. If for some $c > 0$ each of the three series*

$$\sum_{i=1}^{\infty} E\xi_i^c, \qquad \sum_{i=1}^{\infty} \text{Var}(\xi_i^c), \qquad \sum_{i=1}^{\infty} P(|\xi_i| \geq c)$$

converges, then the series $\sum_{i=1}^{\infty} \xi_i$ converges almost surely.

Conversely, if the series $\sum_{i=1}^{\infty} \xi_i$ converges almost surely, then each of the three series above also converges for each $c > 0$.

Proof. We'll only prove the direct statement, leaving the converse as an exercise for the reader.

We first establish the almost sure convergence of the series $\sum_{i=1}^{\infty}(\xi_i^c - E\xi_i^c)$. Let $S_n = \sum_{i=1}^n (\xi_i^c - E\xi_i^c)$. Then, by the Kolmogorov Inequality, for any $\varepsilon > 0$

$$P(\sup_{i \geq 1}|S_{n+i} - S_n| \geq \varepsilon) = \lim_{N \to \infty} P(\max_{1 \leq i \leq N}|S_{n+i} - S_n| \geq \varepsilon) \leq$$

$$\lim_{N \to \infty} \frac{\sum_{i=n+1}^{n+N} E(\xi_i^c)^2}{\varepsilon^2} = \frac{\sum_{i=n+1}^{\infty} E(\xi_i^c)^2}{\varepsilon^2}.$$

The right-hand side can be made arbitrarily small by choosing n large enough. Therefore

$$\lim_{n \to \infty} P(\sup_{i \geq 1}|S_{n+i} - S_n| \geq \varepsilon) = 0.$$

Hence the sequence S_n is fundamental almost surely. Otherwise a set of positive measure would exist where $\sup_{i \geq 1}|S_{n+i} - S_n| \geq \varepsilon$ for some $\varepsilon > 0$. We have therefore proved that the series $\sum_{i=1}^{\infty}(\xi_i^c - E\xi_i^c)$ converges almost surely. By the hypothesis, the series $\sum_{i=1}^{\infty} E\xi_i^c$ converges almost surely. Therefore $\sum_{i=1}^{\infty} \xi_i^c$ converges almost surely.

Since $\sum_{i=1}^{\infty} P(|\xi_i| \geq c) < \infty$ almost surely, the First Borel-Cantelli Lemma implies that $P(\{\omega : |\xi_i| \geq c \text{ for infinitely many } i\}) = 0$. Therefore, $\xi_i^c = \xi_i$ for all but finitely many i with probability one. Thus the series $\sum_{i=1}^{\infty} \xi_i$ also converges almost surely. $\square$

7.3 Problems

1. Let $y_1, y_2, \ldots$ be a sequence such that $0 \leq y_n \leq 1$ for all n, and $\sum_{n=1}^{\infty} y_n = \infty$. Prove that $\prod_{n=1}^{\infty}(1 - y_n) = 0$.

2. Let $\xi_1, \xi_2, \ldots$ be independent identically distributed random variables. Prove that $\sup_n \xi_n = \infty$ almost surely if and only if $P(\xi_1 > A) > 0$ for every A.

3. Let $\xi_1, \xi_2, \ldots$ be a sequence of random variables defined on the same probability space. Prove that there exists a numeric sequence $c_1, c_2, \ldots$ such that $\xi_n / c_n \to 0$ almost surely as $n \to \infty$.

4. For each $\gamma > 2$, define the set $D_\gamma \subset [0, 1]$ as follows: $x \in D_\gamma$ if there is $K_\gamma(x) > 0$ such that for each $q \in \mathbb{N}$

$$\min_{p \in \mathbb{N}} \left| x - \frac{p}{q} \right| \geq \frac{K_\gamma(x)}{q^\gamma}.$$

(The numbers x which satisfy this inequality for some $\gamma > 2$, $K_\gamma(x) > 0$, and all $q \in \mathbb{N}$ are called Diophantine.) Prove that $\lambda(D_\gamma) = 1$, where λ is the Lebesgue measure on $([0, 1], \mathcal{B}([0, 1]))$.

5. Let $\xi_1, \ldots, \xi_n$ be a sequence of n independent random variables, each ξ_i having a symmetric distribution. That is, $P(\xi_i \in A) = P(\xi_i \in -A)$ for any Borel set $A \subseteq \mathbb{R}$. Assume that $E\xi_i^{2m} < \infty$, $i = 1, 2, \ldots, n$. Prove the stronger version of the Kolmogorov Inequality:

$$P\left(\max_{1 \leq k \leq n} |\xi_1 + \ldots + \xi_k| \geq t \right) \leq \frac{E(\xi_1 + \ldots + \xi_n)^{2m}}{t^{2m}}.$$

6. Let $\xi_1, \xi_2, \ldots$ be independent random variables with non-negative values. Prove that the series $\sum_{i=1}^{\infty} \xi_i$ converges almost surely if and only if

$$\sum_{i=1}^{\infty} E \frac{\xi_i}{1 + \xi_i} < \infty.$$

7. Let $\xi_1, \xi_2, \ldots$ be a sequence of independent identically distributed random variables with uniform distribution on $[0, 1]$. Prove that the limit

$$\lim_{n \to \infty} \sqrt[n]{\xi_1 \cdot \ldots \cdot \xi_n}$$

exists with probability one. Find its value.

8. Let $\xi_1, \xi_2, \ldots$ be a sequence of independent random variables, $P(\xi_i = 2^i) = 1/2^i$, $P(\xi_i = 0) = 1 - 1/2^i$, $i \geq 1$. Find the almost sure value of the limit $\lim_{n \to \infty} (\xi_1 + \ldots + \xi_n)/n$.

9. Let $\xi_1, \xi_2, \ldots$ be a sequence of independent identically distributed random variables for which $E\xi_i = 0$ and $E\xi_i^2 = V < \infty$. Prove that for any $\gamma > 1/2$, the series $\sum_{i \geq 1} \xi_i / i^\gamma$ converges almost surely.

10. Let $\xi_1, \xi_2, \ldots$ be independent random variables uniformly distributed on the interval $[-1, 1]$. Let $a_1, a_2, \ldots$ be a sequence of real numbers such that $\sum_{n=1}^{\infty} a_n^2$ converges. Prove that the series $\sum_{n=1}^{\infty} a_n \xi_n$ converges almost surely.

Weak Convergence of Measures

8.1 Definition of Weak Convergence

In this chapter we consider the fundamental concept of weak convergence of probability measures. This will lay the groundwork for the precise formulation of the Central Limit Theorem and other Limit Theorems of probability theory (see Chap. 10).

Let (X, d) be a metric space, $\mathcal{B}(X)$ the σ-algebra of its Borel sets and P_n a sequence of probability measures on $(X, \mathcal{B}(X))$. Recall that $C_b(X)$ denotes the space of bounded continuous functions on X.

Definition 8.1. *The sequence* P_n *converges weakly to the probability measure* P *if, for each* $f \in C_b(X)$,

$$\lim_{n \to \infty} \int_X f(x) d\mathrm{P}_n(x) = \int_X f(x) d\mathrm{P}(x).$$

The weak convergence is sometimes denoted as $\mathrm{P}_n \Rightarrow \mathrm{P}$.

Definition 8.2. *A sequence of real-valued random variables* ξ_n *defined on probability spaces* $(\Omega_n, \mathcal{F}_n, \overline{\mathrm{P}}_n)$ *is said to converge in distribution if the induced measures* P_n, $\mathrm{P}_n(A) = \overline{\mathrm{P}}_n(\xi_n \in A)$, *converge weakly to a probability measure* P.

In Definition 8.1 we could omit the requirement that P_n and P are probability measures. We then obtain the definition of the weak convergence for arbitrary finite measures on $\mathcal{B}(X)$. The following lemma provides a useful criterion for the weak convergence of measures.

Lemma 8.3. *If a sequence of measures* P_n *converges weakly to a measure* P, *then*

$$\limsup_{n \to \infty} \mathrm{P}_n(K) \leq \mathrm{P}(K) \tag{8.1}$$

for any closed set K. *Conversely, if (8.1) holds for any closed set* K, *and* $\mathrm{P}_n(X) = \mathrm{P}(X)$ *for all* n, *then* P_n *converge weakly to* P.

L. Koralov and Y.G. Sinai, *Theory of Probability and Random Processes*,
Universitext, DOI 10.1007/978-3-540-68829-7_8,
© Springer-Verlag Berlin Heidelberg 2012

Proof. First assume that P_n converges to P weakly. Let $\varepsilon > 0$ and select $\delta > 0$ such that $P(K_\delta) < P(K) + \varepsilon$, where K_δ is the δ-neighborhood of the set K. Consider a continuous function f_δ such that $0 \leq f_\delta(x) \leq 1$ for $x \in X$, $f_\delta(x) = 1$ for $x \in K$, and $f_\delta(x) = 0$ for $x \in X \backslash K_\delta$. For example, one can take $f_\delta(x) = \max(1 - \text{dist}(x, K)/\delta, 0)$.

Note that $P_n(K) = \int_K f_\delta dP_n \leq \int_X f_\delta dP_n$ and $\int_X f_\delta dP = \int_{K_\delta} f_\delta dP \leq P(K_\delta) < P(K) + \varepsilon$. Therefore,

$$\lim_{n\to\infty} \sup P_n(K) \leq \lim_{n\to\infty} \int_X f_\delta dP_n = \int_X f_\delta dP < P(K) + \varepsilon,$$

which implies the result since ε was arbitrary.

Let us now assume that $P_n(X) = P(X)$ for all n and $\limsup_{n\to\infty} P_n(K) \leq P(K)$ for any closed set K. Let $f \in C_b(X)$. We can find $a > 0$ and b such that $0 < af + b < 1$. Since $P_n(X) = P(X)$ for all n, if the relation

$$\lim_{n\to\infty} \int_X g(x)dP_n(x) = \int_X g(x)dP(x)$$

is valid for $g = af + b$, then it is also valid for f instead of g. Therefore, without loss of generality, we can assume that $0 < f(x) < 1$ for all x. Define the closed sets $K_i = \{x : f(x) \geq i/k\}$, where $0 \leq i \leq k$. Then

$$\frac{1}{k}\sum_{i=1}^{k} P_n(K_i) \leq \int_X f dP_n \leq \frac{P_n(X)}{k} + \frac{1}{k}\sum_{i=1}^{k} P_n(K_i),$$

$$\frac{1}{k}\sum_{i=1}^{k} P(K_i) \leq \int_X f dP \leq \frac{P(X)}{k} + \frac{1}{k}\sum_{i=1}^{k} P(K_i).$$

Since $\limsup_{n\to\infty} P_n(K_i) \leq P(K)$ for each i, and $P_n(X) = P(X)$, we obtain

$$\limsup_{n\to\infty} \int_X f dP_n \leq \frac{P(X)}{k} + \int_X f dP.$$

Taking the limit as $k \to \infty$, we obtain

$$\limsup_{n\to\infty} \int_X f dP_n \leq \int_X f dP.$$

By considering the function $-f$ instead of f we can obtain

$$\liminf_{n\to\infty} \int_X f dP_n \geq \int_X f dP.$$

This proves the weak convergence of measures.   □

The following lemma will prove useful when proving the Prokhorov Theorem below.

Lemma 8.4. *Let X be a metric space and $\mathcal{B}(X)$ the σ-algebra of its Borel sets. Any finite measure P on $(X, \mathcal{B}(X))$ is regular, that is for any $A \in \mathcal{B}(X)$ and any $\varepsilon > 0$ there are an open set U and a closed set K such that $K \subseteq A \subseteq U$ and $\mathrm{P}(U) - \mathrm{P}(K) < \varepsilon$.*

Proof. If A is a closed set, we can take $K = A$ and consider a sequence of open sets $U_n = \{x : \mathrm{dist}(x, A) < 1/n\}$. Since $\bigcap_n U_n = A$, there is a sufficiently large n such that $\mathrm{P}(U_n) - \mathrm{P}(A) < \varepsilon$. This shows that the statement is true for all closed sets.

Let $\mathcal{K}$ be the collection of sets A such that for any ε there exist K and U with the desired properties. Note that the collection of all closed sets is a π-system. Clearly, $A \in \mathcal{K}$ implies that $X \backslash A \in \mathcal{K}$. Therefore, due to Lemma 4.13, it remains to prove that if $A_1, A_2, \ldots \in \mathcal{K}$ and $A_i \cap A_j = \emptyset$ for $i \neq j$, then $A = \bigcup_n A_n \in \mathcal{K}$.

Let $\varepsilon > 0$. Find n_0 such that $\mathrm{P}(\bigcup_{n=n_0}^{\infty} A_n) < \varepsilon/2$. Find open sets U_n and closed sets K_n such that $K_n \subseteq A_n \subseteq U_n$ and $\mathrm{P}(U_n) - \mathrm{P}(K_n) < \varepsilon/2^{n+1}$ for each n. Then $U = \bigcup_n U_n$ and $K = \bigcup_{n=1}^{n_0} K_n$ have the desired properties, that is $K \subseteq A \subseteq U$ and $\mathrm{P}(U) - \mathrm{P}(K) < \varepsilon$. $\qquad\square$

8.2 Weak Convergence and Distribution Functions

Recall the one-to-one correspondence between the probability measures on $\mathbb{R}$ and the distribution functions. Let F_n and F be the distribution functions corresponding to the measures P_n and P respectively. Note that x is a continuity point of F if and only if $\mathrm{P}(x) = 0$. We now express the condition of weak convergence in terms of the distribution functions.

Theorem 8.5. *The sequence of probability measures P_n converges weakly to the probability measure P if and only if $\lim_{n\to\infty} F_n(x) = F(x)$ for every continuity point x of the function F.*

Proof. Let $\mathrm{P}_n \Rightarrow \mathrm{P}$ and let x be a continuity point of F. We consider the functions f, f_δ^+ and f_δ^-, which are defined as follows:

$$f(y) = \begin{cases} 1, & y \leq x, \\ 0, & y > x, \end{cases}$$

$$f_\delta^+(y) = \begin{cases} 1, & y \leq x, \\ 1 - (y - x)/\delta, & x < y \leq x + \delta, \\ 0, & y > x + \delta, \end{cases}$$

$$f_\delta^-(y) = \begin{cases} 1, & y \leq x - \delta, \\ 1 - (y - x + \delta)/\delta, & x - \delta < y \leq x, \\ 0, & y > x. \end{cases}$$

The functions f_δ^+ and f_δ^- are continuous and $f_\delta^- \leq f \leq f_\delta^+$. Using the fact that x is a continuity point of F we have, for any $\varepsilon > 0$ and $n \geq n_0(\varepsilon)$,

$$F_n(x) = \int_{\mathbb{R}} f(y)dF_n(y) \leq \int_{\mathbb{R}} f_\delta^+(y)dF_n(y)$$

$$\leq \int_{\mathbb{R}} f_\delta^+(y)dF(y) + \frac{\varepsilon}{2} \leq F(x + \delta) + \frac{\varepsilon}{2} \leq F(x) + \varepsilon,$$

if δ is such that $|F(x \pm \delta) - F(x)| \leq \frac{\varepsilon}{2}$. On the other hand, for such n we also have

$$F_n(x) = \int_{\mathbb{R}} f(y)dF_n(y) \geq \int_{\mathbb{R}} f_\delta^-(y)dF_n(y)$$

$$\geq \int_{\mathbb{R}} f_\delta^-(y)dF(y) - \frac{\varepsilon}{2} \geq F(x - \delta) - \frac{\varepsilon}{2} \geq F(x) - \varepsilon.$$

In other words, $|F_n(x) - F(x)| \leq \varepsilon$ for all sufficiently large n.

Now we prove the converse. Let $F_n(x) \to F(x)$ at every continuity point of F. Let f be a bounded continuous function. Let ε be an arbitrary positive constant. We need to prove that

$$|\int_{\mathbb{R}} f(x)dF_n(x) - \int_{\mathbb{R}} f(x)dF(x)| \leq \varepsilon \tag{8.2}$$

for sufficiently large n.

Let $M = \sup |f(x)|$. Since the function F is non-decreasing, it has at most a countable number of points of discontinuity. Select two points of continuity A and B for which $F(A) \leq \frac{\varepsilon}{10M}$ and $F(B) \geq 1 - \frac{\varepsilon}{10M}$. Therefore $F_n(A) \leq \frac{\varepsilon}{5M}$ and $F_n(B) \geq 1 - \frac{\varepsilon}{5M}$ for all sufficiently large n.

Since f is continuous, it is uniformly continuous on $[A, B]$. Therefore we can partition the half-open interval $(A, B]$ into finitely many half-open subintervals $I_1 = (x_0, x_1], I_2 = (x_1, x_2], \ldots, I_n = (x_{n-1}, x_n]$ such that $|f(y) - f(x_i)| \leq \frac{\varepsilon}{10}$ for $y \in I_i$. Moreover, the endpoints x_i can be selected to be continuity points of $F(x)$. Let us define a new function f_ε on $(A, B]$ which is equal to $f(x_i)$ on each of the intervals I_i.

In order to prove (8.2), we write

$$|\int_{\mathbb{R}} f(x)dF_n(x) - \int_{\mathbb{R}} f(x)dF(x)|$$

$$\leq \int_{(-\infty, A]} |f(x)|dF_n(x) + \int_{(-\infty, A]} |f(x)|dF(x)$$

$$+ \int_{(B, \infty)} |f(x)|dF_n(x) + \int_{(B, \infty)} |f(x)|dF(x)$$

$$+ \left| \int_{(A,B]} f(x) dF_n(x) - \int_{(A,B]} f(x) dF(x) \right|.$$

The first term on the right-hand side is estimated from above for large enough n as follows:

$$\int_{(-\infty,A]} |f(x)| dF_n(x) \leq M F_n(A) \leq \frac{\varepsilon}{5}$$

Similarly, the second, third and fourth terms are estimated from above by $\frac{\varepsilon}{10}$, $\frac{\varepsilon}{5}$ and $\frac{\varepsilon}{10}$ respectively.

Since $|f_\varepsilon - f| \leq \frac{\varepsilon}{10}$ on $(A, B]$, the last term can be estimated as follows:

$$\left| \int_{(A,B]} f(x) dF_n(x) - \int_{(A,B]} f(x) dF(x) \right|$$

$$\leq \left| \int_{(A,B]} f_\varepsilon(x) dF_n(x) - \int_{(A,B]} f_\varepsilon(x) dF(x) \right| + \frac{\varepsilon}{5}.$$

Note that

$$\lim_{n \to \infty} \left| \int_{I_i} f_\varepsilon(x) dF_n(x) - \int_{I_i} f_\varepsilon(x) dF(x) \right|$$

$$= \lim_{n \to \infty} \left(|f(x_i)| |F_n(x_i) - F_n(x_{i-1}) - F(x_i) + F(x_{i-1})| \right) = 0,$$

since $F_n(x) \to F(x)$ at the endpoints of the interval I_i. Therefore,

$$\lim_{n \to \infty} \left| \int_{(A,B]} f_\varepsilon(x) dF_n(x) - \int_{(A,B]} f_\varepsilon(x) dF(x) \right| = 0,$$

and thus

$$\left| \int_{(A,B]} f_\varepsilon(x) dF_n(x) - \int_{(A,B]} f_\varepsilon(x) dF(x) \right| \leq \frac{\varepsilon}{5}$$

for large enough n.   □

8.3 Weak Compactness, Tightness, and the Prokhorov Theorem

Let X be a metric space and P_α a family of probability measures on the Borel σ-algebra $\mathcal{B}(X)$. The following two concepts, weak compactness (sometimes also referred to as relative compactness) and tightness, are fundamental in probability theory.

Definition 8.6. *A family of probability measures* $\{P_\alpha\}$ *on* $(X, \mathcal{B}(X))$ *is said to be weakly compact if from any sequence* P_n, $n = 1, 2, \ldots$, *of measures from the family, one can extract a weakly convergent subsequence* P_{n_k}, $k = 1, 2, \ldots$, *that is* $P_{n_k} \Rightarrow P$ *for some probability measure* P.

Remark 8.7. Note that it is not required that $P \in \{P_\alpha\}$.

Definition 8.8. *A family of probability measures $\{P_\alpha\}$ on $(X, \mathcal{B}(X))$ is said to be tight if for any $\varepsilon > 0$ one can find a compact set $K_\varepsilon \subseteq X$ such that $P(K_\varepsilon) \geq 1 - \varepsilon$ for each $P \in \{P_\alpha\}$.*

In the case when $(X, \mathcal{B}(X)) = (\mathbb{R}, \mathcal{B}(\mathbb{R}))$, we have the following theorem.

Theorem 8.9 (Helly Theorem). *A family of probability measures $\{P_\alpha\}$ on $(\mathbb{R}, \mathcal{B}(\mathbb{R}))$ is tight if and only if it is weakly compact.*

The Helly Theorem is a particular case of the following theorem, due to Prokhorov.

Theorem 8.10 (Prokhorov Theorem). *If a family of probability measures $\{P_\alpha\}$ on a metric space X is tight, then it is weakly compact. On a separable complete metric space the two notions are equivalent.*

The proof of the Prokhorov Theorem will be preceded by two lemmas. The first lemma is a general fact from functional analysis, which is a consequence of the Alaoglu Theorem and will not be proved here.

Lemma 8.11. *Let X be a compact metric space. Then from any sequence of measures μ_n on $(X, \mathcal{B}(X))$, such that $\mu_n(X) \leq C < \infty$ for all n, one can extract a weakly convergent subsequence.*

We shall denote an open ball of radius r centered at a point $a \in X$ by $B(a, r)$. The next lemma provides a criterion of tightness for families of probability measures.

Lemma 8.12. *A family $\{P_\alpha\}$ of probability measures on a separable complete metric space X is tight if and only if for any $\varepsilon > 0$ and $r > 0$ there is a finite family of balls $B(a_i, r)$, $i = 1, \ldots, n$, such that*

$$P_\alpha(\bigcup_{i=1}^{n} B(a_i, r)) \geq 1 - \varepsilon$$

for all α.

Proof. Let $\{P_\alpha\}$ be tight, $\varepsilon > 0$, and $r > 0$. Select a compact set K such that $P(K) \geq 1 - \varepsilon$ for all $P \in \{P_\alpha\}$. Since any compact set is totally bounded, there is a finite family of balls $B(a_i, r)$, $i = 1, \ldots, n$, which cover K. Consequently, $P(\bigcup_{i=1}^{n} B(a_i, r)) \geq 1 - \varepsilon$ for all $P \in \{P_\alpha\}$.

Let us prove the converse statement. Fix $\varepsilon > 0$. Then for any integer $k > 0$ there is a family of balls $B^{(k)}(a_i, \frac{1}{k})$, $i = 1, \ldots, n_k$, such that $P(A_k) \geq 1 - 2^{-k}\varepsilon$ for all $P \in \{P_\alpha\}$, where $A_k = \bigcup_{i=1}^{n_k} B^{(k)}(a_i, \frac{1}{k})$. The set $A = \bigcap_{k=1}^{\infty} A_k$ satisfies $P(A) \geq 1 - \varepsilon$ for all $P \in \{P_\alpha\}$ and is totally bounded. Therefore, its closure is compact since X is a complete metric space. $\qquad\square$

Proof of the Prokhorov Theorem. Assume that a family $\{P_\alpha\}$ is weakly compact but not tight. By Lemma 8.12, there exist $\varepsilon > 0$ and $r > 0$ such that for any family $B_1, \ldots, B_n$ of balls of radius r, we have $P(\bigcup_{1 \le i \le n} B_i) \le 1 - \varepsilon$ for some $P \in \{P_\alpha\}$. Since X is separable, it can be represented as a countable union of balls of radius r, that is $X = \bigcup_{i=1}^\infty B_i$. Let $A_n = \bigcup_{1 \le i \le n} B_i$. Then we can select $P_n \in \{P_\alpha\}$ such that $P_n(A_n) \le 1 - \varepsilon$. Assume that a subsequence P_{n_k} converges to a limit P. Since A_m is open, $P(A_m) \le \liminf_{k \to \infty} P_{n_k}(A_m)$ for every fixed m due to Lemma 8.3. Since $A_m \subseteq A_{n_k}$ for large k, we have $P(A_m) \le \liminf_{k \to \infty} P_{n_k}(A_{n_k}) \le 1 - \varepsilon$, which contradicts $\bigcup_{m=1}^\infty A_m = X$. Thus, weak compactness implies tightness.

Now assume that $\{P_\alpha\}$ is tight. Consider a sequence of compact sets K_m such that

$$P(K_m) \ge 1 - \frac{1}{m} \quad \text{for all } P \in \{P_\alpha\}, \quad m = 1, 2, \ldots$$

Consider a sequence of measures $P_n \in \{P_\alpha\}$. By Lemma 8.11, using the diagonalization procedure, we can construct a subsequence P_{n_k} such that, for each m, the restrictions of P_{n_k} to $\widetilde{K}_m = \bigcup_{i=1}^m K_i$ converge weakly to a measure μ_m. Note that $\mu_m(\widetilde{K}_m) \ge 1 - \frac{1}{m}$ since $P_{n_k}(\widetilde{K}_m) \ge 1 - \frac{1}{m}$ for all k.

Let us show that for any Borel set A the sequence $\mu_m(A \bigcap \widetilde{K}_m)$ is non-decreasing. Thus, we need to show that $\mu_{m_1}(A \bigcap \widetilde{K}_{m_1}) \le \mu_{m_2}(A \bigcap \widetilde{K}_{m_2})$ if $m_1 < m_2$. By considering $A \bigcap \widetilde{K}_{m_1}$ instead of A we can assume that $A \subseteq \widetilde{K}_{m_1}$. Fix an arbitrary $\varepsilon > 0$. Due to the regularity of the measures μ_{m_1} and μ_{m_2} (see Lemma 8.4), there exist sets $\overline{U}^i, \overline{K}^i \subseteq \widetilde{K}_{m_i}$, $i = 1, 2$, such that $\overline{U}^i$ $(\overline{K}^i)$ are open (closed) in the topology of $\widetilde{K}_{m_i}$, $\overline{K}^i \subseteq A \subseteq \overline{U}^i$ and

$$\mu_{m_i}(\overline{U}^i) - \varepsilon < \mu_{m_i}(A) < \mu_{m_i}(\overline{K}^i) + \varepsilon, \quad i = 1, 2.$$

Note that $\overline{U}^1 = \overline{\overline{U}} \cap \widetilde{K}_{m_1}$ for some set $\overline{\overline{U}}$ that is open in the topology of $\widetilde{K}_{m_2}$. Let $U = \overline{\overline{U}} \cap \overline{U}^2$ and $K = \overline{K}^1 \cup \overline{K}^2$. Thus $U \subseteq \widetilde{K}_{m_2}$ is open in the topology of $\widetilde{K}_{m_2}$, $K \subseteq \widetilde{K}_{m_1}$ is closed in the topology of $\widetilde{K}_{m_1}$, $K \subseteq A \subseteq U$ and

$$\mu_{m_1}(U \bigcap \widetilde{K}_{m_1}) - \varepsilon < \mu_{m_1}(A) < \mu_{m_1}(K) + \varepsilon, \tag{8.3}$$

$$\mu_{m_2}(U) - \varepsilon < \mu_{m_2}(A) < \mu_{m_2}(K) + \varepsilon. \tag{8.4}$$

Let f be a continuous function on $\widetilde{K}_{m_2}$ such that $0 \le f \le 1$, $f(x) = 1$ if $x \in K$, and $f(x) = 0$ if $x \notin U$. By (8.3) and (8.4),

$$\left| \mu_{m_1}(A) - \int_{\widetilde{K}_{m_1}} f d\mu_{m_1} \right| < \varepsilon,$$

$$\left| \mu_{m_2}(A) - \int_{\widetilde{K}_{m_2}} f d\mu_{m_2} \right| < \varepsilon.$$

Noting that $\int_{\widetilde{K}_{m_i}} f d\mu_{m_i} = \lim_{k\to\infty} \int_{\widetilde{K}_{m_i}} f d\mathrm{P}_{n_k}$, $i = 1, 2$, and $\int_{\widetilde{K}_{m_1}} f d\mathrm{P}_{n_k} \le \int_{\widetilde{K}_{m_2}} f d\mathrm{P}_{n_k}$, we conclude that

$$\mu_{m_1}(A) \le \mu_{m_2}(A) + 2\varepsilon.$$

Since ε was arbitrary, we obtain the desired monotonicity.
Define
$$\mathrm{P}(A) = \lim_{m\to\infty} \mu_m(A \cap \widetilde{K}_m).$$

Note that $\mathrm{P}(X) = \lim_{m\to\infty} \mu_m(\widetilde{K}_m) = 1$. We must show that P is σ-additive in order to conclude that it is a probability measure. If $A = \bigcup_{i=1}^\infty A_i$ is a union of non-intersecting sets, then

$$\mathrm{P}(A) \ge \lim_{m\to\infty} \mu_m(\bigcup_{i=1}^n A_i \cap \widetilde{K}_m) = \sum_{i=1}^n \mathrm{P}(A_i)$$

for each n, and therefore $\mathrm{P}(A) \ge \sum_{i=1}^\infty \mathrm{P}(A_i)$. If $\varepsilon > 0$ is fixed, then for sufficiently large m

$$\mathrm{P}(A) \le \mu_m(A \cap \widetilde{K}_m) + \varepsilon = \sum_{i=1}^\infty \mu_m(A_i \cap \widetilde{K}_m) + \varepsilon \le \sum_{i=1}^\infty \mathrm{P}(A_i) + \varepsilon.$$

Since ε was arbitrary, $\mathrm{P}(A) \le \sum_{i=1}^\infty \mathrm{P}(A_i)$, and thus P is a probability measure.

It remains to show that the measures P_{n_k} converge to the measure P weakly. Let A be a closed set and $\varepsilon > 0$. Then, by the construction of the sets $\widetilde{K}_m$, there is a sufficiently large m such that

$$\limsup_{k\to\infty} \mathrm{P}_{n_k}(A) \le \limsup_{k\to\infty} \mathrm{P}_{n_k}(A \cap \widetilde{K}_m) + \varepsilon \le \mu_m(A) + \varepsilon \le \mathrm{P}(A) + \varepsilon.$$

By Lemma 8.3, this implies the weak convergence of measures. Therefore the family of measures $\{\mathrm{P}_\alpha\}$ is weakly compact.   □

8.4 Problems

1. Let (X, d) be a separable complete metric space. For $x \in X$, let δ_x be the measure on $(X, \mathcal{B}(X))$ which is concentrated at x, that is $\delta(A) = 1$ if $x \in A$, $\delta(A) = 0$ if $x \notin A$, $A \in \mathcal{B}(X)$. Prove that δ_{x_n} converge weakly if and only if there is $x \in X$ such that $x_n \to x$ as $n \to \infty$.
2. Prove that if P_n and P are probability measures, then P_n converges weakly to P if and only if

$$\liminf_{n\to\infty} \mathrm{P}_n(U) \ge \mathrm{P}(U)$$

for any open set U.

3. Prove that if P_n and P are probability measures, then P_n converges to P weakly if and only if
$$\lim_{n \to \infty} P_n(A) = P(A)$$
for all sets A such that $P(\partial A) = 0$, where ∂A is the boundary of the set A.

4. Let X be a metric space and $\mathcal{B}(X)$ the σ-algebra of its Borel sets. Let μ_1 and μ_2 be two probability measures such that $\int_X f d\mu_1 = \int_X f d\mu_2$ for all $f \in C_b(X)$, $f \geq 0$. Prove that $\mu_1 = \mu_2$.

5. Give an example of a family of probability measures P_n on $(\mathbb{R}, \mathcal{B}(\mathbb{R}))$ such that $P_n \Rightarrow P$ (weakly), P_n, P are absolutely continuous with respect to the Lebesgue measure, yet there exists a Borel set A such that $P_n(A)$ does not converge to $P(A)$.

6. Assume that a sequence of random variables ξ_n converges to a random variable ξ in distribution, and a numeric sequence a_n converges to 1. Prove that $a_n \xi_n$ converges to ξ in distribution.

7. Suppose that ξ_n, η_n, $n \geq 1$, and ξ are random variables defined on the same probability space. Prove that if $\xi_n \Rightarrow \xi$ and $\eta_n \Rightarrow c$, where c is a constant, then $\xi_n \eta_n \Rightarrow c\xi$.

8. Prove that if $\xi_n \to \xi$ in probability, then $P_{\xi_n} \Rightarrow P_\xi$, that is the convergence of the random variables in probability implies weak convergence of the corresponding probability measures.

9. Let P_n, P be probability measures on $(\mathbb{R}, \mathcal{B}(\mathbb{R}))$. Suppose that $\int_\mathbb{R} f dP_n \to \int_\mathbb{R} f dP$ as $n \to \infty$ for every infinitely differentiable function f with compact support. Prove that $P_n \Rightarrow P$.

10. Prove that if ξ_n and ξ are defined on the same probability space, ξ is identically equal to a constant, and ξ_n converge to ξ in distribution, then ξ_n converge to ξ in probability.

11. Consider a Markov transition function P on a compact state space X. Prove that the corresponding Markov chain has at least one stationary measure. (Hint: Take an arbitrary initial measure μ and define $\mu_n = (P^*)^n \mu$, $n \geq 0$. Prove that the sequence of measures $\nu_n = (\mu_0 + \ldots + \mu_{n-1})/n$ is weakly compact, and the limit of a subsequence is a stationary measure.)

Characteristic Functions

9.1 Definition and Basic Properties

In this section we introduce the notion of a characteristic function of a probability measure. First we shall formulate the main definitions and theorems
for measures on the real line. Let P be a probability measure on $\mathcal{B}(\mathbb{R})$.

Definition 9.1. *The characteristic function of a measure* P *is the (complex-
valued) function* $\varphi(\lambda)$ *of the variable* $\lambda \in \mathbb{R}$ *given by*

$$\varphi(\lambda) = \int_{-\infty}^{\infty} e^{i\lambda x} d\mathrm{P}(x).$$

If $\mathrm{P} = \mathrm{P}_\xi$, we shall denote the characteristic function by $\varphi_\xi(\lambda)$ and call it
the characteristic function of the random variable ξ. The definition of the
characteristic function means that $\varphi_\xi(\lambda) = \mathrm{E}e^{i\lambda\xi}$. For example, if ξ takes
values $a_1, a_2, \ldots$ with probabilities $p_1, p_2, \ldots$, then

$$\varphi_\xi(\lambda) = \sum_{k=1}^{\infty} p_k e^{i\lambda a_k}.$$

If ξ has a probability density $p_\xi(x)$, then

$$\varphi_\xi(\lambda) = \int_{-\infty}^{\infty} e^{i\lambda x} p_\xi(x) dx.$$

Definition 9.2. *A complex-valued function* $f(\lambda)$ *is said to be non-negative
definite if for any* $\lambda_1, \ldots, \lambda_r$, *the matrix* F *with entries* $F_{kl} = f(\lambda_k - \lambda_l)$ *is
non-negative definite, that is* $(Fv, v) = \sum_{k,l=1}^{r} f(\lambda_k - \lambda_l) v_k \overline{v}_l \geq 0$ *for any
complex vector* $(v_1, \ldots, v_r)$.

L. Koralov and Y.G. Sinai, *Theory of Probability and Random Processes*,
Universitext, DOI 10.1007/978-3-540-68829-7_9,
© Springer-Verlag Berlin Heidelberg 2012

Lemma 9.3 (Properties of Characteristic Functions).

1. $\varphi(0) = 1$.
2. $|\varphi(\lambda)| \leq 1$.
3. If $\eta = a\xi + b$, where a and b are constants, then

$$\varphi_\eta(\lambda) = e^{i\lambda b}\varphi_\xi(a\lambda).$$

4. If $\varphi_\xi(\lambda_0) = e^{2\pi i\alpha}$ for some $\lambda_0 \neq 0$ and some real α, then ξ takes at most a countable number of values. The values of ξ are of the form $\frac{2\pi}{\lambda_0}(\alpha + m)$, where m is an integer.
5. $\varphi(\lambda)$ is uniformly continuous.
6. Any characteristic function $\varphi(\lambda)$ is non-negative definite.
7. Assume that the random variable ξ has an absolute moment of order k, that is $\mathrm{E}|\xi|^k < \infty$. Then $\varphi(\lambda)$ is k times continuously differentiable and

$$\varphi_\xi^{(k)}(0) = i^k \mathrm{E}\xi^k.$$

Proof. The first property is clear from the definition of the characteristic function.

The second property follows from

$$|\varphi(\lambda)| = |\int_{-\infty}^{\infty} e^{i\lambda x} d\mathrm{P}(x)| \leq \int_{-\infty}^{\infty} d\mathrm{P}(x) = 1.$$

The third property follows from

$$\varphi_\eta(\lambda) = \mathrm{E}e^{i\lambda\eta} = \mathrm{E}e^{i\lambda(a\xi+b)} = e^{i\lambda b}\mathrm{E}e^{i\lambda a\xi} = e^{i\lambda b}\varphi_\xi(a\lambda).$$

In order to prove the fourth property, we define $\eta = \xi - \frac{2\pi\alpha}{\lambda_0}$. By the third property,

$$\varphi_\eta(\lambda_0) = e^{-2\pi i\alpha}\varphi_\xi(\lambda_0) = 1.$$

Furthermore,

$$1 = \varphi_\eta(\lambda_0) = \mathrm{E}e^{i\lambda_0\eta} = \mathrm{E}\cos(\lambda_0\eta) + i\mathrm{E}\sin(\lambda_0\eta).$$

Since $\cos(\lambda_0\eta) \leq 1$, the latter equality means that $\cos(\lambda_0\eta) = 1$ with probability one. This is possible only when η takes values of the form $\eta = \frac{2\pi m}{\lambda_0}$, where m is an integer.

The fifth property follows from the Lebesgue Dominated Convergence Theorem, since

$$|\varphi(\lambda) - \varphi(\lambda')| = |\int_{-\infty}^{\infty} (e^{i\lambda x} - e^{i\lambda' x}) d\mathrm{P}(x)| \leq \int_{-\infty}^{\infty} |e^{i(\lambda-\lambda')x} - 1| d\mathrm{P}(x).$$

To prove the sixth property it is enough to note that

$$\sum_{k,l=1}^{r} \varphi(\lambda_k - \lambda_l)v_k\overline{v}_l = \sum_{k,l=1}^{r} \int_{-\infty}^{\infty} e^{i(\lambda_k - \lambda_l)x} v_k\overline{v}_l dP(x)$$

$$= \int_{-\infty}^{\infty} |\sum_{k=1}^{r} v_k e^{i\lambda_k x}|^2 dP(x) \geq 0.$$

The converse is also true. The Bochner Theorem states that any continuous non-negative definite function which satisfies the normalization condition $\varphi(0) = 1$ is the characteristic function of some probability measure (see Sect. 15.3).

The idea of the proof of the seventh property is to use the properties of the Lebesgue integral in order to justify the differentiation in the formal equality

$$\varphi_\xi^{(k)}(\lambda) = \frac{d^k}{d\lambda^k} \int_{-\infty}^{\infty} e^{i\lambda x} dP(x) = i^k \int_{-\infty}^{\infty} x^k e^{i\lambda x} dP(x).$$

The last integral is finite since $E|\xi|^k$ is finite. □

There are more general statements, of which the seventh property is a consequence, relating the existence of various moments of ξ to the smoothness of the characteristic function, with implications going in both directions. Similarly, the rate of decay of $\varphi(\lambda)$ at infinity is responsible for the smoothness class of the distribution. For example, it is not difficult to show that if $\int_{-\infty}^{\infty} |\varphi(\lambda)| d\lambda < \infty$, then the distribution P has a density given by

$$p(x) = \frac{1}{2\pi} \int_{-\infty}^{\infty} e^{-i\lambda x} \varphi(\lambda) d\lambda.$$

The next theorem and its corollary show that one can always recover the measure P from its characteristic function $\varphi(\lambda)$.

Theorem 9.4. *For any interval (a, b)*

$$\lim_{R \to \infty} \frac{1}{2\pi} \int_{-R}^{R} \frac{e^{-i\lambda a} - e^{-i\lambda b}}{i\lambda} \varphi(\lambda) d\lambda = P((a, b)) + \frac{1}{2}P(\{a\}) + \frac{1}{2}P(\{b\}).$$

Proof. By the Fubini Theorem, since the integrand is bounded,

$$\frac{1}{2\pi} \int_{-R}^{R} \frac{e^{-i\lambda a} - e^{-i\lambda b}}{i\lambda} \varphi(\lambda) d\lambda = \frac{1}{2\pi} \int_{-R}^{R} \frac{e^{-i\lambda a} - e^{-i\lambda b}}{i\lambda} \left(\int_{-\infty}^{\infty} e^{i\lambda x} dP(x) \right) d\lambda$$

$$= \frac{1}{2\pi} \int_{-\infty}^{\infty} \left(\int_{-R}^{R} \frac{e^{-i\lambda a} - e^{-i\lambda b}}{i\lambda} e^{i\lambda x} d\lambda \right) dP(x).$$

Furthermore,

$$\int_{-R}^{R} \frac{e^{-i\lambda a} - e^{-i\lambda b}}{i\lambda} e^{i\lambda x} d\lambda = \int_{-R}^{R} \frac{\cos \lambda(x-a) - \cos \lambda(x-b)}{i\lambda} d\lambda$$
$$+ \int_{-R}^{R} \frac{\sin \lambda(x-a) - \sin \lambda(x-b)}{\lambda} d\lambda.$$

The first integral is equal to zero since the integrand is an odd function of λ. The second integrand is even, therefore

$$\int_{-R}^{R} \frac{e^{-i\lambda a} - e^{-i\lambda b}}{i\lambda} e^{i\lambda x} d\lambda = 2 \int_{0}^{R} \frac{\sin \lambda(x-a)}{\lambda} d\lambda - 2 \int_{0}^{R} \frac{\sin \lambda(x-b)}{\lambda} d\lambda.$$

Setting $\mu = \lambda(x-a)$ in the first integral and $\mu = \lambda(x-b)$ in the second integral we obtain

$$2 \int_{0}^{R} \frac{\sin \lambda(x-a)}{\lambda} d\lambda - 2 \int_{0}^{R} \frac{\sin \lambda(x-b)}{\lambda} d\lambda = 2 \int_{R(x-b)}^{R(x-a)} \frac{\sin \mu}{\mu} d\mu.$$

Thus,

$$\frac{1}{2\pi} \int_{-R}^{R} \frac{e^{-i\lambda a} - e^{-i\lambda b}}{i\lambda} \varphi(\lambda) d\lambda = \int_{-\infty}^{\infty} dP(x) \frac{1}{\pi} \int_{R(x-b)}^{R(x-a)} \frac{\sin \mu}{\mu} d\mu.$$

Note that the improper integral $\int_{0}^{\infty} \frac{\sin \mu}{\mu} d\mu$ converges to $\frac{\pi}{2}$ (although it does not converge absolutely). Let us examine the limit

$$\lim_{R \to \infty} \frac{1}{\pi} \int_{R(x-b)}^{R(x-a)} \frac{\sin \mu}{\mu} d\mu$$

for different values of x.

If $x > b$ (or $x < a$) both limits of integration converge to infinity (or minus infinity), and therefore the limit of the integral is equal to zero.

If $a < x < b$,

$$\lim_{R \to \infty} \frac{1}{\pi} \int_{R(x-b)}^{R(x-a)} \frac{\sin \mu}{\mu} d\mu = \frac{1}{\pi} \int_{-\infty}^{\infty} \frac{\sin \mu}{\mu} d\mu = 1.$$

If $x = a$,

$$\lim_{R \to \infty} \frac{1}{\pi} \int_{-R(b-a)}^{0} \frac{\sin \mu}{\mu} d\mu = \frac{1}{\pi} \int_{-\infty}^{0} \frac{\sin \mu}{\mu} d\mu = \frac{1}{2}.$$

If $x = b$,

$$\lim_{R \to \infty} \frac{1}{\pi} \int_{0}^{R(b-a)} \frac{\sin \mu}{\mu} d\mu = \frac{1}{\pi} \int_{0}^{\infty} \frac{\sin \mu}{\mu} d\mu = \frac{1}{2}.$$

Since the integral $\frac{1}{\pi}\int_{R(x-b)}^{R(x-a)}\frac{\sin\mu}{\mu}d\mu$ is bounded in x and R, we can apply the Lebesgue Dominated Convergence Theorem to obtain

$$\lim_{R\to\infty}\int_{-\infty}^{\infty}dP(x)\frac{1}{\pi}\int_{R(x-b)}^{R(x-a)}\frac{\sin\mu}{\mu}d\mu$$

$$=\int_{-\infty}^{\infty}dP(x)\lim_{R\to\infty}\frac{1}{\pi}\int_{R(x-b)}^{R(x-a)}\frac{\sin\mu}{\mu}d\mu=P((a,b))+\frac{1}{2}P(\{a\})+\frac{1}{2}P(\{b\}).$$

□

Corollary 9.5. *If two probability measures have equal characteristic functions, then they are equal.*

Proof. By Theorem 9.4, the distribution functions must coincide at all common continuity points. The set of discontinuity points for each of the distribution functions is at most countable, therefore the distribution functions coincide on a complement to a countable set, which implies that they coincide at all points, since they are both right-continuous. □

Definition 9.6. *The characteristic function of a measure P on $\mathbb{R}^n$ is the (complex-valued) function $\varphi(\lambda)$ of the variable $\lambda\in\mathbb{R}^n$ given by*

$$\varphi(\lambda)=\int_{\mathbb{R}^n}e^{i(\lambda,x)}dP(x),$$

where (λ,x) is the inner product of vectors λ and x in $\mathbb{R}^n$.

The above properties of the characteristic function of a measure on $\mathbb{R}$ can be appropriately re-formulated and remain true for measures on $\mathbb{R}^n$. In particular, if two probability measures on $\mathbb{R}^n$ have equal characteristic functions, then they are equal.

9.2 Characteristic Functions and Weak Convergence

One of the reasons why characteristic functions are helpful in probability theory is the following criterion of weak convergence of probability measures.

Theorem 9.7. *Let P_n be a sequence of probability measures on $\mathbb{R}$ with characteristic functions $\varphi_n(\lambda)$ and let P be a probability measure on $\mathbb{R}$ with characteristic function $\varphi(\lambda)$. Then $P_n\Rightarrow P$ if and only if $\lim_{n\to\infty}\varphi_n(\lambda)=\varphi(\lambda)$ for every λ.*

Proof. The weak convergence $P_n \Rightarrow P$ implies that

$$\varphi_n(\lambda) = \int_{-\infty}^{\infty} e^{i\lambda x} dP_n(x) = \int_{-\infty}^{\infty} \cos \lambda x dP_n(x) + i \int_{-\infty}^{\infty} \sin \lambda x dP_n(x) \rightarrow$$

$$\int_{-\infty}^{\infty} \cos \lambda x dP(x) + i \int_{-\infty}^{\infty} \sin \lambda x dP(x) = \int_{-\infty}^{\infty} e^{i\lambda x} dP(x) = \varphi(\lambda),$$

so the implication in one direction is trivial.

To prove the converse statement we need the following lemma.

Lemma 9.8. *Let* P *be a probability measure on the line and* φ *be its characteristic function. Then for every* $\tau > 0$ *we have*

$$P([-\frac{2}{\tau}, \frac{2}{\tau}]) \geq |\frac{1}{\tau} \int_{-\tau}^{\tau} \varphi(\lambda) d\lambda| - 1.$$

Proof. By the Fubini Theorem,

$$\frac{1}{2\tau} \int_{-\tau}^{\tau} \varphi(\lambda) d\lambda = \frac{1}{2\tau} \int_{-\tau}^{\tau} \left(\int_{-\infty}^{\infty} e^{i\lambda x} dP(x) \right) d\lambda$$

$$= \frac{1}{2\tau} \int_{-\infty}^{\infty} \left(\int_{-\tau}^{\tau} e^{i\lambda x} d\lambda \right) dP(x) = \int_{-\infty}^{\infty} \frac{e^{ix\tau} - e^{-ix\tau}}{2ix\tau} dP(x) = \int_{-\infty}^{\infty} \frac{\sin x\tau}{x\tau} dP(x).$$

Therefore,

$$|\frac{1}{2\tau} \int_{-\tau}^{\tau} \varphi(\lambda) d\lambda| = |\int_{-\infty}^{\infty} \frac{\sin x\tau}{x\tau} dP(x)|$$

$$\leq |\int_{|x| \leq \frac{2}{\tau}} \frac{\sin x\tau}{x\tau} dP(x)| + |\int_{|x| > \frac{2}{\tau}} \frac{\sin x\tau}{x\tau} dP(x)|$$

$$\leq \int_{|x| \leq \frac{2}{\tau}} |\frac{\sin x\tau}{x\tau}| dP(x) + \int_{|x| > \frac{2}{\tau}} |\frac{\sin x\tau}{x\tau}| dP(x).$$

Since $|\sin x\tau / x\tau| \leq 1$ for all x and $|\sin x\tau / x\tau| \leq 1/2$ for $|x| > 2/\tau$, the last expression is estimated from above by

$$\int_{|x| \leq \frac{2}{\tau}} dP(x) + \frac{1}{2} \int_{|x| > \frac{2}{\tau}} dP(x)$$

$$= P([-\frac{2}{\tau}, \frac{2}{\tau}]) + \frac{1}{2}(1 - P([-\frac{2}{\tau}, \frac{2}{\tau}])) = \frac{1}{2} P([-\frac{2}{\tau}, \frac{2}{\tau}]) + \frac{1}{2},$$

which implies the statement of the lemma.   □

We now return to the proof of the theorem. Let $\varepsilon > 0$ and $\varphi_n(\lambda) \to \varphi(\lambda)$ for each λ. Since $\varphi(0) = 1$ and $\varphi(\lambda)$ is a continuous function, there exists $\tau > 0$ such that $|\varphi(\lambda) - 1| < \frac{\varepsilon}{4}$ when $|\lambda| < \tau$. Thus

$$\left| \int_{-\tau}^{\tau} \varphi(\lambda) d\lambda \right| = \left| \int_{-\tau}^{\tau} (\varphi(\lambda) - 1) d\lambda + 2\tau \right| \geq 2\tau - \left| \int_{-\tau}^{\tau} (\varphi(\lambda) - 1) d\lambda \right|$$

$$\geq 2\tau - \int_{-\tau}^{\tau} |(\varphi(\lambda) - 1)| d\lambda \geq 2\tau - 2\tau \frac{\varepsilon}{4} = 2\tau(1 - \frac{\varepsilon}{4}).$$

Therefore,

$$\left| \frac{1}{\tau} \int_{-\tau}^{\tau} \varphi(\lambda) d\lambda \right| \geq 2 - \frac{\varepsilon}{2}.$$

Since $\varphi_n(\lambda) \to \varphi(\lambda)$ and $|\varphi_n(\lambda)| \leq 1$, by the Lebesgue Dominated Convergence Theorem,

$$\lim_{n \to \infty} \left| \frac{1}{\tau} \int_{-\tau}^{\tau} \varphi_n(\lambda) d\lambda \right| = \left| \frac{1}{\tau} \int_{-\tau}^{\tau} \varphi(\lambda) d\lambda \right| \geq 2 - \frac{\varepsilon}{2}.$$

Thus there exists an N such that for all $n \geq N$ we have

$$\left| \frac{1}{\tau} \int_{-\tau}^{\tau} \varphi_n(\lambda) d\lambda \right| \geq 2 - \varepsilon.$$

By Lemma 9.8, for such n

$$P_n([-\frac{2}{\tau}, \frac{2}{\tau}]) \geq \left| \frac{1}{\tau} \int_{-\tau}^{\tau} \varphi_n(\lambda) d\lambda \right| - 1 \geq 1 - \varepsilon.$$

For each $n < N$ we choose $t_n > 0$ such that $P_n([-t_n, t_n]) \geq 1 - \varepsilon$. If we set $K = \max(\frac{2}{\tau}, \max_{1 \leq n < N} t_n)$, we find that $P_n([-K, K]) \geq 1 - \varepsilon$ for all n. Thus the sequence of measures P_n is tight and, by the Prokhorov Theorem, is weakly compact.

Let P_{n_i} be a weakly convergent subsequence, $P_{n_i} \Rightarrow \tilde{P}$. We now show that $\tilde{P} = P$. Let us denote the characteristic function of $\tilde{P}$ by $\tilde{\varphi}(\lambda)$. By the first part of our theorem, $\varphi_n(\lambda) \to \tilde{\varphi}(\lambda)$ for all λ. On the other hand, by assumption $\varphi_n(\lambda) \to \varphi(\lambda)$ for all λ. Therefore $\tilde{\varphi}(\lambda) = \varphi(\lambda)$. By Corollary 9.5, $\tilde{P} = P$.

It remains to establish that the entire sequence P_n converges to P. Assume that this is not true. Then for some bounded continuous function f there exist a subsequence $\{n_i\}$ and $\varepsilon > 0$ such that

$$\left| \int_{-\infty}^{\infty} f(x) dP_{n_i}(x) - \int_{-\infty}^{\infty} f(x) dP(x) \right| > \varepsilon.$$

We extract a weakly convergent subsequence $P_{n_j'}$ from the sequence P_{n_i}, that is $P_{n_j'} \Rightarrow \overline{P}$. The same argument as before shows that $\overline{P} = P$, and therefore

$$\lim_{j\to\infty} \int_{-\infty}^{\infty} f(x)dP_{n_j'}(x) = \int_{-\infty}^{\infty} f(x)dP(x).$$

Hence the contradiction.   □

Remark 9.9. Theorem 9.7 remains true for measures and characteristic functions on $\mathbb{R}^n$. In this case the characteristic functions depend on n variables $\lambda = (\lambda_1, \ldots, \lambda_n)$, and the weak convergence is equivalent to the convergence of $\varphi_n(\lambda)$ to $\varphi(\lambda)$ for every λ.

Remark 9.10. One can show that if $\varphi_n(\lambda) \to \varphi(\lambda)$ for every λ, where φ is a characteristic function, then this convergence is uniform on any compact set of values of λ.

Remark 9.11. One can show that if a sequence of characteristic functions $\varphi_n(\lambda)$ converges to a continuous function $\varphi(\lambda)$, then the sequence of probability measures P_n converges weakly to some probability measure P.

Let us consider a collection of n random variables $\xi_1, \ldots, \xi_n$ with characteristic functions $\varphi_1, \ldots, \varphi_n$. Let φ be the characteristic function of the random vector $(\xi_1, \ldots, \xi_n)$. The condition of independence of $\xi_1, \ldots, \xi_n$ is easily expressed in terms of characteristic functions.

Lemma 9.12. *The random variables $\xi_1, \ldots, \xi_n$ are independent if and only if* $\varphi(\lambda_1, \ldots, \lambda_n) = \varphi_1(\lambda_1) \cdot \ldots \cdot \varphi_n(\lambda_n)$ *for all $(\lambda_1, \ldots, \lambda_n)$.*

Proof. If $\xi_1, \ldots, \xi_n$ are independent, then by Theorem 4.8

$$\varphi(\lambda_1, \ldots, \lambda_n) = \mathrm{E}e^{i(\lambda_1\xi_1 + \ldots + \lambda_n\xi_n)} = \mathrm{E}e^{i\lambda_1\xi_1} \cdot \ldots \cdot \mathrm{E}e^{i\lambda_n\xi_n} = \varphi_1(\lambda_1) \cdot \ldots \cdot \varphi_n(\lambda_n).$$

Conversely, assume that $\varphi(\lambda_1, \ldots, \lambda_n) = \varphi_1(\lambda_1) \cdot \ldots \cdot \varphi_n(\lambda_n)$. Let $\widetilde{\xi}_1, \ldots, \widetilde{\xi}_n$ be independent random variables, which have the same distributions as $\xi_1, \ldots, \xi_n$ respectively, and therefore have the same characteristic functions. Then the characteristic function of the vector $(\widetilde{\xi}_1, \ldots, \widetilde{\xi}_n)$ is equal to $\varphi_1(\lambda_1) \cdot \ldots \cdot \varphi_n(\lambda_n)$ by the first part of the lemma. Therefore, by Remark 9.9, the measure on $\mathbb{R}^n$ induced by the vector $(\widetilde{\xi}_1, \ldots, \widetilde{\xi}_n)$ is the same as the measure induced by the vector $(\xi_1, \ldots, \xi_n)$. Thus the random variables $\xi_1, \ldots, \xi_n$ are also independent.   □

9.3 Gaussian Random Vectors

Gaussian random vectors appear in a large variety of problems, both in pure mathematics and in applications. Their distributions are limits of distributions of normalized sums of independent or weakly correlated random variables.

Recall that a random variable is called Gaussian with parameters $(0,1)$ if it has the density $p(x) = \frac{1}{\sqrt{2\pi}}e^{-\frac{x^2}{2}}$. Let $\eta = (\eta_1, \ldots, \eta_n)$ be a random vector defined on a probability space $(\Omega, \mathcal{F}, \mathrm{P})$. Note that the Gaussian property of the vector is defined in terms of the distribution of the vector (the measure on $\mathbb{R}^n$, which is induced by η).

Definition 9.13. *A random vector* $\eta = (\eta_1, \ldots, \eta_n)$ *on* $(\Omega, \mathcal{F}, \mathrm{P})$ *is called Gaussian if there is a vector* $\xi = (\xi_1, \ldots, \xi_n)$ *of independent Gaussian random variables with parameters* $(0,1)$ *which may be defined on a different probability space* $(\widetilde{\Omega}, \widetilde{\mathcal{F}}, \widetilde{\mathrm{P}})$, *an* $n \times n$ *matrix* A, *and a vector* $a = (a_1, \ldots, a_n)$ *such that the vectors* η *and* $A\xi + a$ *have the same distribution.*

Remark 9.14. It does not follow from this definition that the random vector ξ can be defined on the same probability space, or that η can be represented in the form $\eta = A\xi + a$. Indeed, as a pathological example we can consider the space Ω which consists of one element ω, and define $\eta(\omega) = 0$. This is a Gaussian random variable, since we can take a Gaussian random variable ξ with parameters $(0,1)$ defined on a different probability space, and take $A = 0$, $a = 0$. On the other hand, a Gaussian random variable with parameters $(0,1)$ cannot be defined on the space Ω itself.

The covariance matrix of a random vector, its density, and the characteristic function can be expressed in terms of the distribution that the vector induces on $\mathbb{R}^n$. Therefore, in the calculations below, we can assume without loss of generality that $\eta = A\xi + a$.

Remark 9.15. Here we discuss only real-valued Gaussian vectors. Some of the formulas below need to be modified if we allow complex matrices A and vectors a. Besides, the distribution of a complex-valued Gaussian vector is not determined uniquely by its covariance matrix.

Let us examine expectations and the covariances of different components of a Gaussian random vector $\eta = A\xi + a$. Since $\mathrm{E}\xi_i = 0$ for all i, it is clear that $\mathrm{E}\eta_i = a_i$. Regarding the covariance,

$$\mathrm{Cov}(\eta_i, \eta_j) = \mathrm{E}(\eta_i - a_i)(\eta_j - a_j) = \mathrm{E}(\sum_k A_{ik}\xi_k \sum_l A_{jl}\xi_l)$$

$$= \sum_k \sum_l A_{ik} A_{jl} \mathrm{E}(\xi_k \xi_l) = \sum_k A_{ik} A_{jk} = (AA^*)_{ij}.$$

We shall refer to the matrix $B = AA^*$ as the covariance matrix of the Gaussian vector η.

Note that the matrix A is not determined uniquely by the covariance matrix, that is, there are pairs of square matrices A_1, A_2 such that $A_1 A_1^* = A_2 A_2^*$. The distribution of a Gaussian random vector, however, is determined by the expectation and the covariance matrix (see below). The distribution of a Gaussian random vector with expectation a and covariance matrix B will be denoted by $N(a, B)$.

If $\det B \neq 0$, then there is a density corresponding to the distribution of η in $\mathbb{R}^n$. Indeed, the multi-dimensional density corresponding to the vector ξ is equal to

$$p_\xi(x_1, \ldots, x_n) = (2\pi)^{-\frac{n}{2}} e^{-\frac{||x||^2}{2}}.$$

Let μ_ξ and μ_η be the measures on $\mathbb{R}^n$ induced by the random vectors ξ and η respectively. The random vector η can be obtained from ξ by a composition with an affine transformation of the space $\mathbb{R}^n$, that is $\eta = L\xi$, where $Lx = Ax + a$. Therefore μ_η is the same as the push-forward of the measure μ_ξ by the map $L : \mathbb{R}^n \to \mathbb{R}^n$ (see Sect. 3.2). The Jacobian $J(x)$ of L is equal to $(\det B)^{\frac{1}{2}}$ for all x. Therefore the density corresponding to the random vector η is equal to

$$p_\eta(x) = J^{-1}(x)p_\xi(L^{-1}x) = (\det B)^{-\frac{1}{2}}(2\pi)^{-\frac{n}{2}} e^{-\frac{||A^{-1}(x-a)||^2}{2}}$$

$$= (\det B)^{-\frac{1}{2}}(2\pi)^{-\frac{n}{2}} e^{-\frac{(B^{-1}(x-a),(x-a))}{2}}.$$

Let us now examine the characteristic function of a Gaussian random vector. For a Gaussian variable ξ with parameters $(0, 1)$,

$$\varphi(\lambda) = \mathrm{E}e^{i\lambda\xi} = \frac{1}{\sqrt{2\pi}} \int_{-\infty}^\infty e^{i\lambda x} e^{-\frac{x^2}{2}} dx = \frac{1}{\sqrt{2\pi}} \int_{-\infty}^\infty e^{-\frac{(x-i\lambda)^2}{2} - \frac{\lambda^2}{2}} dx$$

$$= e^{-\frac{\lambda^2}{2}} \frac{1}{\sqrt{2\pi}} \int_{-\infty}^\infty e^{-\frac{u^2}{2}} du = e^{-\frac{\lambda^2}{2}}.$$

Therefore, in the multi-dimensional case,

$$\varphi(\lambda) = \mathrm{E}e^{i(\lambda,\eta)} = \mathrm{E}e^{i\sum_i \lambda_i(\sum_k A_{ik}\xi_k + a_i)} = e^{i(\lambda,a)} \prod_k \mathrm{E}e^{i(\sum_i \lambda_i A_{ik})\xi_k} =$$

$$e^{i(\lambda,a)} \prod_k e^{-\frac{(\sum_i \lambda_i A_{ik})^2}{2}} = e^{i(\lambda,a)} e^{-\frac{\sum_k(\sum_i \lambda_i A_{ik})^2}{2}} = e^{i(\lambda,a) - \frac{1}{2}(B\lambda,\lambda)}.$$

Since the characteristic function determines the distribution of the random vector uniquely, this calculation shows that the distribution of a Gaussian random vector is uniquely determined by its expectation and covariance matrix.

The property that the characteristic function of a random vector is

$$\varphi(\lambda) = e^{i(\lambda,a) - \frac{1}{2}(B\lambda,\lambda)}$$

for some vector a and a non-negative definite matrix B can be taken as a definition of a Gaussian random vector equivalent to Definition 9.13 (see Problem 11).

Recall that for two independent random variables with finite variances the covariance is equal to zero. For random variables which are components of a Gaussian vector the converse implication is also valid.

Lemma 9.16. *If* $(\eta_1, \ldots, \eta_n)$ *is a Gaussian vector, and* $\mathrm{Cov}(\eta_i, \eta_j) = 0$ *for* $i \neq j$, *then the random variables* $\eta_1, \ldots, \eta_n$ *are independent.*

Proof. Let e_i denote the vector whose i-th component is equal to one and the rest of the components are equal to zero. If $\mathrm{Cov}(\eta_i, \eta_j) = 0$ for $i \neq j$, then the covariance matrix B is diagonal, while $\mathrm{Cov}(\eta_i, \eta_i) = B_{ii}$. Therefore the characteristic function of the random vector η is $\varphi(\lambda) = e^{i(\lambda, a) - \frac{1}{2} \sum_i B_{ii} \lambda_i^2}$, while the characteristic function of η_i is equal to

$$\varphi_i(\lambda_i) = \mathrm{E}e^{i \lambda_i \eta_i} = \mathrm{E}e^{i(\lambda_i e_i, \eta)} = \varphi(\lambda_i e_i) = e^{i \lambda_i a_i - \frac{1}{2} B_{ii} \lambda_i^2}.$$

This implies the independence of the random variables by Lemma 9.12. $\quad\square$

9.4 Problems

1. Is $\varphi(\lambda) = \cos(\lambda^2)$ a characteristic function of some distribution?
2. Find the characteristic functions of the following distributions: (1) $\xi = \pm 1$ with probabilities $\frac{1}{2}$; (2) binomial distribution; (3) Poisson distribution with parameter λ; (4) exponential distribution; (5) uniform distribution on $[a, b]$.
3. Let $\xi_1, \xi_2, \ldots$ be a sequence of random variables on a probability space $(\Omega, \mathcal{F}, \mathrm{P})$, and $\mathcal{G}$ be a σ-subalgebra of $\mathcal{F}$. Assume that ξ_n is independent of $\mathcal{G}$ for each n, and that $\lim_{n \to \infty} \xi_n = \xi$ almost surely. Prove that ξ is independent of $\mathcal{G}$.
4. Prove that if the measure P is discrete, then its characteristic function $\varphi(\lambda)$ does not tend to zero as $\lambda \to \infty$.
5. Prove that if the characteristic function $\varphi(\lambda)$ is analytic in a neighborhood of $\lambda = 0$, then there exist constants $c_1, c_2 > 0$ such that for every $x > 0$ we have $\mathrm{P}((-\infty, -x)) \leq c_1 e^{-c_2 x}$, and $\mathrm{P}((x, \infty)) \leq c_1 e^{-c_2 x}$.
6. Assume that ξ_1 and ξ_2 are Gaussian random variables. Does this imply that (ξ_1, ξ_2) is a Gaussian random vector?
7. Prove that if $(\xi_1, \ldots, \xi_n)$ is a Gaussian vector, then $\xi = a_1 \xi_1 + \ldots + a_n \xi_n$ is a Gaussian random variable. Find its expectation and variance.
8. Let ξ be a Gaussian random variable and $a_0, a_1, \ldots, a_n$ some real numbers. Prove that the characteristic function of the random variable $\eta = a_0 + a_1 \xi + \ldots + a_n \xi^n$ is infinitely differentiable.
9. Let $\xi_1, \ldots, \xi_n$ be independent Gaussian random variables with $N(0, 1)$ distribution. Find the density and the characteristic function of $\xi_1^2 + \ldots + \xi_n^2$.
10. Let $\xi_1, \xi_2, \ldots$ be a sequence of independent identically distributed random variables with $N(0, 1)$ distribution. Let $0 < \lambda < 1$ and η_0 be independent of $\xi_1, \xi_2, \ldots$. Let η_n, $n \geq 1$, be defined by $\eta_n = \lambda \eta_{n-1} + \xi_n$. Show that η_n is a Markov chain. Find its stationary distribution.

11. Let a random vector η have the characteristic function

$$\varphi(\lambda) = e^{i(\lambda,a)-\frac{1}{2}(B\lambda,\lambda)}$$

for some vector a and a non-negative definite matrix B. Prove that there are a vector $\xi = (\xi_1,\ldots,\xi_n)$ of independent Gaussian random variables with parameters $(0,1)$ defined on some probability space $(\widetilde{\Omega}, \widetilde{\mathcal{F}}, \widetilde{P})$, and an $n \times n$ matrix A such that the vectors η and $A\xi + a$ have the same distribution.

12. Prove that, for Gaussian vectors, convergence of covariance matrices implies convergence in distribution.

13. Let $(\xi_1,\ldots,\xi_{2n})$ be a Gaussian vector. Assuming that $E\xi_i = 0$, $1 \le i \le 2n$, prove that

$$E(\xi_1\ldots\xi_{2n}) = \sum_\sigma E(\xi_{\sigma_1}\xi_{\sigma_2})\ldots E(\xi_{\sigma_{2n-1}}\xi_{\sigma_{2n}}),$$

where $\sigma = ((\sigma_1,\sigma_2),\ldots,(\sigma_{2n-1},\sigma_{2n}))$, $1 \le \sigma_i \le 2n$, is a partition of the set $\{1,\ldots,2n\}$ into n pairs, and the summation extends over all the partitions. (The permutation of elements of a pair is considered to yield the same partition.)

14. Let $(\xi_1,\ldots,\xi_{2n-1})$ be a random vector with the density

$$p(x_1,\ldots,x_{2n-1}) = c_n \exp(-\frac{1}{2}(x_1^2 + \sum_{i=1}^{2n-2}(x_{i+1}-x_i)^2 + x_{2n-1}^2)),$$

where c_n is a normalizing constant. Prove that this is a Gaussian vector and find the value of the normalizing constant. Prove that there is a constant a which does not depend on n such that $\mathrm{Var}(\xi_n) \ge an$ for all $n \ge 1$.

15. Let $(\xi_1,\ldots,\xi_n)$ be a random vector uniformly distributed on the ball

$$\xi_1^2 + \ldots + \xi_n^2 \le n.$$

Prove that the joint distribution of (ξ_1,ξ_2,ξ_3) converges to a three-dimensional Gaussian distribution.

Limit Theorems

10.1 Central Limit Theorem, the Lindeberg Condition

Limit Theorems describe limiting distributions of appropriately scaled sums of a large number of random variables. It is usually assumed that the random variables are either independent, or almost independent, in some sense. In the case of the Central Limit Theorem that we prove in this section, the random variables are independent and the limiting distribution is Gaussian. We first introduce the definitions.

Let $\xi_1, \xi_2, \ldots$ be a sequence of independent random variables with finite variances, $m_i = E\xi_i$, $\sigma_i^2 = \text{Var}(\xi_i)$, $\zeta_n = \sum_{i=1}^{n} \xi_i$, $M_n = E\zeta_n = \sum_{i=1}^{n} m_i$, $D_n^2 = \text{Var}(\zeta_n) = \sum_{i=1}^{n} \sigma_i^2$. Let $F_i = F_{\xi_i}$ be the distribution function of the random variable ξ_i.

Definition 10.1. *The Lindeberg condition is said to be satisfied if*

$$\lim_{n \to \infty} \frac{1}{D_n^2} \sum_{i=1}^{n} \int_{\{x : |x - m_i| \geq \varepsilon D_n\}} (x - m_i)^2 dF_i(x) = 0$$

for every $\varepsilon > 0$.

Remark 10.2. The Lindeberg condition easily implies that $\lim_{n \to \infty} D_n = \infty$ (see formula (10.5) below).

Theorem 10.3 (Central Limit Theorem, Lindeberg Condition). *Let $\xi_1, \xi_2, \ldots$ be a sequence of independent random variables with finite variances. If the Lindeberg condition is satisfied, then the distributions of $(\zeta_n - M_n)/D_n$ converge weakly to $N(0, 1)$ distribution as $n \to \infty$.*

Proof. We may assume that $m_i = 0$ for all i. Otherwise we can consider a new sequence of random variables $\tilde{\xi}_i = \xi_i - m_i$, which have zero expectations, and

L. Koralov and Y.G. Sinai, *Theory of Probability and Random Processes*,   131
Universitext, DOI 10.1007/978-3-540-68829-7_10,
© Springer-Verlag Berlin Heidelberg 2012

for which the Lindeberg condition is also satisfied. Let $\varphi_i(\lambda)$ and $\varphi_{\tau_n}(\lambda)$ be the characteristic functions of the random variables ξ_i and $\tau_n = \frac{\zeta_n}{D_n}$ respectively. By Theorem 9.7, it is sufficient to prove that for all $\lambda \in \mathbb{R}$

$$\varphi_{\tau_n}(\lambda) \to e^{-\frac{\lambda^2}{2}} \quad \text{as} \quad n \to \infty. \tag{10.1}$$

Fix $\lambda \in \mathbb{R}$ and note that the left-hand side of (10.1) can be written as follows:

$$\varphi_{\tau_n}(\lambda) = \mathrm{E}e^{i\lambda\tau_n} = \mathrm{E}e^{i(\frac{\lambda}{D_n})(\xi_1+\dots+\xi_n)} = \prod_{i=1}^n \varphi_i(\frac{\lambda}{D_n}).$$

We shall prove that

$$\varphi_i(\frac{\lambda}{D_n}) = 1 - \frac{\lambda^2\sigma_i^2}{2D_n^2} + a_i^n \tag{10.2}$$

for some $a_i^n = a_i^n(\lambda)$ such that for any λ

$$\lim_{n\to\infty} \sum_{i=1}^n |a_i^n| = 0. \tag{10.3}$$

Assuming (10.2) for now, let us prove the theorem. By Taylor's formula, for any complex number z with $|z| < \frac{1}{4}$

$$\ln(1 + z) = z + \theta(z)|z|^2, \tag{10.4}$$

with $|\theta(z)| \leq 1$, where ln denotes the principal value of the logarithm (the analytic continuation of the logarithm from the positive real semi-axis to the half-plane $\mathrm{Re}(z) > 0$).

We next show that

$$\lim_{n\to\infty} \max_{1\leq i\leq n} \frac{\sigma_i^2}{D_n^2} = 0. \tag{10.5}$$

Indeed, for any $\varepsilon > 0$,

$$\max_{1\leq i\leq n} \frac{\sigma_i^2}{D_n^2} \leq \max_{1\leq i\leq n} \frac{\int_{\{x:|x|\geq\varepsilon D_n\}} x^2 dF_i(x)}{D_n^2} + \max_{1\leq i\leq n} \frac{\int_{\{x:|x|\leq\varepsilon D_n\}} x^2 dF_i(x)}{D_n^2}.$$

The first term on the right-hand side of this inequality tends to zero by the Lindeberg condition. The second term does not exceed ε^2, since the integrand does not exceed $\varepsilon^2 D_n^2$ on the domain of integration. This proves (10.5), since ε was arbitrary.

Therefore, when n is large enough, we can put $z = -\frac{\lambda^2\sigma_i^2}{2D_n^2} + a_i^n$ in (10.4) and obtain

$$\sum_{i=1}^n \ln\varphi_i(\frac{\lambda}{D_n}) = \sum_{i=1}^n \frac{-\lambda^2\sigma_i^2}{2D_n^2} + \sum_{i=1}^n a_i^n + \sum_{i=1}^n \theta_i|\frac{-\lambda^2\sigma_i^2}{2D_n^2} + a_i^n|^2$$

with $|\theta_i| \leq 1$. The first term on the right-hand side of this expression is equal to $-\frac{\lambda^2}{2}$. The second term tends to zero due to (10.3). The third term tends to zero since

$$\sum_{i=1}^{n} \theta_i |\frac{-\lambda^2 \sigma_i^2}{2D_n^2} + a_i^n|^2 \leq \max_{1 \leq i \leq n} \{\frac{\lambda^2 \sigma_i^2}{2D_n^2} + |a_i^n|\} \sum_{i=1}^{n} (\frac{\lambda^2 \sigma_i^2}{2D_n^2} + |a_i^n|)$$

$$\leq c(\lambda) \max_{1 \leq i \leq n} \{\frac{\lambda^2 \sigma_i^2}{2D_n^2} + |a_i^n|\},$$

where $c(\lambda)$ is a constant, while the second factor converges to zero by (10.3) and (10.5). We have thus demonstrated that

$$\lim_{n \to \infty} \sum_{i=1}^{n} \ln \varphi_i(\frac{\lambda}{D_n}) = -\frac{\lambda^2}{2},$$

which clearly implies (10.1). It remains to prove (10.2). We use the following simple relations:

$$e^{ix} = 1 + ix + \frac{\theta_1(x)x^2}{2},$$

$$e^{ix} = 1 + ix - \frac{x^2}{2} + \frac{\theta_2(x)x^3}{6},$$

which are valid for all real x, with $|\theta_1(x)| \leq 1$ and $|\theta_2(x)| \leq 1$. Then

$$\varphi_i(\frac{\lambda}{D_n}) = \int_{-\infty}^{\infty} e^{\frac{i\lambda}{D_n}x} dF_i(x) = \int_{|x| \geq \varepsilon D_n} (1 + \frac{i\lambda}{D_n}x + \frac{\theta_1(x)(\lambda x)^2}{2D_n^2}) dF_i(x)$$

$$+ \int_{|x| < \varepsilon D_n} (1 + \frac{i\lambda x}{D_n} - \frac{\lambda^2 x^2}{2D_n^2} + \frac{\theta_2(x)|\lambda x|^3}{6D_n^3}) dF_i(x)$$

$$= 1 - \frac{\lambda^2 \sigma_i^2}{2D_n^2} + \frac{\lambda^2}{2D_n^2} \int_{|x| \geq \varepsilon D_n} (1 + \theta_1(x)) x^2 dF_i(x)$$

$$+ \frac{|\lambda|^3}{6D_n^3} \int_{|x| < \varepsilon D_n} \theta_2(x) |x|^3 dF_i(x).$$

Here we have used that

$$\int_{-\infty}^{\infty} x dF_i(x) = \mathrm{E}\xi_i = 0.$$

In order to prove (10.2), we need to show that

$$\sum_{i=1}^{n} \frac{\lambda^2}{2D_n^2} \int_{|x| \geq \varepsilon D_n} (1+\theta_1(x)) x^2 dF_i(x) + \sum_{i=1}^{n} \frac{|\lambda|^3}{6D_n^3} \int_{|x| < \varepsilon D_n} \theta_2(x)|x|^3 dF_i(x) \to 0.$$

$$(10.6)$$

The second sum in (10.6) can be estimated as

$$\left| \sum_{i=1}^{n} \frac{|\lambda|^3}{6D_n^3} \int_{|x|<\varepsilon D_n} \theta_2(x)|x|^3 dF_i(x) \right|$$

$$\leq \left| \sum_{i=1}^{n} \frac{|\lambda|^3\varepsilon}{6D_n^3} \int_{|x|<\varepsilon D_n} \theta_2(x)x^2 D_n dF_i(x) \right| \leq \sum_{i=1}^{n} \frac{|\lambda|^3\varepsilon\sigma_i^2}{6D_n^2} = \frac{\varepsilon|\lambda|^3}{6},$$

which can be made arbitrarily small by selecting a sufficiently small ε. The first sum in (10.6) tends to zero by the Lindeberg condition.   □

Remark 10.4. The proof can be easily modified to demonstrate that the convergence in (10.1) is uniform on any compact set of values of λ. We shall need this fact in the next section.

The Lindeberg condition is clearly satisfied for every sequence of independent identically distributed random variables with finite variances. We therefore have the following Central Limit Theorem for independent identically distributed random variables.

Theorem 10.5. *Let $\xi_1, \xi_2, \ldots$ be a sequence of independent identically distributed random variables with $m = \mathrm{E}\xi_1$ and $0 < \sigma^2 = \mathrm{Var}(\xi_1) < \infty$. Then the distributions of $(\zeta_n - nm)/\sqrt{n}\sigma$ converge weakly to $N(0,1)$ distribution as $n \to \infty$.*

Theorem 10.3 also implies the Central Limit Theorem under the following Lyapunov condition.

Definition 10.6. *The Lyapunov condition is said to be satisfied if there is a $\delta > 0$ such that*

$$\lim_{n\to\infty} \frac{1}{D_n^{2+\delta}} \sum_{i=1}^{n} \mathrm{E}(|\xi_i - m_i|^{2+\delta}) = 0.$$

Theorem 10.7 (Central Limit Theorem, Lyapunov Condition). *Let $\xi_1, \xi_2, \ldots$ be a sequence of independent random variables with finite variances. If the Lyapunov condition is satisfied, then the distributions of $(\zeta_n - M_n)/D_n$ converge weakly to $N(0,1)$ distribution as $n \to \infty$.*

Proof. Let $\varepsilon, \delta > 0$. Then,

$$\frac{\int_{\{x:|x-m_i|\geq\varepsilon D_n\}} (x - m_i)^2 dF_i(x)}{D_n^2}$$

$$\leq \frac{\int_{\{x:|x-m_i|\geq\varepsilon D_n\}} (x - m_i)^{2+\delta} dF_i(x)}{D_n^2(\varepsilon D_n)^{\delta}} \leq \varepsilon^{-\delta} \frac{\mathrm{E}(|\xi_i - m_i|^{2+\delta})}{D_n^{2+\delta}}.$$

Therefore, a sequence of random variables satisfying the Lyapunov condition also satisfies the Lindeberg condition.   □

If condition (10.5) is satisfied, then the Lindeberg condition is not only sufficient, but also necessary for the Central Limit Theorem to hold. We state the following theorem without providing a proof.

Theorem 10.8 (Lindeberg-Feller). *Let $\xi_1, \xi_2, \ldots$ be a sequence of independent random variables with finite variances such that the condition (10.5) is satisfied. Then the Lindeberg condition is satisfied if and only if the Central Limit Theorem holds, that is the distributions of $(\zeta_n - M_n)/D_n$ converge weakly to $N(0,1)$ distribution as $n \to \infty$.*

There are various generalizations of the Central Limit Theorem, not presented here, where the condition of independence of random variables is replaced by conditions of weak dependence in some sense. Other important generalizations concern vector-valued random variables.

10.2 Local Limit Theorem

The Central Limit Theorem proved in the previous section states that the measures on $\mathbb{R}$ induced by normalized sums of independent random variables converge weakly to the Gaussian measure $N(0,1)$. Under certain additional conditions this statement can be strengthened to include the point-wise convergence of the densities. In the case of integer-valued random variables (where no densities exist) the corresponding statement is the following Local Central Limit Theorem, which is a generalization of the de Moivre-Laplace Theorem.

Let ξ be an integer-valued random variable. Let $X = \{x_1, x_2, \ldots\}$ be the finite or countable set consisting of those values of ξ for which $p_j = \mathrm{P}(\xi = x_j) \neq 0$. Let $Y = \{x - y : x, y \in X\}$ be the set of all the pair-wise differences between the elements of X. We shall say that ξ spans the set of integers $\mathbb{Z}$ if the greatest common divisor of all the elements of Y equals 1.

Lemma 10.9. *If ξ spans $\mathbb{Z}$, and $\varphi(\lambda) = \mathrm{E}e^{i\xi\lambda}$ is the characteristic function of the variable ξ, then for any $\delta > 0$*

$$\sup_{\delta \leq |\lambda| \leq \pi} |\varphi(\lambda)| < 1. \tag{10.7}$$

Proof. Suppose that $|\varphi(\lambda)| = 1$ for some $\lambda \neq 0$. From Lemma 9.3 (property 4) it follows that the values of ξ are of the form $2\pi(\alpha + m)/\lambda$ for some fixed α, where m can take arbitrary integer values. This is clearly impossible if $\lambda \in [\delta, \pi]$ and ξ spans $\mathbb{Z}$, and therefore $|\varphi(\lambda)| < 1$. Since $|\varphi(\lambda)|$ is continuous, (10.7) holds. $\qquad\square$

Let $\xi_1, \xi_2, \ldots$ be a sequence of integer-valued independent identically distributed random variables. Let $m = \mathrm{E}\xi_1$, $\sigma^2 = \mathrm{Var}(\xi_1) < \infty$, $\zeta_n = \sum_{i=1}^{n} \xi_i$, $M_n = \mathrm{E}\zeta_n = nm$, $D_n^2 = \mathrm{Var}(\zeta_n) = n\sigma^2$. We shall be interested in the probability of the event that ζ_n takes an integer value k. Let $\mathrm{P}_n(k) = \mathrm{P}(\zeta_n = k)$, $z = z(n, k) = \frac{k - M_n}{D_n}$.

Theorem 10.10 (Local Limit Theorem). *Let $\xi_1, \xi_2, \ldots$ be a sequence of independent identically distributed integer-valued random variables with finite variances such that ξ_1 spans $\mathbb{Z}$. Then*

$$\lim_{n\to\infty} \left(D_n P_n(k) - \frac{1}{\sqrt{2\pi}} e^{-\frac{z^2}{2}} \right) = 0 \qquad (10.8)$$

uniformly in k.

Proof. We shall prove the theorem for the case $m = 0$, since the general case requires only trivial modifications. Let $\varphi(\lambda)$ be the characteristic function of each of the variables ξ_i. Then the characteristic function of the random variable ζ_n is

$$\varphi_{\zeta_n}(\lambda) = \varphi^n(\lambda) = \sum_{k=-\infty}^{\infty} P_n(k) e^{i\lambda k}.$$

Thus $\varphi^n(\lambda)$ is the Fourier series with coefficients $P_n(k)$, and we can use the formula for Fourier coefficients to find $P_n(k)$:

$$2\pi P_n(k) = \int_{-\pi}^{\pi} \varphi^n(\lambda) e^{-i\lambda k} d\lambda = \int_{-\pi}^{\pi} \varphi^n(\lambda) e^{-i\lambda z D_n} d\lambda.$$

Therefore, after a change of variables we obtain

$$2\pi D_n P_n(k) = \int_{-\pi D_n}^{\pi D_n} e^{-i\lambda z} \varphi^n\left(\frac{\lambda}{D_n}\right) d\lambda.$$

From the formula for the characteristic function of the Gaussian distribution

$$\frac{1}{\sqrt{2\pi}} e^{-\frac{z^2}{2}} = \frac{1}{2\pi} \int_{-\infty}^{\infty} e^{i\lambda z - \frac{\lambda^2}{2}} d\lambda = \frac{1}{2\pi} \int_{-\infty}^{\infty} e^{-i\lambda z - \frac{\lambda^2}{2}} d\lambda.$$

We can write the difference in (10.8) multiplied by 2π as a sum of four integrals:

$$2\pi\left(D_n P_n(k) - \frac{1}{\sqrt{2\pi}} e^{-\frac{z^2}{2}} \right) = I_1 + I_2 + I_3 + I_4,$$

where

$$I_1 = \int_{-T}^{T} e^{-i\lambda z} \left(\varphi^n\left(\frac{\lambda}{D_n}\right) - e^{-\frac{\lambda^2}{2}} \right) d\lambda,$$

$$I_2 = -\int_{|\lambda|>T} e^{-i\lambda z - \frac{\lambda^2}{2}} d\lambda,$$

$$I_3 = \int_{\delta D_n \leq |\lambda| \leq \pi D_n} e^{-i\lambda z} \varphi^n\left(\frac{\lambda}{D_n}\right) d\lambda,$$

$$I_4 = \int_{T \leq |\lambda| < \delta D_n} e^{-i\lambda z} \varphi^n\left(\frac{\lambda}{D_n}\right) d\lambda,$$

where the positive constants $T < \delta D_n$ and $\delta < \pi$ will be selected later. By Remark 10.4, the convergence $\lim_{n\to\infty} \varphi^n(\frac{\lambda}{D_n}) = e^{-\frac{\lambda^2}{2}}$ is uniform on the interval $[-T, T]$. Therefore $\lim_{n\to\infty} I_1 = 0$ for any T.

The second integral can be estimated as follows:

$$|I_2| \leq \int_{|\lambda|>T} |e^{-i\lambda z - \frac{\lambda^2}{2}}| d\lambda = \int_{|\lambda|>T} e^{-\frac{\lambda^2}{2}} d\lambda,$$

which can be made arbitrarily small by selecting T large enough, since the improper integral $\int_{-\infty}^{\infty} e^{-\frac{\lambda^2}{2}} d\lambda$ converges.

The third integral is estimated as follows:

$$|I_3| \leq \int_{\delta D_n \leq |\lambda| \leq \pi D_n} |e^{-i\lambda z} \varphi^n(\frac{\lambda}{D_n})| d\lambda \leq 2\pi\sigma\sqrt{n} (\sup_{\delta \leq |\lambda| \leq \pi} |\varphi(\lambda)|)^n,$$

which tends to zero as $n \to \infty$ due to (10.7).

In order to estimate the fourth integral, we note that the existence of the variance implies that the characteristic function is a twice continuously differentiable complex-valued function with $\varphi'(0) = im = 0$ and $\varphi''(0) = -\sigma^2$. Therefore, applying the Taylor formula to the real and imaginary parts of φ, we obtain

$$\varphi(\lambda) = 1 - \frac{\sigma^2 \lambda^2}{2} + o(\lambda^2) \quad \text{as} \quad \lambda \to 0.$$

For $|\lambda| \leq \delta$ and δ sufficiently small, we obtain

$$|\varphi(\lambda)| \leq 1 - \frac{\sigma^2 \lambda^2}{4} \leq e^{-\frac{\sigma^2 \lambda^2}{4}}.$$

If $|\lambda| \leq \delta D_n$, then

$$|\varphi(\frac{\lambda}{D_n})|^n \leq e^{\frac{-n\sigma^2 \lambda^2}{4D_n^2}} = e^{-\frac{\lambda^2}{4}}.$$

Therefore,

$$|I_4| \leq 2 \int_T^{\delta D_n} e^{-\frac{\lambda^2}{4}} d\lambda \leq 2 \int_T^{\infty} e^{-\frac{\lambda^2}{4}} d\lambda.$$

This can be made arbitrarily small by selecting sufficiently large T. This completes the proof of the theorem.   □

Whenwe studied the recurrence and transience of random walks on $\mathbb{Z}^d$ (Chap. 6) we needed to estimate the probability that a path returns to the origin after $2n$ steps:

$$u_{2n} = P(\sum_{j=1}^{2n} \omega_j = 0).$$

Here ω_j are independent identically distributed random variables with values in $\mathbb{Z}^d$ with the distribution p_y, $y \in \mathbb{Z}^d$, where $p_y = \frac{1}{2d}$ if $y = \pm e_s$, $1 \leq s \leq d$, and 0 otherwise.

Let us use the characteristic functions to study the asymptotics of u_{2n} as $n \to \infty$. The characteristic function of ω_j is equal to

$$\mathrm{E}e^{i(\lambda,\omega_i)} = \frac{1}{2d}(e^{i\lambda_1} + e^{-i\lambda_1} + \ldots + e^{i\lambda_d} + e^{-i\lambda_d}) = \frac{1}{d}(\cos(\lambda_1) + \ldots + \cos(\lambda_d)),$$

where $\lambda = (\lambda_1, \ldots, \lambda_d) \in \mathbb{R}^d$. Therefore, the characteristic function of the sum $\sum_{j=1}^{2n} \omega_j$ is equal to $\varphi_{2n}(\lambda) = \frac{1}{d^{2n}}(\cos(\lambda_1) + \ldots + \cos(\lambda_d))^{2n}$. On the other hand,

$$\varphi_{2n}(\lambda) = \sum_{k \in \mathbb{Z}^d} \mathrm{P}_n(k)e^{i(\lambda,k)},$$

where $\mathrm{P}_n(k) = \mathrm{P}(\sum_{j=1}^{2n} \omega_j = k)$. Integrating both sides of the equality

$$\sum_{k \in \mathbb{Z}^d} \mathrm{P}_n(k)e^{i(\lambda,k)} = \frac{1}{d^{2n}}(\cos(\lambda_1) + \ldots + \cos(\lambda_d))^{2n}$$

over λ, we obtain

$$(2\pi)^d u_{2n} = \frac{1}{d^{2n}} \int_{-\pi}^{\pi} \ldots \int_{-\pi}^{\pi} (\cos(\lambda_1) + \ldots + \cos(\lambda_d))^{2n} d\lambda_1 \ldots d\lambda_d.$$

The asymptotics of the latter integral can be treated with the help of the so-called Laplace asymptotic method. The Laplace method is used to describe the asymptotic behavior of integrals of the form

$$\int_D f(\lambda)e^{sg(\lambda)} d\lambda,$$

where D is a domain in $\mathbb{R}^d$, f and g are smooth functions, and $s \to \infty$ is a large parameter. The idea is that if $f(\lambda) > 0$ for $\lambda \in D$, then the main contribution to the integral comes from arbitrarily small neighborhoods of the maxima of the function g. Then the Taylor formula can be used to approximate the function g in small neighborhoods of its maxima. In our case the points of the maxima are $\lambda_1 = \ldots = \lambda_d = 0$ and $\lambda_1 = \ldots = \lambda_d = \pm\pi$. We state the result for the problem at hand without going into further detail:

$$\int_{-\pi}^{\pi} \ldots \int_{-\pi}^{\pi} (\cos(\lambda_1) + \ldots + \cos(\lambda_d))^{2n} d\lambda_1 \ldots d\lambda_d$$

$$= \int_{-\pi}^{\pi} \ldots \int_{-\pi}^{\pi} e^{2n \ln|\cos(\lambda_1) + \ldots + \cos(\lambda_d)|} d\lambda_1 \ldots d\lambda_d$$

$$\sim c \sup(|\cos(\lambda_1) + \ldots + \cos(\lambda_d)|)^{2n} n^{-\frac{d}{2}} = cd^{2n} n^{-\frac{d}{2}},$$

which implies that $u_{2n} \sim cn^{-\frac{d}{2}}$ as $n \to \infty$ with another constant c.

10.3 Central Limit Theorem and Renormalization Group Theory

The Central Limit Theorem states that Gaussian distributions can be obtained as limits of distributions of properly normalized sums of independent random variables. If the random variables $\xi_1, \xi_2, \ldots$ forming the sum are independent and identically distributed, then it is enough to assume that they have a finite second moment.

In this section we shall take another look at the mechanism of convergence of normalized sums, which may help explain why the class of distributions of ξ_i, for which the central limit theorem holds, is so large. We shall view the densities (assuming that they exist) of the normalized sums as iterations of a certain non-linear transformation applied to the common density of ξ_i. The method presented below is called the renormalization group method. It can be generalized in several ways (for example, to allow the variables to be weakly dependent). We do not strive for maximal generality, however. Instead, we consider again the case of independent random variables.

Let $\xi_1, \xi_2, \ldots$ be a sequence of independent identically distributed random variables with zero expectation and finite variance. We define the random variables

$$\zeta_n = 2^{-\frac{n}{2}} \sum_{i=1}^{2^n} \xi_i, \quad n \geq 0.$$

Then

$$\zeta_{n+1} = \frac{1}{\sqrt{2}} (\zeta_n' + \zeta_n''),$$

where

$$\zeta_n' = 2^{-\frac{n}{2}} \sum_{i=1}^{2^n} \xi_i, \quad \zeta_n'' = 2^{-\frac{n}{2}} \sum_{i=2^n+1}^{2^{n+1}} \xi_i.$$

Clearly, ζ_n' and ζ_n'' are independent identically distributed random variables. Let us assume that ξ_i have a density, which will be denoted by p_0. Note that $\zeta_0 = \xi_1$, and thus the density of ζ_0 is also p_0. Let us denote the density of ζ_n by p_n and its distribution by P_n. Then

$$p_{n+1}(x) = \sqrt{2} \int_{-\infty}^{\infty} p_n(\sqrt{2}x - u) p_n(u) du.$$

Thus the sequence p_n can be obtained from p_0 by iterating the non-linear operator T, which acts on the space of densities according to the formula

$$Tp(x) = \sqrt{2} \int_{-\infty}^{\infty} p(\sqrt{2}x - u) p(u) du, \tag{10.9}$$

that is $p_{n+1} = Tp_n$ and $p_n = T^n p_0$. Note that if p is the density of a random variable with zero expectation, then so is Tp. In other words,

$$\int_{-\infty}^{\infty} x(Tp)(x)dx = 0 \quad \text{if} \quad \int_{-\infty}^{\infty} xp(x)dx = 0. \qquad (10.10)$$

Indeed, if ζ' and ζ'' are independent identically distributed random variables with zero mean and density p, then $\frac{1}{\sqrt{2}}(\zeta'+\zeta'')$ has zero mean and density Tp. Similarly, for a density p such that $\int_{-\infty}^{\infty} xp(x)dx = 0$, the operator T preserves the variance, that is

$$\int_{-\infty}^{\infty} x^2(Tp)(x)dx = \int_{-\infty}^{\infty} x^2 p(x)dx. \qquad (10.11)$$

Let $p_G(x) = \frac{1}{\sqrt{2\pi}}e^{-\frac{x^2}{2}}$ be the density of the Gaussian distribution and μ_G the Gaussian measure on the real line (the measure with the density p_G). It is easy to check that p_G is a fixed point of T, that is $p_G = Tp_G$. The fact that the convergence $P_n \Rightarrow \mu_G$ holds for a wide class of initial densities is related to the stability of this fixed point.

In the general theory of non-linear operators the investigation of the stability of a fixed point starts with an investigation of its stability with respect to the linear approximation. In our case it is convenient to linearize not the operator T itself, but a related operator, as explained below.

Let $H = L^2(\mathbb{R}, \mathcal{B}, \mu_G)$ be the Hilbert space with the inner product

$$(f,g) = \frac{1}{\sqrt{2\pi}} \int_{-\infty}^{\infty} f(x)\overline{g}(x)\exp(-\frac{x^2}{2})dx.$$

Let h be an element of H, that is a measurable function such that

$$\|h\|^2 = \frac{1}{\sqrt{2\pi}} \int_{-\infty}^{\infty} h^2(x)\exp(-\frac{x^2}{2})dx < \infty.$$

Assume that $\|h\|$ is small. We perturb the Gaussian density as follows:

$$p_h(x) = p_G(x) + \frac{h(x)}{\sqrt{2\pi}}\exp(-\frac{x^2}{2}) = \frac{1}{\sqrt{2\pi}}(1+h(x))\exp(-\frac{x^2}{2}).$$

In order for p_h to be a density of a probability measure, we need to assume that

$$\int_{-\infty}^{\infty} h(x)\exp(-\frac{x^2}{2})dx = 0. \qquad (10.12)$$

Moreover, in order for p_h to correspond to a random variable with zero expectation, we assume that

$$\int_{-\infty}^{\infty} xh(x)\exp(-\frac{x^2}{2})dx = 0. \qquad (10.13)$$

Let us define a non-linear operator $\widetilde{L}$ by the implicit relation

$$Tp_h(x) = \frac{1}{\sqrt{2\pi}} \exp(-\frac{x^2}{2})(1 + (\widetilde{L}h)(x)). \tag{10.14}$$

Thus,

$$T^n p_h(x) = \frac{1}{\sqrt{2\pi}} \exp(-\frac{x^2}{2})(1 + (\widetilde{L}^n h)(x)).$$

This formula shows that in order to study the behavior of $T^n p_h(x)$ for large n, it is sufficient to study the behavior of $\widetilde{L}^n h$ for large n. We can write

$$Tp_h(x) =$$

$$\frac{1}{\sqrt{2\pi}} \int_{-\infty}^{\infty} (1 + h(\sqrt{2}x - u)) \exp(-\frac{(\sqrt{2}x - u)^2}{2})(1 + h(u)) \exp(-\frac{u^2}{2})du$$

$$= \frac{1}{\sqrt{2\pi}} \int_{-\infty}^{\infty} \exp(-\frac{(\sqrt{2}x - u)^2}{2} - \frac{u^2}{2})du$$

$$+ \frac{1}{\sqrt{2\pi}} \int_{-\infty}^{\infty} \exp(-\frac{(\sqrt{2}x - u)^2}{2} - \frac{u^2}{2})(h(\sqrt{2}x - u) + h(u))du + O(||h||^2)$$

$$= \frac{1}{\sqrt{2\pi}} \exp(-\frac{x^2}{2}) + \frac{\sqrt{2}}{\pi} \int_{-\infty}^{\infty} \exp(-x^2 + \sqrt{2}xu - u^2)h(u)du + O(||h||^2)$$

$$= \frac{1}{\sqrt{2\pi}} \exp(-\frac{x^2}{2})(1 + (Lh)(x)) + O(||h||^2),$$

where the linear operator L is given by the formula

$$(Lh)(x) = \frac{2}{\sqrt{\pi}} \int_{-\infty}^{\infty} \exp(-\frac{x^2}{2} + \sqrt{2}xu - u^2)h(u)du. \tag{10.15}$$

It is referred to as the Gaussian integral operator. Comparing two expressions for $Tp_h(x)$, the one above and the one given by (10.14), we see that

$$\widetilde{L}h = Lh + O(||h||^2),$$

that is L is the linearization of $\widetilde{L}$ at zero.

It is not difficult to show that (10.15) defines a bounded self-adjoint operator on H. It has a complete set of eigenvectors, which are the Hermite polynomials

$$h_k(x) = \exp(\frac{x^2}{2})(\frac{d}{dx})^k \exp(-\frac{x^2}{2}), \quad k \geq 0.$$

The corresponding eigenvalues are $\lambda_k = 2^{1-\frac{k}{2}}$, $k \geq 0$. We see that $\lambda_0, \lambda_1 > 1$, $\lambda_2 = 1$, while $0 < \lambda_k \leq 1/\sqrt{2}$ for $k \geq 3$. Let H_k, $k \geq 0$, be one-dimensional subspaces of H spanned by h_k. By (10.12) and (10.13) the initial vector h is orthogonal to H_0 and H_1, and thus $h \in H \ominus (H_0 \oplus H_1)$.

If $h \perp H_0$, then $\widetilde{L}(h) \perp H_0$ follows from (10.14), since (10.12) holds and p_h is a density. Similarly, if $h \perp H_0 \oplus H_1$, then $\widetilde{L}(h) \perp H_0 \oplus H_1$ follows from (10.10) and (10.14). Thus the subspace $H \ominus (H_0 \oplus H_1)$ is invariant not only for L, but also for $\widetilde{L}$. Therefore we can restrict both operators to this subspace, which can be further decomposed as follows:

$$H \ominus (H_0 \oplus H_1) = H_2 \oplus [H \ominus (H_0 \oplus H_1 \oplus H_2)].$$

Note that for an initial vector $h \in H \ominus (H_0 \oplus H_1)$, by (10.11) the operator $\widetilde{L}$ preserves its projection to H_2, that is

$$\int_{-\infty}^{\infty} (x^2 - 1)h(x) \exp(-\frac{x^2}{2}) = \int_{-\infty}^{\infty} (x^2 - 1)(\widetilde{L}h)(x) \exp(-\frac{x^2}{2}).$$

Let U be a small neighborhood of zero in H, and H^h the set of vectors whose projection to H_2 is equal to the projection of h onto H_2. Let $U^h = U \cap H^h$. It is not difficult to show that one can choose U such that $\widetilde{L}$ leaves U^h invariant for all sufficiently small h. Note that $\widetilde{L}$ is contracting on U^h for small h, since L is contracting on $H \ominus (H_0 \oplus H_1 \oplus H_2)$. Therefore it has a unique fixed point. It is easy to verify that this fixed point is the function

$$f_h(x) = \frac{1}{\sigma(p_h)} \exp(\frac{x^2}{2} - \frac{x^2}{2\sigma^2(p_h)}) - 1,$$

where $\sigma^2(p_h)$ is the variance of a random variable with density p_h,

$$\sigma^2(p_h) = \frac{1}{\sqrt{2\pi}} \int_{-\infty}^{\infty} x^2(1 + h(x)) \exp(-\frac{x^2}{2})dx.$$

Therefore, by the contracting mapping principle,

$$\widetilde{L}^n h \to f_h \quad \text{as } n \to \infty,$$

and consequently

$$T^n p_h(x) = \frac{1}{\sqrt{2\pi}} \exp(-\frac{x^2}{2})(1 + (\widetilde{L}^n h)(x)) \to$$

$$\frac{1}{\sqrt{2\pi}} \exp(-\frac{x^2}{2})(1 + f_h(x)) = \frac{1}{\sqrt{2\pi}\sigma(p_h)} \exp(-\frac{x^2}{2\sigma^2(p_h)}).$$

We see that $T^n p_h(x)$ converges in the space H to the density of the Gaussian distribution with variance $\sigma^2(p_h)$. This easily implies the convergence of distributions.

It is worth stressing again that the arguments presented in this section were based on the assumption that h is small, thus allowing us to state the convergence of the normalized sums ζ_n to the Gaussian distribution, provided the distribution of ξ_i is a small perturbation of the Gaussian distribution. The proof of the Central Limit Theorem in Sect. 10.1 went through regardless of this assumption.

10.4 Probabilities of Large Deviations

In the previous chapters we considered the probabilities

$$P(|\sum_{i=1}^{n} \xi_i - \sum_{i=1}^{n} m_i| \geq t)$$

with $m_i = \mathrm{E}\xi_i$ for sequences of independent random variables $\xi_1, \xi_2, \ldots$, and we estimated these probabilities using the Chebyshev Inequality

$$P(|\sum_{i=1}^{n} \xi_i - \sum_{i=1}^{n} m_i| \geq t) \leq \frac{\sum_{i=1}^{n} d_i}{t^2}, \quad d_i = \mathrm{Var}(\xi_i).$$

In particular, if the random variables ξ_i are identically distributed, then for some constant c which does not depend on n, and with $d = d_1$:

(a) For $t = c\sqrt{n}$ we have $\frac{d}{c^2}$ on the right-hand side of the inequality;
(b) For $t = cn$ we have $\frac{d}{c^2 n}$ on the right-hand side of the inequality.

We know from the Central Limit Theorem that in the case (a) the corresponding probability converges to a positive limit as $n \to \infty$. This limit can be calculated using the Gaussian distribution. This means that in the case (a) the order of magnitude of the estimate obtained from the Chebyshev Inequality is correct. On the other hand, in the case (b) the estimate given by the Chebyshev Inequality is very crude. In this section we obtain more precise estimates in the case (b).

Let us consider a sequence of independent identically distributed random variables. We denote their common distribution function by F. We make the following assumption about F

$$R(\lambda) = \int_{-\infty}^{\infty} e^{\lambda x} dF(x) < \infty \tag{10.16}$$

for all λ, $-\infty < \lambda < \infty$. This condition is automatically satisfied if all the ξ_i are bounded. It is also satisfied if the probabilities of large values of ξ_i decay faster than exponentially.

We now note several properties of the function $R(\lambda)$. From the finiteness of the integral in (10.16) for all λ, it follows that the derivatives

$$R'(\lambda) = \int_{-\infty}^{\infty} x e^{\lambda x} dF(x), \quad R''(\lambda) = \int_{-\infty}^{\infty} x^2 e^{\lambda x} dF(x)$$

exist for all λ. Let us consider $m(\lambda) = \frac{R'(\lambda)}{R(\lambda)}$. Then

$$m'(\lambda) = \frac{R''(\lambda)}{R(\lambda)} - \left(\frac{R'(\lambda)}{R(\lambda)}\right)^2 = \int_{-\infty}^{\infty} \frac{x^2}{R(\lambda)} e^{\lambda x} dF(x) - \left(\int_{-\infty}^{\infty} \frac{x}{R(\lambda)} e^{\lambda x} dF(x)\right)^2.$$

We define a new distribution function $F_\lambda(x) = \frac{1}{R(\lambda)} \int_{(-\infty,x]} e^{\lambda t} dF(t)$ for each λ. Then $m(\lambda) = \int_{-\infty}^{\infty} x dF_\lambda(x)$ is the expectation of a random variable with this distribution, and $m'(\lambda)$ is the variance. Therefore $m'(\lambda) > 0$ if F is a non-trivial distribution, that is it is not concentrated at a point. We exclude the latter case from further consideration. Since $m'(\lambda) > 0$, $m(\lambda)$ is a monotonically increasing function.

We say that M^+ is an upper limit in probability for a random variable ξ if $P(\xi > M^+) = 0$, and $P(M^+ - \varepsilon \le \xi \le M^+) > 0$ for every $\varepsilon > 0$. One can define the lower limit in probability in the same way. If $P(\xi > M) > 0$ ($P(\xi < M) > 0$) for any M, then $M^+ = \infty$ ($M^- = -\infty$). In all the remaining cases M^+ and M^- are finite. The notion of the upper (lower) limit in probability can be recast in terms of the distribution function as follows:

$$M^+ = \sup\{x : F(x) < 1\}, \quad M^- = \inf\{x : F(x) > 0\}.$$

Lemma 10.11. *Under the assumption (10.16) on the distribution function, the limits for $m(\lambda)$ are as follows:*

$$\lim_{\lambda \to \infty} m(\lambda) = M^+, \quad \lim_{\lambda \to -\infty} m(\lambda) = M^-.$$

Proof. We shall only prove the first statement since the second one is proved analogously. If $M^+ < \infty$, then from the definition of F_λ

$$\int_{(M^+,\infty)} dF_\lambda(x) = \frac{1}{R(\lambda)} \int_{(M^+,\infty)} e^{\lambda x} dF(x) = 0$$

for each λ. Note that $\int_{(M^+,\infty)} dF_\lambda(x) = 0$ implies that

$$m(\lambda) = \int_{(-\infty,M^+]} x dF_\lambda(x) \le M^+,$$

and therefore $\lim_{\lambda \to \infty} m(\lambda) \le M^+$. It remains to prove the opposite inequality.

Let $M^+ \le \infty$. If $M^+ = 0$, then $m(\lambda) \le 0$ for all λ. Therefore, we can assume that $M^+ \neq 0$. Take $M \in (0, M^+)$ if $M^+ > 0$ and $M \in (-\infty, M^+)$ if $M^+ < 0$. Choose a finite segment $[A, B]$ such that $M < A < B \le M^+$ and $\int_{[A,B]} dF(x) > 0$. Then

$$\int_{(-\infty,M]} e^{\lambda x} dF(x) \le e^{\lambda M},$$

while

$$\int_{(M,\infty)} e^{\lambda x} dF(x) \ge e^{\lambda A} \int_{[A,B]} dF(x),$$

which implies that

$$\int_{(-\infty,M]} e^{\lambda x}dF(x) = o(\int_{(M,\infty)} e^{\lambda x}dF(x)) \quad \text{as} \quad \lambda \to \infty.$$

Similarly,

$$\int_{(-\infty,M]} xe^{\lambda x}dF(x) = O(e^{\lambda M}),$$

while

$$|\int_{(M,\infty)} xe^{\lambda x}dF(x)| = |\int_{(M,M^+]} xe^{\lambda x}dF(x)| \geq \min(|A|,|B|)e^{\lambda A}\int_{[A,B]} dF(x),$$

which implies that

$$\int_{(-\infty,M]} xe^{\lambda x}dF(x) = o(\int_{(M,\infty)} xe^{\lambda x}dF(x)) \quad \text{as} \quad \lambda \to \infty.$$

Therefore,

$$\lim_{\lambda \to \infty} m(\lambda) = \lim_{\lambda \to \infty} \frac{\int_{(-\infty,\infty)} xe^{\lambda x}dF(x)}{\int_{(-\infty,\infty)} e^{\lambda x}dF(x)} = \lim_{\lambda \to \infty} \frac{\int_{(M,\infty)} xe^{\lambda x}dF(x)}{\int_{(M,\infty)} e^{\lambda x}dF(x)} \geq M.$$

Since M can be taken to be arbitrary close to M^+, we conclude that $\lim_{\lambda \to \infty} m(\lambda) = M^+$. $\qquad\square$

We now return to considering the probabilities of the deviations of sums of independent identically distributed random variables from the sums of their expectations. Consider c such that $m = \mathrm{E}\xi_i < c < M^+$. We shall be interested in the probability $\mathrm{P}_{n,c} = \mathrm{P}(\xi_1 + \ldots + \xi_n > cn)$. Since $c > m$, this is the probability of the event that the sum of the random variables takes values which are far away from the mathematical expectation of the sum. Such values are called large deviations (from the expectation). We shall describe a method for calculating the asymptotics of these probabilities which is usually called Cramer's method.

Let λ_0 be such that $m(\lambda_0) = c$. Such λ_0 exists by Lemma 10.11 and is unique since $m(\lambda)$ is strictly monotonic. Note that $m = m(0) < c$. Therefore $\lambda_0 > 0$ by the monotonicity of $m(\lambda)$.

Theorem 10.12. $\mathrm{P}_{n,c} \leq B_n(R(\lambda_0)e^{-\lambda_0 c})^n$, where $\lim_{n\to\infty} B_n = \frac{1}{2}$.

Proof. We have

$$\mathrm{P}_{n,c} = \int \ldots \int_{x_1+\ldots+x_n>cn} dF(x_1)\ldots dF(x_n)$$

$$\leq (R(\lambda_0))^n e^{-\lambda_0 cn} \int \ldots \int_{x_1+\ldots+x_n>cn} \frac{e^{\lambda_0(x_1+\ldots+x_n)}}{(R(\lambda_0))^n} dF(x_1)\ldots dF(x_n)$$

$$= (R(\lambda_0)e^{-\lambda_0 c})^n \int \ldots \int_{x_1+\ldots+x_n>cn} dF_{\lambda_0}(x_1)\ldots dF_{\lambda_0}(x_n).$$

To estimate the latter integral, we can consider independent identically distributed random variables $\tilde{\xi}_1,\ldots,\tilde{\xi}_n$ with distribution F_{λ_0}. The expectation of such random variables is equal to $\int_{\mathbb{R}} x dF_{\lambda_0}(x) = m(\lambda_0) = c$. Therefore

$$\int \ldots \int_{x_1+\ldots+x_n>cn} dF_{\lambda_0}(x_1)\ldots dF_{\lambda_0}(x_n) = P(\tilde{\xi}_1+\ldots+\tilde{\xi}_n > cn)$$

$$= P(\tilde{\xi}_1+\ldots+\tilde{\xi}_n - nm(\lambda_0) > 0)$$

$$= P(\frac{\tilde{\xi}_1+\ldots+\tilde{\xi}_n - nm(\lambda_0)}{\sqrt{nd(\lambda_0)}} > 0) \to \frac{1}{2}$$

as $n \to \infty$. Here $d(\lambda_0)$ is the variance of the random variables $\tilde{\xi}_i$, and the convergence of the probability to $\frac{1}{2}$ follows from the Central Limit Theorem. $\square$

The lower estimate turns out to be somewhat less elegant.

Theorem 10.13. *For any $b > 0$ there exists $p(b, \lambda_0) > 0$ such that*

$$P_{n,c} \geq (R(\lambda_0)e^{-\lambda_0 c})^n e^{-\lambda_0 b\sqrt{n}} p_n,$$

with $\lim_{n\to\infty} p_n = p(b, \lambda_0) > 0$.

Proof. As in Theorem 10.12,

$$P_{n,c} \geq \int \ldots \int_{cn<x_1+\ldots+x_n<cn+b\sqrt{n}} dF(x_1)\ldots dF(x_n)$$

$$\geq (R(\lambda_0))^n e^{-\lambda_0(cn+b\sqrt{n})} \int \ldots \int_{cn<x_1+\ldots+x_n<cn+b\sqrt{n}} dF_{\lambda_0}(x_1)\ldots dF_{\lambda_0}(x_n).$$

The latter integral, as in the case of Theorem 10.12, converges to a positive limit by the Central Limit Theorem. $\square$

In Theorems 10.12 and 10.13 the number $R(\lambda_0)e^{-\lambda_0 c} = r(\lambda_0)$ is involved. It is clear that $r(0) = 1$. Let us show that $r(\lambda_0) < 1$. We have

$$\ln r(\lambda_0) = \ln R(\lambda_0) - \lambda_0 c = \ln R(\lambda_0) - \ln R(0) - \lambda_0 c.$$

By Taylor's formula,

$$\ln R(\lambda_0) - \ln R(0) = \lambda_0 (\ln R)'(\lambda_0) - \frac{\lambda_0^2}{2}(\ln R)''(\lambda_1),$$

where λ_1 is an intermediate point between 0 and λ_0. Furthermore,

$$(\ln R)'(\lambda_0) = \frac{R'(\lambda_0)}{R(\lambda_0)} = m(\lambda_0) = c, \quad \text{and} \quad (\ln R)''(\lambda_1) > 0,$$

since it is the variance of the distribution F_{λ_1}. Thus

$$\ln r(\lambda_0) = -\frac{\lambda_0^2}{2}(\ln R)''(\lambda_1) < 0.$$

From Theorems 10.12 and 10.13 we obtain the following corollary.

Corollary 10.14.

$$\lim_{n \to \infty} \frac{1}{n} \ln P_{n,c} = \ln r(\lambda_0) < 0.$$

Proof. Indeed, let $b = 1$ in Theorem 10.13. Then

$$\ln r(\lambda_0) - \frac{\lambda_0}{\sqrt{n}} - \frac{\ln p_n}{n} \leq \frac{\ln P_{n,c}}{n} \leq \ln r(\lambda_0) + \frac{1}{n} \ln B_n.$$

We complete the proof by taking the limit as $n \to \infty$. $\qquad\square$

This corollary shows that the probabilities $P_{n,c}$ decay exponentially in n. In other words, they decay much faster than suggested by the Chebyshev Inequality.

10.5 Other Limit Theorems

The Central Limit Theorem applies to sums of independent identically distributed random variables when the variances of these variables are finite. When the variances are infinite, different Limit Theorems may apply, giving different limiting distributions.

As an example, we consider a sequence of independent identically distributed random variables $\xi_1, \xi_2, \ldots$, whose distribution is given by a symmetric density $p(x)$, $p(x) = p(-x)$, such that

$$p(x) \sim \frac{c}{|x|^{\alpha+1}} \quad \text{as } |x| \to \infty, \tag{10.17}$$

where $0 < \alpha < 2$ and c is a constant. The condition of symmetry is imposed for the sake of simplicity. Consider the normalized sum

$$\eta_n = \frac{\xi_1 + \ldots + \xi_n}{n^{\frac{1}{\alpha}}}.$$

Theorem 10.15. *As $n \to \infty$, the distributions of η_n converge weakly to a limiting distribution whose characteristic function is $\psi(\lambda) = e^{-c_1|\lambda|^\alpha}$, where c_1 is a function of c.*

Remark 10.16. For $\alpha = 2$, the convergence to the Gaussian distribution is also true, but the normalization of the sum is different:

$$\eta_n = \frac{\xi_1 + \ldots + \xi_n}{(n \ln n)^{\frac{1}{2}}}.$$

Remark 10.17. For $\alpha = 1$ we have the convergence to the Cauchy distribution.

In order to prove Theorem 10.15, we shall need the following lemma.

Lemma 10.18. *Let* $\varphi(\lambda)$ *be the characteristic function of the random variables* $\xi_1, \xi_2, \ldots$. *Then,*

$$\varphi(\lambda) = 1 - c_1 |\lambda|^\alpha + o(|\lambda|^\alpha) \quad as \quad \lambda \to 0.$$

Remark 10.19. This is a particular case of the so-called Tauberian Theorems, which relate the behavior of a distribution at infinity to the behavior of the characteristic function near $\lambda = 0$.

Proof. Take a constant M large enough, so that the density $p(x)$ can be represented as $p(x) = \frac{c(1+g(x))}{|x|^{\alpha+1}}$ for $|x| \geq M$, where $g(x)$ is a bounded function, $g(x) \to 0$ as $|x| \to \infty$. For simplicity of notation, assume that $\lambda \to 0+$. For $\lambda < 1/M$ we break the integral defining $\varphi(\lambda)$ into five parts:

$$\varphi(\lambda) = \int_{-\infty}^{-\frac{1}{\lambda}} p(x)e^{i\lambda x}dx + \int_{-\frac{1}{\lambda}}^{-M} p(x)e^{i\lambda x}dx + \int_{-M}^{M} p(x)e^{i\lambda x}dx$$

$$+ \int_{M}^{\frac{1}{\lambda}} p(x)e^{i\lambda x}dx + \int_{\frac{1}{\lambda}}^{\infty} p(x)e^{i\lambda x}dx$$

$$= I_1(\lambda) + I_2(\lambda) + I_3(\lambda) + I_4(\lambda) + I_5(\lambda).$$

The integral $I_3(\lambda)$ is a holomorphic function of λ equal to $\int_{-M}^{M} p(x)dx$ at $\lambda = 0$. The derivative $I_3'(0)$ is equal to $\int_{-M}^{M} p(x)ixdx = 0$, since $p(x)$ is an even function. Therefore, for any fixed M

$$I_3(\lambda) = \int_{-M}^{M} p(x)dx + O(\lambda^2) \quad as \quad \lambda \to 0.$$

Using a change of variables and the Dominated Convergence Theorem, we obtain

$$I_1(\lambda) = \int_{-\infty}^{-\frac{1}{\lambda}} p(x)e^{i\lambda x}dx = \int_{-\infty}^{-\frac{1}{\lambda}} \frac{c(1+g(x))}{|x|^{\alpha+1}}e^{i\lambda x}dx$$

$$= c\lambda^\alpha \int_{-\infty}^{-1} \frac{(1+g(\frac{y}{\lambda}))}{|y|^{\alpha+1}}e^{iy}dy \sim c\lambda^\alpha \int_{-\infty}^{-1} \frac{e^{iy}}{|y|^{\alpha+1}}dy.$$

Similarly,

$$I_5(\lambda) \sim c\lambda^\alpha \int_1^\infty \frac{e^{iy}}{|y|^{\alpha+1}} dy.$$

Next, since $p(x)$ is an even function,

$$I_2(\lambda) + I_4(\lambda) = \int_{-\frac{1}{\lambda}}^{-M} p(x)(e^{i\lambda x} - 1 - i\lambda x)dx + \int_M^{\frac{1}{\lambda}} p(x)(e^{i\lambda x} - 1 - i\lambda x)dx$$

$$+ \int_{-\frac{1}{\lambda}}^{-M} p(x)dx + \int_M^{\frac{1}{\lambda}} p(x)dx. \tag{10.18}$$

The third term on the right-hand side is equal to

$$\int_{-\frac{1}{\lambda}}^{-M} p(x)dx = \int_{-\infty}^{-M} p(x)dx - \int_{-\infty}^{-\frac{1}{\lambda}} \frac{c(1+g(x))}{|x|^{\alpha+1}} dx$$

$$= \int_{-\infty}^{-M} p(x)dx + c_0\lambda^\alpha + o(\lambda^\alpha),$$

where c_0 is some constant. Similarly,

$$\int_M^{\frac{1}{\lambda}} p(x)dx = \int_M^\infty p(x)dx + c_0\lambda^\alpha + o(\lambda^\alpha).$$

The first two terms on the right-hand side of (10.18) can be treated with the help of the same change of variables that was used to find the asymptotics of $I_1(\lambda)$. Therefore, taking into account the asymptotic behavior of each term, we obtain

$$I_1(\lambda) + I_2(\lambda) + I_3(\lambda) + I_4(\lambda) + I_5(\lambda)$$

$$= \int_{-\infty}^\infty p(x)dx - c_1\lambda^\alpha + o(\lambda^\alpha) = 1 - c_1\lambda^\alpha + o(\lambda^\alpha),$$

where c_1 is another constant. $\qquad\square$

Proof of Theorem 10.15. The characteristic function of η_n has the form

$$\varphi_{\eta_n}(\lambda) = \mathrm{E}e^{i\lambda \frac{\xi_1 + \ldots + \xi_n}{n^{1/\alpha}}} = (\varphi(\frac{\lambda}{n^{1/\alpha}}))^n.$$

In our case, λ is fixed and $n \to \infty$. Therefore we can use Lemma 10.18 to conclude

$$(\varphi(\frac{\lambda}{n^{1/\alpha}}))^n = (1 - \frac{c_1|\lambda|^\alpha}{n} + o(\frac{1}{n}))^n \to e^{-c_1|\lambda|^\alpha}.$$

By Remark 9.11, the function $e^{-c_1|\lambda|^\alpha}$ is a characteristic function of some distribution. $\qquad\square$

Consider a sequence of independent identically distributed random variables $\xi_1, \xi_2, \ldots$ with zero expectation. While both Theorem 10.15 and the Central Limit Theorem state that the normalized sums of the random variables converge weakly, there is a crucial difference in the mechanisms of convergence. Let us show that, in the case of the Central Limit Theorem, the contribution of each individual term to the sum is negligible. This is not so in the situation described by Theorem 10.15. For random variables with distributions of the form (10.17), the largest term of the sum is commensurate with the entire sum.

First consider the situation described by the Central Limit Theorem. Let $F(x)$ be the distribution function of each of the random variables $\xi_1, \xi_2, \ldots$, which have finite variance. Then, for each $a > 0$, we have

$$nP(|\xi_1| \geq a\sqrt{n}) = n \int_{|x| \geq a\sqrt{n}} dF(x) \leq \frac{1}{a^2} \int_{|x| \geq a\sqrt{n}} x^2 dF(x).$$

The last integral converges to zero as $n \to \infty$ since $\int_{\mathbb{R}} x^2 dF(x)$ is finite.

The Central Limit Theorem states that the sum $\xi_1 + \ldots + \xi_n$ is of order $\sqrt{n}$ for large n. We can estimate the probability that the largest term in the sum is greater than $a\sqrt{n}$ for $a > 0$. Due to the independence of the random variables,

$$P(\max_{1 \leq i \leq n} |\xi_i| \geq a\sqrt{n}) \leq nP(|\xi_1| \geq a\sqrt{n}) \to 0 \quad \text{as} \quad n \to \infty.$$

Let us now assume that the distribution of each random variable is given by a symmetric density $p(x)$ for which (10.17) holds. Theorem 10.15 states that the sum $\xi_1 + \ldots + \xi_n$ is of order $n^{\frac{1}{\alpha}}$ for large n. For $a > 0$ we can estimate from below the probability that the largest term in the sum is greater than $an^{\frac{1}{\alpha}}$. Namely,

$$P(\max_{1 \leq i \leq n} |\xi_i| \geq an^{\frac{1}{\alpha}}) = 1 - P(\max_{1 \leq i \leq n} |\xi_i| < an^{\frac{1}{\alpha}}) = 1 - (P(|\xi_1| < an^{\frac{1}{\alpha}}))^n$$

$$= 1 - (1 - P(|\xi_1| \geq an^{\frac{1}{\alpha}}))^n.$$

By (10.17),

$$P(|\xi_1| \geq an^{\frac{1}{\alpha}}) \sim \int_{|x| \geq an^{\frac{1}{\alpha}}} \frac{c}{|x|^{\alpha+1}} dx = \frac{2c}{\alpha a^\alpha n}.$$

Therefore,

$$\lim_{n \to \infty} P(\max_{1 \leq i \leq n} |\xi_i| \geq an^{\frac{1}{\alpha}}) = \lim_{n \to \infty} (1 - (1 - \frac{2c}{\alpha a^\alpha n})^n) = 1 - \exp(-\frac{2c}{\alpha a^\alpha}) > 0.$$

This justifies our remarks on the mechanism of convergence of sums of random variables with densities satisfying (10.17).

Consider an arbitrary sequence of independent identically distributed random variables $\xi_1, \xi_2, \ldots$. Assume that for some A_n, B_n the distributions of the normalized sums

$$\frac{\xi_1 + \ldots + \xi_n - A_n}{B_n} \qquad (10.19)$$

converge weakly to a non-trivial limit.

Definition 10.20. *A distribution which can appear as a limit of normalized sums (10.19) for some sequence of independent identically distributed random variables $\xi_1, \xi_2, \ldots$ and some sequences A_n, B_n is called a stable distribution.*

There is a general formula for characteristic functions of stable distributions. It is possible to show that the sequences A_n, B_n cannot be arbitrary. They are always products of power functions and the so-called "slowly changing" functions, for which a typical example is any power of the logarithm.

Finally, we consider a Limit Theorem for a particular problem in one-dimensional random walks. It provides another example of a proof of a Limit Theorem with the help of characteristic functions. Let $\xi_1, \xi_2, \ldots$ be the consecutive moments of return of a simple symmetric one-dimensional random walk to the origin. In this case ξ_1, $\xi_2 - \xi_1$, $\xi_3 - \xi_2, \ldots$ are independent identically distributed random variables. We shall prove that the distributions of ξ_n/n^2 converge weakly to a non-trivial distribution.

Let us examine the characteristic function of the random variable ξ_1. Recall that in Sect. 6.2 we showed that the generating function of ξ_1 is equal to

$$F(z) = \mathrm{E} z^{\xi_1} = 1 - \sqrt{1 - z^2}.$$

This formula holds for $|z| < 1$, and can be extended by continuity to the unit circle $|z| = 1$. Here, the branch of the square root with the non-negative real part is selected. Now

$$\varphi(\lambda) = \mathrm{E} e^{i\lambda\xi_1} = \mathrm{E}(e^{i\lambda})^{\xi_1} = 1 - \sqrt{1 - e^{2i\lambda}}.$$

Since ξ_n is a sum of independent identically distributed random variables, the characteristic function of $\frac{\xi_n}{n^2}$ is equal to

$$(\varphi(\frac{\lambda}{n^2}))^n = (1 - \sqrt{1 - e^{\frac{2i\lambda}{n^2}}})^n = (1 - \frac{\sqrt{-2i\lambda}}{n} + o(\frac{1}{n}))^n \sim e^{\sqrt{-2i\lambda}}.$$

By Remark 9.11, this implies that the distribution of $\frac{\xi_n}{n^2}$ converges weakly to the distribution with the characteristic function $e^{\sqrt{-2i\lambda}}$.

10.6 Problems

1. Prove the following Central Limit Theorem for independent identically distributed random vectors. Let $\xi_1 = (\xi_1^{(1)}, \ldots, \xi_1^{(k)}), \xi_2 = (\xi_2^{(1)}, \ldots, \xi_2^{(k)}), \ldots$ be a sequence of independent identically distributed random vectors in $\mathbb{R}^k$. Let m and D be the expectation and the covariance matrix, respectively, of the random vector ξ_1. That is,

$$m = (m^1, \ldots, m^k), \quad m^i = \mathrm{E}\xi_1^{(i)}, \quad \text{and} \quad D = (d^{ij})_{1 \leq i, j \leq k}, \quad d^{ij} = \mathrm{Cov}(\xi_1^{(i)}, \xi_1^{(j)}).$$

Assume that $|d^{ij}| < \infty$ for all i, j. Prove that the distributions of

$$(\xi_1 + .. + \xi_n - nm)/\sqrt{n}$$

converge weakly to $N(0, D)$ distribution as $n \to \infty$.

2. Two people are playing a series of games against each other. In each game each player either wins a certain amount of money or loses the same amount of money, both with probability $1/2$. With each new game the stake increases by a dollar. Let S_n denote the change of the fortune of the first player by the end of the first n games.

   (a) Find a function $f(n)$ such that the random variables $S_n/f(n)$ converge in distribution to some limit which is not a distribution concentrated at zero and identify the limiting distribution.

   (b) If R_n denotes the change of the fortune of the second player by the end of the first n games, what is the limit, in distribution, of the random vectors $(S_n/f(n), R_n/f(n))$?

3. Let $\xi_1, \xi_2, \ldots$ be a sequence of independent identically distributed random variables with $\mathrm{E}\xi_1 = 0$ and $0 < \sigma^2 = \mathrm{Var}(\xi_1) < \infty$. Prove that the distributions of $(\sum_{i=1}^n \xi_i)/(\sum_{i=1}^n \xi_i^2)^{1/2}$ converge weakly to $N(0, 1)$ distribution as $n \to \infty$.

4. Let $\xi_1, \xi_2, \ldots$ be independent identically distributed random variables such that $\mathrm{P}(\xi_n = -1) = \mathrm{P}(\xi_n = 1) = 1/2$. Let $\zeta_n = \sum_{i=1}^n \xi_i$. Prove that

$$\lim_{n \to \infty} \mathrm{P}(\zeta_n = k^2 \text{ for some } k \in \mathbb{N}) = 0.$$

5. Let $\omega = (\omega_0, \omega_1, \ldots)$ be a trajectory of a simple symmetric random walk on $\mathbb{Z}^3$. Prove that for any $\varepsilon > 0$

$$\mathrm{P}(\lim_{n \to \infty} (n^{\varepsilon - \frac{1}{6}} \|\omega_n\|) = \infty) = 1.$$

6. Let $\xi_1, \xi_2, \ldots$ be independent identically distributed random variables such that $\mathrm{P}(\xi_n = -1) = \mathrm{P}(\xi_n = 1) = 1/2$. Let $\zeta_n = \sum_{i=1}^n \xi_i$. Find the limit

$$\lim_{n \to \infty} \frac{\ln \mathrm{P}((\zeta_n/n) > \varepsilon)}{n}.$$

7. Let $\xi_1, \xi_2, \ldots$ be independent identically distributed random variables with the Cauchy distribution. Prove that

$$\liminf_{n\to\infty} P(\max(\xi_1, \ldots, \xi_n) > xn) \geq \exp(-\pi x).$$

for any $x \geq 0$.

8. Let $\xi_1, \xi_2, \ldots$ be independent identically distributed random variables with the uniform distribution on the interval $[-1/2, 1/2]$. What is the limit (in distribution) of the sequence

$$\zeta_n = (\sum_{i=1}^{n} 1/\xi_i)/n.$$

9. Let $\xi_1, \xi_2, \ldots$ be independent random variables with uniform distribution on $[0, 1]$. Given $\alpha \in \mathbb{R}$, find a_n and b_n such that the sequence

$$(\sum_{i=1}^{n} i^\alpha \xi_i - a_n)/b_n$$

converges in distribution to a limit which is different from zero.

10. Let $\xi_1, \xi_2, \ldots$ be independent random variables with uniform distribution on $[0,1]$. Show that for any continuous function $f(x, y, z)$ on $[0, 1]^3$

$$\frac{1}{\sqrt{n}}(\sum_{j=1}^{n} f(\xi_j, \xi_{j+1}, \xi_{j+2}) - n\int_0^1 \int_0^1 \int_0^1 f(x, y, z)dxdydz)$$

converges in distribution.

11

Several Interesting Problems

In this chapter we describe three applications of probability theory. [1]The exposition is more difficult and more concise than in previous chapters.

11.1 Wigner Semicircle Law for Symmetric Random Matrices

There are many mathematical problems which are related to eigenvalues of large matrices. When the matrix elements are random, the eigenvalues are also random. Let A be a real $n \times n$ symmetric matrix with eigenvalues $\lambda_1^{(n)}, \ldots, \lambda_n^{(n)}$. Since the matrix is symmetric, all the eigenvalues are real. We can consider the discrete measure μ^n (we shall call it the eigenvalue measure) which assigns the weight $\frac{1}{n}$ to each of the eigenvalues, that is for any Borel set $B \in \mathcal{B}(\mathbb{R})$, the measure μ^n is defined by

$$\mu^n(B) = \frac{1}{n} \sharp \{1 \le i \le n : \lambda_i^{(n)} \in B\}.$$

In this section we shall study the asymptotic behavior of the measures μ^n or, more precisely, their moments, as the size n of the matrix goes to infinity. The following informal discussion serves to justify our interest in this problem.

If, for a sequence of measures η^n, all their moments converge to the corresponding moments of some measure η, then (under certain additional conditions on the growth rate of the moments of the measures η^n) the measures themselves converge weakly to η. In our case, the k-th moment of the eigenvalue measure $M_k^n = \int_{-\infty}^{\infty} \lambda^k d\mu^n(\lambda)$ is a random variable. We shall demonstrate that for a certain class of random matrices, the moments M_k^n converge in probability to the k-th moment of the measure whose density is given by the semicircle law

---

[1] This chapter can be omitted during the first reading.

L. Koralov and Y.G. Sinai, *Theory of Probability and Random Processes*, 155
Universitext, DOI 10.1007/978-3-540-68829-7_11,
© Springer-Verlag Berlin Heidelberg 2012

$$p(\lambda) = \begin{cases} \frac{2}{\pi}\sqrt{1-\lambda^2} & \text{if } -1 \le \lambda \le 1, \\ 0 & \text{otherwise.} \end{cases}$$

Thus the eigenvalue measures converge, in a certain sense, to a non-random measure on the real line with the density given by the semicircle law. This is a part of the statement proved by E. Wigner in 1951.

Let us introduce the appropriate notations. Let ξ_{ij}^n, $1 \le i,j \le n$, $n = 1,2,\ldots$, be a collection of identically distributed random variables (with distributions independent of i, j, and n) such that:

1. For each n the random variables ξ_{ij}^n, $1 \le i \le j \le n$, are independent.
2. The matrix $(A^n)_{ij} = \frac{1}{2\sqrt{n}}(\xi_{ij}^n)$ is symmetric, that is $\xi_{ij}^n = \xi_{ji}^n$.
3. The random variables ξ_{ij}^n have symmetric distributions, that is for any Borel set $B \in \mathcal{B}(\mathbb{R})$ we have $P(\xi_{ij}^n \in B) = P(\xi_{ij}^n \in -B)$.
4. All the moments of ξ_{ij}^n are finite, that is $E(\xi_{ij}^n)^k < \infty$ for all $k \ge 1$, while the variance is equals one, $\text{Var}(\xi_{ij}^n) = 1$.

Let $m_k = \frac{2}{\pi}\int_{-1}^1 \lambda^k \sqrt{1-\lambda^2}d\lambda$ be the moments of the measure with the density given by the semicircle law. In this section we prove the following.

Theorem 11.1. *Let ξ_{ij}^n, $1 \le i,j \le n$, $n = 1,2,\ldots$, be a collection of identically distributed random variables for which the conditions 1–4 above are satisfied. Let M_k^n be the k-th moment of the eigenvalue measure μ^n for the matrix A^n, that is*

$$M_k^n = \frac{1}{n}\sum_{i=1}^n (\lambda_i^{(n)})^k.$$

Then

$$\lim_{n\to\infty} EM_k^n = m_k$$

and

$$\lim_{n\to\infty} \text{Var}(M_k^n) = 0.$$

Proof. We shall prove the first statement by reducing it to a combinatorial problem. The second one can be proved similarly and we shall not discuss it in detail.

Our first observation is

$$M_k^n = \frac{1}{n}\sum_{i=1}^n (\lambda_i^{(n)})^k = \frac{1}{n}\text{Tr}((A^n)^k), \tag{11.1}$$

where $\lambda_i^{(n)}$ are the eigenvalues of the matrix A^n, and $\text{Tr}((A^n)^k)$ is the trace of its k-th power. Thus we need to analyze the quantity

$$EM_k^n = \frac{1}{n}E\text{Tr}((A^n)^k) = \frac{1}{n}(\frac{1}{2\sqrt{n}})^k E(\sum_{i_1,\ldots,i_k=1}^n \xi_{i_1 i_2}^n \xi_{i_2 i_3}^n \cdots \xi_{i_k i_1}^n).$$

Recall that $\xi_{ij}^n = \xi_{ji}^n$, and that the random variables $\xi_{ij}^n, 1 \leq i \leq j \leq n$, are independent for each n. Therefore each of the terms $E\xi_{i_1 i_2}^n \xi_{i_2 i_3}^n \cdots \xi_{i_k i_1}^n$ is equal to the product of factors of the form $E(\xi_{ij}^n)^{p(i,j)}$, where $1 \leq i \leq j \leq n$, $1 \leq p(i,j) \leq k$, and $\sum p(i,j) = k$. If k is odd, then at least one of the factors is the expectation of an odd power of ξ_{ij}^n, which is equal to zero, since the distributions of ξ_{ij}^n are symmetric. Thus $EM_k^n = 0$ if k is odd. The fact that $m_k = 0$ if k is odd is obvious.

Let $k = 2r$ be even. Then

$$EM_{2r}^n = \frac{1}{2^{2r} n^{r+1}} E\left(\sum_{i_1,\ldots,i_{2r}=1}^{n} \xi_{i_1 i_2}^n \xi_{i_2 i_3}^n \cdots \xi_{i_{2r} i_1}^n \right).$$

Observe that we can identify an expression of the form $\xi_{i_1 i_2}^n \xi_{i_2 i_3}^n \cdots \xi_{i_{2r} i_1}^n$ with a closed path of length $2r$ on the set of n points $\{1, 2, \ldots, n\}$, which starts at i_1, next goes to i_2, etc., and finishes at i_1. The transitions from a point to itself are allowed, for example if $i_1 = i_2$.

We shall say that a path $(i_1, \ldots, i_{2r}, i_{2r+1})$ goes through a pair (i, j), if for some s either $i_s = i$ and $i_{s+1} = j$, or $i_s = j$ and $i_{s+1} = i$. Here $1 \leq i \leq j \leq n$, and $1 \leq s \leq 2r$. Note that i and j are not required to be distinct.

As in the case of odd k, the expectation $E\xi_{i_1 i_2}^n \xi_{i_2 i_3}^n \cdots \xi_{i_{2r} i_1}^n$ is equal to zero, unless the path passes through each pair (i, j), $1 \leq i \leq j \leq n$, an even number of times. Otherwise the expectation $E\xi_{i_1 i_2}^n \xi_{i_2 i_3}^n \cdots \xi_{i_{2r} i_1}^n$ would contain a factor $E(\xi_{ij}^n)^{p(i,j)}$ with an odd $p(i, j)$.

There are four different types of closed paths of length $2r$ which pass through each pair (i, j), $1 \leq i \leq j \leq n$, an even number of times:

1. A path contains an elementary loop ($i_s = i_{s+1}$ for some s).
2. A path does not contain elementary loops, but passes through some pair (i, j) at least four times.
3. A path does not contain elementary loops, nor passes through any pair more than two times, but forms at least one loop. That is the sequence $(i_1, \ldots, i_{2r})$ contains a subsequence $(i_{s_1}, i_{s_2}, \ldots, i_{s_q}, i_{s_1})$ made out of consecutive elements of the original sequence, where $q > 2$, and all the elements of the subsequence other than the first and the last one are different.
4. A path spans a tree with r edges, and every edge is passed twice.

Note that any closed path of length $2r$ of types 1, 2 or 3 passes through at most r points. This is easily seen by induction on r, once we recall that the path passes through each pair an even number of times.

The total number of ways to select $q \leq r$ points out of a set of n points is bounded by n^q. With q points fixed, there are at most $c_{q,r}$ ways to select a path of length $2r$ on the set of these q points, where $c_{q,r}$ is some constant. Therefore, the number of paths of types 1–3 is bounded by $\sum_{q=1}^{r} c_{q,r} n^q \leq C_r n^r$, where C_r is some constant.

Since we assumed that all the moments of the random variables ξ_{ij}^n are finite, and r is fixed, the expressions $E\xi_{i_1 i_2}^n \xi_{i_2 i_3}^n \dots \xi_{i_{2r} i_1}^n$ are bounded uniformly in n,

$$|E\xi_{i_1 i_2}^n \xi_{i_2 i_3}^n \dots \xi_{i_{2r} i_1}^n| < k_r.$$

Therefore, the contribution to EM_{2r}^n from all the paths of types 1–3 is bounded from above by $\frac{1}{2^{2r}n^{r+1}}C_r n^r k_r$, which tends to zero as $n \to \infty$.

For a path which has the property that for each pair it either passes through the pair twice or does not pass through it at all, the expression $E\xi_{i_1 i_2}^n \xi_{i_2 i_3}^n \dots \xi_{i_{2r} i_1}^n$ is equal to a product of r expressions of the form $E(\xi_{ij}^n)^2$. Since the expectation of each of the variables ξ_{ij}^n is zero and the variance is equal to one, we obtain $E\xi_{i_1 i_2}^n \xi_{i_2 i_3}^n \dots \xi_{i_{2r} i_1}^n = 1$. It remains to estimate the number of paths of length $2r$ which span a tree whose every edge is passed twice. We shall call them eligible paths.

With each eligible path we can associate a trajectory of one-dimensional simple symmetric random walk $\omega = (\omega_0, \dots, \omega_{2r})$, where $\omega_0 = \omega_{2r} = 0$, and ω_i is the number of edges that the path went through only once during the first i steps. The trajectory ω has the property that $\omega_i \geq 0$ for all $0 \leq i \leq 2r$. Note that if the trajectory $(\omega_0, \dots, \omega_{2r})$ is fixed, there are exactly $n(n-1)\dots(n-r)$ corresponding eligible paths. Indeed, the starting point for the path can be selected in n different ways. The first step can be taken to any of the $n-1$ remaining points, the next time when the path does not need to retract along its route (that is $\omega_i > \omega_{i-1}$) there will be $n-2$ points where the path can jump, etc.

We now need to calculate the number of trajectories $(\omega_0, \dots, \omega_{2r})$ for which $\omega_0 = \omega_{2r} = 0$ and $\omega_i \geq 0$ for all $0 \leq i \leq 2r$. The proof of Lemma 6.7 contains an argument based on the Reflection Principle which shows that the number of such trajectories is $\frac{(2r)!}{r!(r+1)!}$. Thus, there are $\frac{n!(2r)!}{(n-r-1)!r!(r+1)!}$ eligible paths of length $2r$. We conclude that

$$\lim_{n \to \infty} EM_{2r}^n = \lim_{n \to \infty} \frac{1}{2^{2r}n^{r+1}} \frac{n!(2r)!}{(n-r-1)!r!(r+1)!} = \frac{(2r)!}{2^{2r}r!(r+1)!}. \quad (11.2)$$

The integral defining m_{2r} can be calculated explicitly, and the value of m_{2r} is seen to be equal to the right-hand side of (11.2). This completes the proof of the theorem. □

Remark 11.2. The Chebyshev Inequality implies that for any $\varepsilon > 0$

$$P(|M_k^n - EM_k^n| \geq \varepsilon) \leq (\text{Var}(M_k^n)/\varepsilon^2) \to 0 \quad \text{as } n \to \infty.$$

Since $EM_k^n \to m_k$ as $n \to \infty$, this implies that

$$\lim_{n \to \infty} M_k^n = m_k \quad \text{in probability}$$

for any $k \geq 1$.

11.2 Products of Random Matrices

In this section we consider the limiting behavior of products of random matrices $g \in \mathrm{SL}(2, \mathbb{R})$, where $\mathrm{SL}(2, \mathbb{R})$ is the group of two-dimensional matrices with determinant 1. Each matrix $g = \begin{pmatrix} a & b \\ c & d \end{pmatrix}$ satisfies the relation $ad - bc = 1$ and therefore $\mathrm{SL}(2, \mathbb{R})$ can be considered as a three-dimensional submanifold in $\mathbb{R}^4$. Assume that a probability distribution on $\mathrm{SL}(2, \mathbb{R})$ is given. We define

$$g^{(n)} = g(n)g(n-1)\ldots g(2)g(1),$$

where $g(k)$ are independent elements of $\mathrm{SL}(2, \mathbb{R})$ with distribution P. Denote the distribution of $g^{(n)}$ by $\mathrm{P}^{(n)}$. We shall discuss statements of the type of the Law of Large Numbers and the Central Limit Theorem for the distribution $\mathrm{P}^{(n)}$. We shall see that for the products of random matrices, the corresponding statements differ from the statements for sums of independent identically distributed random variables.

A detailed treatment would require some notions from hyperbolic geometry, which would be too specific for this book. We shall use a more elementary approach and obtain the main conclusions from the "first order approximation". We assume that P has a density (in natural coordinates) which is a continuous function with compact support.

The subgroup O of orthogonal matrices

$$o = o(\varphi) = \begin{pmatrix} \cos\varphi & \sin\varphi \\ -\sin\varphi & \cos\varphi \end{pmatrix}, \quad 0 \leq \varphi < 2\pi,$$

will play a special role. It is clear that

$$o(\varphi_1)o(\varphi_2) = o((\varphi_1 + \varphi_2) \pmod{2\pi}).$$

Lemma 11.3. *Each matrix $g \in \mathrm{SL}(2, \mathbb{R})$ can be represented as $g = o(\varphi)d(\lambda)o(\psi)$, where $o(\varphi), o(\psi) \in O$ and $d(\lambda) = \begin{pmatrix} \lambda & 0 \\ 0 & \lambda^{-1} \end{pmatrix}$ is a diagonal matrix for which $\lambda \geq 1$. Such a representation is unique if $\lambda \neq 1$.*

Proof. If φ and ψ are the needed values of the parameters, then $o(-\varphi)go(-\psi)$ is a diagonal matrix. Since

$$o(-\varphi)go(-\psi) = \begin{pmatrix} \cos\varphi & -\sin\varphi \\ \sin\varphi & \cos\varphi \end{pmatrix} \begin{pmatrix} a & b \\ c & d \end{pmatrix} \begin{pmatrix} \cos\psi & -\sin\psi \\ \sin\psi & \cos\psi \end{pmatrix}, \quad (11.3)$$

we have the equations

$$a\tan\varphi + b\tan\varphi\tan\psi + c + d\tan\psi = 0, \quad (11.4)$$

$$a\tan\psi - b - c\tan\varphi\tan\psi + d\tan\varphi = 0. \quad (11.5)$$

Multiplying (11.4) by c, (11.5) by b, and summing up the results, we obtain the following expression for $\tan\varphi$:

$$\tan\varphi = -\frac{ab+cd}{ac+bd}\tan\psi + \frac{b^2-c^2}{ac+bd}.$$

The substitution of this expression into (11.4) gives us a quadratic equation for $\tan\psi$. It is easy to check that it always has two solutions, one of which corresponds to $\lambda \geq 1$.   □

We can now write

$$g^{(n)} = o(\varphi^{(n)})d(\lambda^{(n)})o(\psi^{(n)}).$$

We shall derive, in some approximation, the recurrent relations for $\varphi^{(n)}, \psi^{(n)}$, and $\lambda^{(n)}$, which will imply that $\psi^{(n)}$ converges with probability one to a random limit, $\varphi^{(n)}$ is a Markov chain with compact state space, and $\frac{\ln\lambda^{(n)}}{n}$ converges with probability one to a non-random positive limit a such that the distribution of $\frac{\ln\lambda^{(n)}-na}{\sqrt{n}}$ converges to a Gaussian distribution. We have

$$g(n+1) = o(\varphi(n+1))d(\lambda(n+1))o(\psi(n+1))$$

and

$$\begin{aligned}
g^{(n+1)} &= g(n+1)g^{(n)}\\
&= o(\varphi(n+1))d(\lambda(n+1))o(\psi(n+1))o(\varphi^{(n)})d(\lambda^{(n)})o(\psi^{(n)})\\
&= o(\varphi(n+1))d(\lambda(n+1))o(\overline{\varphi}^{(n)})d(\lambda^{(n)})o(\psi^{(n)}),
\end{aligned}$$

where $o(\overline{\varphi}^{(n)}) = o(\psi(n+1))o(\varphi^{(n)})$, $\overline{\varphi}^{(n)} = \psi(n+1) + \varphi^{(n)} (\mathrm{mod}\ 2\pi)$. Note that $\{\varphi(n+1), \psi(n+1), \lambda(n+1)\}$ for different n are independent identically distributed random variables whose joint probability distribution has a density with compact support. By Lemma 11.3, we can write

$$d(\lambda(n+1))o(\overline{\varphi}^{(n)})d(\lambda^{(n)}) = o(\overline{\overline{\varphi}}^{(n)})d(\lambda^{(n+1)})o(\overline{\overline{\psi}}^{(n)}).$$

This shows that

$$\varphi^{(n+1)} = \varphi(n+1) + \overline{\overline{\varphi}}^{(n)} \ (\mathrm{mod}\ 2\pi), \quad \psi^{(n+1)} = \psi(n+1) + \overline{\overline{\psi}}^{(n)} \ (\mathrm{mod}\ 2\pi).$$

Our next step is to derive more explicit expressions for $o(\overline{\overline{\varphi}}^{(n)})$, $d(\lambda^{(n+1)})$, and $o(\overline{\overline{\psi}}^{(n)})$. We have

$$\begin{aligned}
&d(\lambda(n))o(\overline{\varphi}^{(n)})d(\lambda^{(n)})\\
&= \begin{pmatrix} \lambda(n) & 0 \\ 0 & \lambda^{-1}(n) \end{pmatrix} \begin{pmatrix} \cos\overline{\varphi}^{(n)} & \sin\overline{\varphi}^{(n)} \\ -\sin\overline{\varphi}^{(n)} & \cos\overline{\varphi}^{(n)} \end{pmatrix} \begin{pmatrix} \lambda^{(n)} & 0 \\ 0 & (\lambda^{(n)})^{-1} \end{pmatrix} \qquad (11.6)\\
&= \begin{pmatrix} \lambda(n)\lambda^{(n)}\cos\overline{\varphi}^{(n)} & \lambda(n)(\lambda^{(n)})^{-1}\sin\overline{\varphi}^{(n)} \\ -\lambda^{-1}(n)\lambda^{(n)}\sin\overline{\varphi}^{(n)} & \lambda^{-1}(n)(\lambda^{(n)})^{-1}\cos\overline{\varphi}^{(n)} \end{pmatrix}.
\end{aligned}$$

As was previously mentioned, all $\lambda(n)$ are bounded from above. Therefore, all $\lambda^{-1}(n)$ are bounded from below. Assume now that $\lambda^{(n)} \gg 1$. Then from (11.4), and with a, b, c, d taken from (11.3)

$$\tan \overline{\overline{\varphi}}^{(n)} = -\frac{c}{a} + O(\frac{1}{\lambda^{(n)}}) = \lambda^{-2}(n) \tan \overline{\varphi}^{(n)} + O(\frac{1}{\lambda^{(n)}}),$$

where $\tan \varphi = O(1)$, $\tan \psi = O(1)$. Therefore, in the main order of magnitude

$$\varphi^{(n+1)} = \varphi(n+1) + \overline{\overline{\varphi}}^{(n)} = \varphi(n+1) + f(g^{(n+1)}, \varphi^{(n)}),$$

which shows that, with the same precision, $\{\varphi^{(n)}\}$ is a Markov chain with compact state space. Since the transition probabilities have densities, this Markov chain has a stationary distribution. From (11.5)

$$\tan \overline{\overline{\psi}}^{(n)} (1 - \frac{c}{a} \tan \overline{\overline{\varphi}}^{(n)}) = \frac{b}{a} - \frac{d}{a} \tan \overline{\overline{\varphi}}^{(n)}$$

or

$$\tan \overline{\overline{\psi}}^{(n)} (1 + \frac{c^2}{a^2} + O((\lambda^{(n)})^{-1})) = \frac{b}{a} + \frac{dc}{a^2} + O((\lambda^{(n)})^{-2}),$$

which shows that $\tan \overline{\overline{\psi}}^{(n)} = O((\lambda^{(n)})^{-1})$, that is $\overline{\overline{\psi}}^{(n)} = O((\lambda^{(n)})^{-1})$. Therefore $\psi^{(n+1)} = \psi^{(n)} + \overline{\overline{\psi}}^{(n)}$, and the limit $\lim_{n \to \infty} \psi^{(n)}$ exists with probability one, since we will show that $\lambda^{(n)}$ grows exponentially with probability one.

From (11.3) and (11.6) it follows easily that $\lambda^{(n+1)} = \lambda^{(n)} \lambda(n)(1 + O((\lambda^{(n)})^{-1}))$. Since $\lambda(n) > 1$ and $\lambda(n)$ are independent random variables, $\lambda^{(n)}$ grow exponentially with n. As previously mentioned, we do not provide accurate estimates of all the remainders.

11.3 Statistics of Convex Polygons

In this section we consider a combinatorial problem with an unusual space Ω and a quite unexpected "Law of Large Numbers". The problem was first studied in the works of A. Vershik and I. Baranyi.

For each $n \geq 1$, introduce the space $\Omega_n(1, 1)$ of convex polygons ω which go out from $(0, 0)$, end up at $(1, 1)$, are contained inside the unit square $\{(x_1, x_2) : 0 \leq x_1 \leq 1, 0 \leq x_2 \leq 1\}$ and have the vertices of the form $(\frac{n_1}{n}, \frac{n_2}{n})$. Here n_1 and n_2 are integers such that $0 \leq n_i \leq n$, $i = 1, 2$. The vertices belong to the lattice $\frac{1}{n} \mathbb{Z}^2$. The space $\Omega_n(1, 1)$ is finite and we can consider the uniform probability distribution P_n on $\Omega_n(1, 1)$, for which

$$P_n(\omega) = \frac{1}{|\Omega_n(1, 1)|},$$

$\omega \in \Omega_n(1,1)$. Let $\mathcal{L}$ be the curve on the (x_1, x_2)-plane given by the equation

$$\mathcal{L} = \{(x_1, x_2) : (x_1 + x_2)^2 = 4x_2\}.$$

Clearly, $\mathcal{L}$ is invariant under the map

$$(x_1, x_2) \mapsto (1 - x_2, 1 - x_1).$$

For $\varepsilon > 0$ let $\mathcal{U}_\varepsilon$ be the ε-neighborhood around $\mathcal{L}$. The main result of this section is the following theorem.

Theorem 11.4. *For each $\varepsilon > 0$,*

$$\mathbf{P}_n \{\omega \in \mathcal{U}_\varepsilon\} \to 1$$

as $n \to \infty$.

In other words, the majority (in the sense of probability distribution $\mathbf{P}_n$) of convex polygons $\omega \in \Omega_n(1,1)$ is concentrated in a small neighborhood $\mathcal{U}_\varepsilon$. We shall provide only a sketch of the proof.

Proof. We enlarge the space $\Omega_n(1,1)$ by introducing a countable space Ω_n of all convex polygons ω which go out from $(0,0)$ and belong to the half-plane $x_2 \geq 0$. Now it is not necessary for polygons to end up at $(1,1)$, but the number of vertices must be finite, and the vertices must be of the form $(\frac{n_1}{n}, \frac{n_2}{n})$.

Let M be the set of pairs of mutually coprime positive integers $m = (m_1, m_2)$. It is convenient to include the pairs $(1,0)$ and $(0,1)$ in M. Set $\tau(m) = \frac{m_2}{m_1}$ so that $\tau(1,0) = 0$ and $\tau(0,1) = \infty$. If $m \neq m'$, then $\tau(m) \neq \tau(m')$.

Lemma 11.5. *The space Ω_n can be represented as the space $C_0(M)$ of non-negative integer-valued functions defined on M which are different from zero only on a finite non-empty subset of M.*

Proof. For any $\nu \in C_0(M)$ take $m^{(j)} = (m_1^{(j)}, m_2^{(j)})$ with $\nu(m^{(j)}) > 0$. Choose the ordering so that $\tau(m^{(j)}) > \tau(m^{(j+1)})$. Thus the polygon ω whose consecutive sides are made of the vectors $\frac{1}{n}\nu(m^{(j)})m^{(j)}$ is convex. The converse can be proved in the same way. □

It is clear that the coordinates of the last point of ω are

$$\overline{x}_1 = \frac{1}{n}\sum_j \nu(m^{(j)})m_1^{(j)} = \frac{1}{n}\sum_{m \in M} \nu(m)m_1,$$

$$\overline{x}_2 = \frac{1}{n}\sum_j \nu(m^{(j)})m_2^{(j)} = \frac{1}{n}\sum_{m \in M} \nu(m)m_2.$$

Denote by $\Omega_n(\overline{x}_1, \overline{x}_2)$ the set of $\omega \in \Omega_n$ with given $(\overline{x}_1, \overline{x}_2)$ and $N_n(\overline{x}_1, \overline{x}_2) = |\Omega_n(\overline{x}_1, \overline{x}_2)|$.

We shall need a probability distribution Q_n on Ω_n for which

$$q_n(\omega) = \prod_{m=(m_1, m_2) \in M} (z_1^{m_1} z_2^{m_2})^{\nu(m)} (1 - z_1^{m_1} z_2^{m_2}),$$

where $\nu(m) \in C_0(M)$ is the function corresponding to the polygon ω. Here $0 < z_i < 1$, $i = 1, 2$, are parameters which can depend on n and will be chosen later. It is clear that, with respect to Q_n, each $\nu(m)$ has the exponential distribution with parameter $z_1^{m_1} z_2^{m_2}$, and the random variables $\nu(m)$ are independent. We can write

$$Q_n(\Omega_n(\overline{x}_1, \overline{x}_2)) = \sum_{\omega \in \Omega_n(\overline{x}_1, \overline{x}_2)} q(\omega)$$

$$= z_1^{n\overline{x}_1} z_2^{n\overline{x}_2} N_n(\overline{x}_1, \overline{x}_2) \prod_{(m_1, m_2) \in M} (1 - z_1^{m_1} z_2^{m_2}). \tag{11.7}$$

Theorem 11.6.

$$\ln N_n(1, 1) = n^{\frac{2}{3}} \left[3 \left(\frac{\zeta(3)}{\zeta(2)} \right)^{\frac{1}{3}} + o(1) \right]$$

as $n \to \infty$. Here $\zeta(r)$ is the Riemann zeta-function, $\zeta(r) = \sum_{k \geq 1} \frac{1}{k^r}$.

Proof. By (11.7), for any $0 < z_1, z_2 < 1$,

$$N_n(1, 1) = z_1^{-n} z_2^{-n} \prod_{m=(m_1, m_2) \in M} (1 - z_1^{m_1} z_2^{m_2})^{-1} Q_n(\Omega_n(1, 1)). \tag{11.8}$$

The main step in the proof is the choice of z_1, z_2, so that

$$E_{z_1, z_2} \left(\frac{1}{n} \sum_{m \in M} \nu(m) m_1 \right) = E_{z_1, z_2} \left(\frac{1}{n} \sum_{m \in M} \nu(m) m_2 \right) = 1. \tag{11.9}$$

The expectations with respect to the exponential distribution can be written explicitly:

$$E_{z_1, z_2} \nu(m) = \frac{z_1^{m_1} z_2^{m_2}}{1 - z_1^{m_1} z_2^{m_2}}.$$

Therefore (11.9) takes the form

$$E_{z_1, z_2} \left(\frac{1}{n} \sum_{m \in M} \nu(m) m_1 \right) = \sum_{m \in M} \frac{m_1 z_1^{m_1} z_2^{m_2}}{n(1 - z_1^{m_1} z_2^{m_2})} = 1, \tag{11.10}$$

$$E_{z_1, z_2} \left(\frac{1}{n} \sum_{m \in M} \nu(m) m_2 \right) = \sum_{m \in M} \frac{m_2 z_1^{m_1} z_2^{m_2}}{n(1 - z_1^{m_1} z_2^{m_2})} = 1. \tag{11.11}$$

The expressions (11.10) and (11.11) can be considered as equations for $z_1 = z_1(n)$, $z_2 = z_2(n)$.

We shall look for the solutions z_1, z_2 in the form $z_1 = 1 - \frac{\alpha_1}{n^{1/3}}$, $z_2 = 1 - \frac{\alpha_2}{n^{1/3}}$, where α_1 and α_2 vary within fixed boundaries, $0 < \text{const} \le \alpha_1, \alpha_2 \le \text{const}$. The fact that such solutions exist needs a separate justification, which we do not provide. For z_1 and z_2 as above, we have

$$z_1^n = \left(1 - \frac{\alpha_1}{n^{1/3}}\right)^n = \exp\{-\alpha_1 n^{2/3}(1 + o_1(1))\},$$

$$z_2^n = \left(1 - \frac{\alpha_2}{n^{1/3}}\right)^n = \exp\{-\alpha_2 n^{2/3}(1 + o_2(1))\},$$

where $o_1(1)$ and $o_2(1)$ tend to zero as $n \to \infty$ uniformly over all considered values of α_1, α_2.

Set $m_i = n^{1/3} t_i$, for $i = 1, 2$. Then t_i belongs to the lattice $\frac{1}{n^{1/3}}\mathbb{Z}$, $i = 1, 2$, and it should not be forgotten that m_1, m_2 are coprime. Thus,

$$\prod_{m \in M}(1 - z_1^{m_1} z_2^{m_2}) = \exp\left\{\sum_{(t_1, t_2)} \ln\left(1 - \left(1 - \frac{\alpha_1}{n^{1/3}}\right)^{t_1 n^{1/3}}\left(1 - \frac{\alpha_2}{n^{1/3}}\right)^{t_2 n^{1/3}}\right)\right\}$$

$$= \exp\left\{\frac{n^{2/3}}{\zeta(2)}\int_0^\infty\int_0^\infty \ln\left(1 - e^{-\alpha_1 t_1 - \alpha_2 t_2}\right)(1 + o(1))\, dt_1 dt_2\right\},$$

where $o(1)$ tends to zero as $n \to \infty$ uniformly for all considered values of α_1, α_2.

The factor $\frac{1}{\zeta(2)}$ enters the above expression due to the fact that the density of coprime pairs $m = (m_1, m_2)$ among all pairs $m = (m_1, m_2)$ equals exactly $\frac{1}{\zeta(2)}$ (see Sect. 1.3).

The integral $\int_0^\infty \int_0^\infty \ln(1 - e^{-\alpha_1 t_1 - \alpha_2 t_2})\, dt_1 dt_2$ can be computed explicitly. The change of variables $\alpha_i t_i = t_i'$, $i = 1, 2$, shows that it is equal to

$$\frac{1}{\alpha_1 \alpha_2}\int_0^\infty\int_0^\infty \ln\left(1 - e^{-t_1 - t_2}\right)\, dt_1 dt_2.$$

The last integral equals $-\zeta(3)$. To see this, one should write down the Taylor expansion for the logarithm and integrate each term separately. Returning to (11.10) and (11.11), we obtain

$$E_{z_1, z_2}\left(\frac{1}{n}\sum_{m \in M}\nu(m)m_1\right) = \sum_{m \in M}\frac{m_1 z_1^{m_1} z_2^{m_2}}{n(1 - z_1^{m_1} z_2^{m_2})}$$

$$= \sum_{t_1, t_2}\frac{t_1\left(1 - \frac{\alpha_1}{n^{1/3}}\right)^{t_1 n^{1/3}}\left(1 - \frac{\alpha_2}{n^{1/3}}\right)^{t_2 n^{1/3}} n^{-2/3}}{1 - \left(1 - \frac{\alpha_1}{n^{1/3}}\right)^{t_1 n^{1/3}}\left(1 - \frac{\alpha_2}{n^{1/3}}\right)^{t_2 n^{1/3}}}$$

$$= \frac{1}{\zeta(2)}\int_0^\infty\int_0^\infty\frac{t_1 e^{-\alpha_1 t_1 - \alpha_2 t_2}}{1 - e^{-\alpha_1 t_1 - \alpha_2 t_2}}\, dt_1 dt_2 \,(1 + o(1))$$

$$= \frac{1}{\zeta(2)} \frac{\partial}{\partial \alpha_1} \left(\int_0^\infty \int_0^\infty \ln(1 - e^{-\alpha_1 t_1 - \alpha_2 t_2}) \ dt_1 dt_2 \right)$$
$$(1 + o(1))$$
$$= \frac{\zeta(3)}{\zeta(2)\alpha_1^2 \alpha_2}(1 + o(1)),$$

or

$$\alpha_1^2 \alpha_2 = \frac{\zeta(3)}{\zeta(2)}(1 + o(1)).$$

In an analogous way, from (11.11) we obtain

$$\alpha_1 \alpha_2^2 = \frac{\zeta(3)}{\zeta(2)}(1 + o(1)).$$

This gives

$$\alpha_1 = \left(\frac{\zeta(3)}{\zeta(2)}\right)^{1/3}(1 + o(1)), \quad \alpha_2 = \left(\frac{\zeta(3)}{\zeta(2)}\right)^{1/3}(1 + o(1)),$$

and

$$z_1 = 1 - \left(\frac{\zeta(3)}{\zeta(2)n}\right)^{1/3}(1 + o(1)), \quad z_2 = 1 - \left(\frac{\zeta(3)}{\zeta(2)}\right)^{1/3}(1 + o(1)). \quad (11.12)$$

The sums

$$\eta_1 = \sum_{m \in M} \nu(m)m_1, \quad \eta_2 = \sum_{m \in M} \nu(m)m_2,$$

with respect to the probability distribution Q_n, are sums of independent random variables which are not identically distributed. It is possible to check that their variances $\mathcal{D}\eta_1, \mathcal{D}\eta_2$ grow as $n^{4/3}$. The same method as in the proof of the Local Central Limit Theorem in Sect. 10.2 can be used to prove that

$$Q_n(\eta_1 = n, \eta_2 = n) \sim \frac{\text{const}}{\sqrt{\mathcal{D}\eta_1 \mathcal{D}\eta_2}}$$

as $n \to \infty$. Returning to (11.8), we see that

$$\ln N_n(1,1) \sim n^{2/3}\left[\alpha_1 + \alpha_2 - \frac{1}{\zeta(2)} \int_0^\infty \int_0^\infty \ln(1 - e^{-\alpha_1 t_1 - \alpha_2 t_2}) \ dt_1 dt_2\right]$$

$$\sim n^{2/3} \, 3 \left(\frac{\zeta(3)}{\zeta(2)}\right)^{1/3}.$$

This completes the proof of Theorem 11.6. $\qquad\qquad\qquad\qquad\qquad$ □

Now we shall prove Theorem 11.4. We assume that the values of z_1 and z_2 are chosen as in the proof of Theorem 11.6, and we shall find the "mathematical expectation" of a convex polygon in the limit $n \to \infty$. The statement of the theorem will follow from the usual arguments in the Law of Large Numbers based on the Chebyshev Inequality.

For the convex polygons we consider, it is convenient to use the parametrization $x_1 = f_1(\tau)$, $x_2 = f_2(\tau)$, where τ is the slope of $m^{(j)}$ which is considered as an independent parameter, $0 \le \tau \le \infty$. Clearly, $\tau = \frac{dx_2}{dx_1}$. Let us fix two numbers τ', τ'', so that $\tau'' - \tau'$ is small. The coordinates of the increment of a random curve on the interval $[\tau', \tau'']$ on the τ axis have the form

$$\overline{\eta}_k = \sum_{m:\tau' \le \frac{m_2}{m_1} \le \tau''} \frac{m_k}{n} \nu(m), \qquad k = 1, 2,$$

and the expectation

$$\mathbf{E}\overline{\eta}_1 = \sum_{\tau' \le \frac{t_2}{t_1} \le \tau''} t_1 \frac{\left(1 - \frac{\alpha_1}{n^{1/3}}\right)^{t_1 n^{1/3}} \left(1 - \frac{\alpha_2}{n^{1/3}}\right)^{t_2 n^{1/3}} n^{-2/3}}{1 - \left(1 - \frac{\alpha_1}{n^{1/3}}\right)^{t_1 n^{1/3}} \left(1 - \frac{\alpha_2}{n^{1/3}}\right)^{t_2 n^{1/3}}}$$

$$\sim \frac{1}{\zeta(2)} \int \int_{\tau' \le \frac{t_2}{t_1} \le \tau''} \frac{t_1 e^{-\alpha_1 t_1 - \alpha_2 t_2}}{1 - e^{-\alpha_1 t_1 - \alpha_2 t_2}} \, dt_1 dt_2$$

$$\sim \frac{\tau'' - \tau'}{\zeta(2)} \int \frac{t_1^2 e^{-\alpha_1 t_1 - \alpha_2 \tau' t_1}}{1 - e^{-\alpha_1 t_1 - \alpha_2 \tau' t_1}} \, dt_1.$$

One can compute $\mathbf{E}\overline{\eta}_2$ in an analogous way.

The last integral can be computed explicitly as before, and it equals $\frac{C_1}{(\alpha_1 + \alpha_2 \tau')^3}$, where C_1 is an absolute constant whose exact value plays no role. When $n \to \infty$, we have $\alpha_1 = \alpha_2 = \alpha = \frac{\zeta(3)}{\zeta(2)}$. As $\tau'' - \tau' \to 0$, we get the differential equation

$$\frac{dx_1}{d\left(\frac{dx_2}{dx_1}\right)} = \frac{C_1}{\alpha^3 \left(1 + \frac{dx_2}{dx_1}\right)^3},$$

or equivalently

$$\frac{d}{dx_1}\left(\frac{dx_2}{dx_1}\right) = \alpha^3 C_1^{-1} \left(1 + \frac{dx_2}{dx_1}\right)^3,$$

or

$$\frac{1}{\left(1 + \frac{dx_2}{dx_1}\right)^2} = \frac{1}{\left(\frac{d}{dx_1}(x_1 + x_2)\right)^2} = C_2^{-1}(x_1 + C_3),$$

for some constants C_2, C_3. Thus

$$\frac{d}{dx_1}(x_1 + x_2) = \sqrt{\frac{C_2}{x_1 + C_3}},$$

or

$$x_1 + x_2 = 2\sqrt{C_2}\sqrt{x_1 + C_3},$$

so that

$$(x_1 + x_2)^2 = 4C_2(x_1 + C_3).$$

The value of C_3 must be zero, since our curve goes through $(0,0)$, while the value of C_2 must be one, since our curve goes through $(1,1)$. Therefore, $(x_1 + x_2)^2 = 4x_1$. □

Part II

Random Processes

Basic Concepts

12.1 Definitions of a Random Process and a Random Field

Consider a family of random variables X_t defined on a common probability space $(\Omega, \mathcal{F}, P)$ and indexed by a parameter $t \in T$. If the parameter set T is a subset of the real line (most commonly $\mathbb{Z}$, $\mathbb{Z}^+$, $\mathbb{R}$, or $\mathbb{R}^+$), we refer to the parameter t as time, and to X_t as a random process. If T is a subset of a multi-dimensional space, then X_t called a random field.

All the random variables X_t are assumed to take values in a common measurable space, which will be referred to as the state space of the random process or field. We shall always assume that the state space is a metric space with the σ-algebra of Borel sets. In particular, we shall encounter real and complex-valued processes, processes with values in $\mathbb{R}^d$, and others with values in a finite or countable set.

Let us discuss the relationship between random processes with values in a metric space S and probability measures on the space of S-valued functions defined on the parameter set T. For simplicity of notation, we shall assume that $S = \mathbb{R}$. Consider the set $\widetilde{\Omega}$ of all functions $\widetilde{\omega} : T \to \mathbb{R}$. Given a finite collection of points $t_1, \dots, t_k \in T$ and a Borel set $A \in \mathcal{B}(\mathbb{R}^k)$, we define a finite-dimensional cylinder (or simply a cylindrical set or a cylinder) as

$$\{\widetilde{\omega} : (\widetilde{\omega}(t_1), \dots, \widetilde{\omega}(t_k)) \in A\}.$$

The collection of all cylindrical sets, for which $t_1, \dots, t_k$ are fixed and $A \in \mathcal{B}(\mathbb{R}^k)$ is allowed to vary, is a σ-algebra, which will be denoted by $\mathcal{B}_{t_1,\dots,t_k}$. Let $\mathcal{B}$ be the smallest σ-algebra containing all $\mathcal{B}_{t_1,\dots,t_k}$ for all possible choices of $k, t_1, \dots, t_k$. Thus $(\widetilde{\Omega}, \mathcal{B})$ is a measurable space.

If we fix $\omega \in \Omega$, and consider $X_t(\omega)$ as a function of t, then we get a realization (also called a sample path) of a random process. The mapping $\omega \to X_t(\omega)$ from Ω to $\widetilde{\Omega}$ is measurable, since the pre-image of any cylindrical set is measurable,

L. Koralov and Y.G. Sinai, *Theory of Probability and Random Processes*,   171
Universitext, DOI 10.1007/978-3-540-68829-7_12,
© Springer-Verlag Berlin Heidelberg 2012

$$\{\omega : (X_{t_1}(\omega), \ldots, X_{t_k}(\omega)) \in A\} \in \mathcal{F}.$$

Therefore, a random process induces a probability measure $\widetilde{\mathrm{P}}$ on the space $(\widetilde{\Omega}, \mathcal{B})$.

Definition 12.1. *Two processes X_t and Y_t, which need not be defined on the same probability space, are said to have the same finite-dimensional distributions if the vectors $(X_{t_1}, \ldots, X_{t_k})$ and $(Y_{t_1}, \ldots, Y_{t_k})$ have the same distributions for any k and any $t_1, \ldots, t_k \in T$.*

By Lemma 4.14, if two processes X_t and Y_t have the same finite-dimensional distributions, then they induce the same measure on $(\widetilde{\Omega}, \mathcal{B})$.

If we are given a probability measure $\widetilde{\mathrm{P}}$ on $(\widetilde{\Omega}, \mathcal{B})$, we can consider the process $\widetilde{X}_t$ on $(\widetilde{\Omega}, \mathcal{B}, \widetilde{\mathrm{P}})$ defined via $\widetilde{X}_t(\widetilde{\omega}) = \widetilde{\omega}(t)$. If $\widetilde{\mathrm{P}}$ is induced by a random process X_t, then the processes X_t and $\widetilde{X}_t$ clearly have the same finite-dimensional distributions.

We shall use the notations X_t and $X_t(\omega)$ for a random process and a realization of a process, respectively, often without specifying explicitly the underlying probability space or probability measure.

Let X_t be a random process defined on a probability space $(\Omega, \mathcal{F}, \mathrm{P})$ with parameter set either $\mathbb{R}$ or $\mathbb{R}^+$.

Definition 12.2. *A random process X_t is said to be measurable if $X_t(\omega)$, considered as a function of the two variables ω and t, is measurable with respect to the product σ-algebra $\mathcal{F} \times \mathcal{B}(T)$, where $\mathcal{B}(T)$ is the σ-algebra of Borel subsets of T.*

Lemma 12.3. *If every realization of a process is right-continuous, or every realization of a process is left-continuous, then the process is measurable.*

Proof. Let every realization of a process X_t be right-continuous. (The left-continuous case is treated similarly.) Define a sequence of processes Y_t^n by

$$Y_t^n(\omega) = X_{(k+1)/2^n}(\omega)$$

for $k/2^n < t \leq (k+1)/2^n$, where $t \in T$, $k \in \mathbb{Z}$. The mapping $(\omega, t) \to Y_t^n(\omega)$ is clearly measurable with respect to the product σ-algebra $\mathcal{F} \times \mathcal{B}(T)$. Furthermore, due to right-continuity of X_t, we have $\lim_{n \to \infty} Y_t^n(\omega) = X_t(\omega)$ for all $\omega \in \Omega$, $t \in T$. By Theorem 3.1, the mapping $(\omega, t) \to X_t(\omega)$ is measurable. $\square$

Definition 12.4. *Let X_t and Y_t be two random processes defined on the same probability space $(\Omega, \mathcal{F}, \mathrm{P})$. A process Y_t is said to be a modification of X_t if $\mathrm{P}(X_t = Y_t) = 1$ for every $t \in T$.*

It is clear that if Y_t is a modification of X_t, then they have the same finite-dimensional distributions.

Definition 12.5. *Two processes X_t and Y_t, $t \in T$, are indistinguishable if there is a set Ω' of full measure such that*

$$X_t(\omega) = Y_t(\omega) \quad \text{for all} \quad t \in T, \omega \in \Omega'.$$

If the parameter set is countable, then two processes are indistinguishable if and only if they are modifications of one another. If the parameter set is uncountable, then two processes may be modifications of one another, yet fail to be indistinguishable (see Problem 4).

Lemma 12.6. *Let the parameter set for the processes X_t and Y_t be either $\mathbb{R}$ or $\mathbb{R}^+$. If Y_t is a modification of X_t and both processes have right-continuous realizations (or both processes have left-continuous realizations), then they are indistinguishable.*

Proof. Let S be a dense countable subset in the parameter set T. Then there is a set Ω' of full measure such that

$$X_t(\omega) = Y_t(\omega) \quad \text{for all} \quad t \in S, \omega \in \Omega',$$

since Y_t is a modification of X_t. Due to right-continuity (or left-continuity), we then have

$$X_t(\omega) = Y_t(\omega) \quad \text{for all} \quad t \in T, \omega \in \Omega'.$$

$\square$

Let X_t be a random process defined on a probability space $(\Omega, \mathcal{F}, \mathrm{P})$. Then $\mathcal{F}^X = \sigma(X_t, t \in T)$ is called the σ-algebra generated by the process.

Definition 12.7. *The processes $X_t^1, \ldots, X_t^d$ defined on a common probability space are said to be independent, if the σ-algebras $\mathcal{F}^{X^1}, \ldots, \mathcal{F}^{X^d}$ are independent.*

12.2 Kolmogorov Consistency Theorem

The correspondence between random processes and probability measures on $(\widetilde{\Omega}, \mathcal{B})$ is helpful when studying the existence of random processes with prescribed finite-dimensional distributions. Namely, given a probability measure $\mathrm{P}_{t_1,\ldots,t_k}$ on each of the σ-algebras $\mathcal{B}_{t_1,\ldots,t_k}$, we would like to check whether there exists a measure $\widetilde{\mathrm{P}}$ on $(\widetilde{\Omega}, \mathcal{B})$ whose restriction to $\mathcal{B}_{t_1,\ldots,t_k}$ coincides with $\mathrm{P}_{t_1,\ldots,t_k}$. If such a measure exists, then the process $\widetilde{X}_t(\widetilde{\omega}) = \widetilde{\omega}(t)$ defined on $(\widetilde{\Omega}, \mathcal{B}, \widetilde{\mathrm{P}})$ has the prescribed finite-dimensional distributions.

We shall say that a collection of probability measures $\{\mathrm{P}_{t_1,\ldots,t_k}\}$ satisfies the consistency conditions if it has the following two properties:

(a) For every permutation π, every $t_1,\ldots,t_k$ and $A \in \mathcal{B}(\mathbb{R}^k)$,

$$\mathrm{P}_{t_1,\ldots,t_k}((\widetilde{\omega}(t_1),\ldots,\widetilde{\omega}(t_k)) \in A) = \mathrm{P}_{\pi(t_1,\ldots,t_k)}((\widetilde{\omega}(t_1),\ldots,\widetilde{\omega}(t_k)) \in A).$$

(b) For every $t_1,\ldots,t_k,t_{k+1}$ and $A \in \mathcal{B}(\mathbb{R}^k)$, we have

$$\mathrm{P}_{t_1,\ldots,t_k}((\widetilde{\omega}(t_1),\ldots,\widetilde{\omega}(t_k)) \in A)=\mathrm{P}_{t_1,\ldots,t_{k+1}}((\widetilde{\omega}(t_1),\ldots,\widetilde{\omega}(t_{k+1})) \in A\times\mathbb{R}).$$

Note that if the measures $\{\mathrm{P}_{t_1,\ldots,t_k}\}$ are induced by a common probability measure $\widetilde{\mathrm{P}}$ on $(\widetilde{\Omega},\mathcal{B})$, then they automatically satisfy the consistency conditions. The converse is also true.

Theorem 12.8 (Kolmogorov). *Assume that we are given a family of finite-dimensional probability measures $\{\mathrm{P}_{t_1,\ldots,t_k}\}$ satisfying the consistency conditions. Then there exists a unique σ-additive probability measure $\widetilde{\mathrm{P}}$ on $\mathcal{B}$ whose restriction to each $\mathcal{B}_{t_1,\ldots,t_k}$ coincides with $\mathrm{P}_{t_1,\ldots,t_k}$.*

Proof. The collection of all cylindrical sets is an algebra. Given a cylindrical set $B \in \mathcal{B}_{t_1,\ldots,t_k}$, we denote $m(B) = \mathrm{P}_{t_1,\ldots,t_k}(B)$. While the same set B may belong to different σ-algebras $\mathcal{B}_{t_1,\ldots,t_k}$ and $\mathcal{B}_{s_1,\ldots,s_{k'}}$, the consistency conditions guarantee that $m(B)$ is defined correctly. We would like to apply the Caratheodory Theorem (Theorem 3.19) to show that m can be extended in a unique way, as a measure, to the σ-algebra $\mathcal{B}$. Thus, in order to satisfy the assumptions of the Caratheodory Theorem, we need to show that m is a σ-additive function on the algebra of all cylindrical sets.

First, note that m is additive. Indeed, if $B, B_1,\ldots,B_n$ are cylindrical sets, $B \in \mathcal{B}_{t_1^0,\ldots,t_{k_0}^0}, B_1 \in \mathcal{B}_{t_1^1,\ldots,t_{k_1}^1},\ldots,B_n \in \mathcal{B}_{t_1^n,\ldots,t_{k_n}^n}$, then we can find a σ-algebra $\mathcal{B}_{t_1,\ldots,t_k}$ such that all of these sets belong to $\mathcal{B}_{t_1,\ldots,t_k}$ (it is sufficient to take $t_1 = t_1^0,\ldots,t_k = t_{k_n}^n$). If, in addition, $B = B_1 \cup \ldots \cup B_n$, where $B_i \cap B_j = \emptyset$ for $i \neq j$, then the relation

$$m(B) = \mathrm{P}_{t_1,\ldots,t_k}(B) = \sum_{i=1}^n \mathrm{P}_{t_1,\ldots,t_k}(B_i) = \sum_{i=1}^n m(B_i)$$

holds since $\mathrm{P}_{t_1,\ldots,t_k}$ is a measure.

Next, let us show that m is σ-subadditive. That is, for any cylindrical sets $B, B_1, B_2,\ldots$, the relation $B \subseteq \cup_{i=1}^\infty B_i$ implies that $m(B) \leq \sum_{i=1}^\infty m(B_i)$. This, together with the finite additivity of m, will immediately imply that m is σ-additive (see Remark 1.19). Assume that m is not σ-subadditive, that is, there are cylindrical sets $B, B_1, B_2,\ldots$ and a positive ε such that $B \subseteq \cup_{i=1}^\infty B_i$, and at the same time $m(B) = \sum_{i=1}^\infty m(B_i) + \varepsilon$. Let $A, A_1,\ldots$ be Borel sets such that

$$B = \{\widetilde{\omega} : (\widetilde{\omega}(t_1^0),\ldots,\widetilde{\omega}(t_{k_0}^0)) \in A\}, \quad B_i = \{\widetilde{\omega} : (\widetilde{\omega}(t_1^i),\ldots,\widetilde{\omega}(t_{k_i}^i)) \in A_i\}, i \geq 1.$$

For each set of indices $t_1, \ldots, t_k$, we can define the measure $\mathrm{P}'_{t_1, \ldots, t_k}$ on $\mathbb{R}^k$ via

$$\mathrm{P}'_{t_1, \ldots, t_k}(A) = \mathrm{P}_{t_1, \ldots, t_k}(\{\widetilde{\omega} : (\widetilde{\omega}(t_1), \ldots, \widetilde{\omega}(t_k)) \in A\}), \quad A \in \mathcal{B}(\mathbb{R}^k).$$

By Lemma 8.4, each of the measures $\mathrm{P}'_{t_1, \ldots, t_k}$ is regular. Therefore, we can find a closed set A' and open sets $A'_1, A'_2, \ldots$ such that $A' \subseteq A$, $A_i \subseteq A'_i$, $i \geq 1$, and

$$\mathrm{P}'_{t^0_1, \ldots, t^0_{k_0}}(A \setminus A') < \varepsilon/4, \quad \mathrm{P}'_{t^i_1, \ldots, t^i_{k_i}}(A'_i \setminus A_i) < \varepsilon/2^{i+1} \text{ for } i \geq 1.$$

By taking the intersection of A' with a large enough closed ball, we can ensure that A' is compact and $\mathrm{P}'_{t^0_1, \ldots, t^0_{k_0}}(A \setminus A') < \varepsilon/2$. Let us define

$$B' = \{\widetilde{\omega} : (\widetilde{\omega}(t^0_1), \ldots, \widetilde{\omega}(t^0_{k_0})) \in A'\}, B'_i = \{\widetilde{\omega} : (\widetilde{\omega}(t^i_1), \ldots, \widetilde{\omega}(t^i_{k_i})) \in A'_1\}, i \geq 1.$$

Therefore, $B' \subseteq \cup_{i=1}^{\infty} B'_i$ with $m(B') > \sum_{i=1}^{\infty} m(B'_i)$.

We can consider $\widetilde{\Omega}$ as a topological space with product topology (the weakest topology for which all the projections $\pi(t) : \widetilde{\omega} \to \widetilde{\omega}(t)$ are continuous). Then the sets B'_i, $i \geq 1$ are open, and B' is closed in the product topology. Furthermore, we can use Tychonoff's Theorem to show that B' is compact. Tychonoff's Theorem can be formulated as follows.

Theorem 12.9. *Let $\{K_t\}_{t \in T}$ be a family of compact spaces. Let $\widetilde{K}$ be the product space, that is the family of all $\{k_t\}_{t \in T}$ with $k_t \in K_t$. Then $\widetilde{K}$ is compact in the product topology.*

The proof of Tychonoff's Theorem can be found in the book "Functional Analysis", volume I, by Reed and Simon. In order to apply it, we define $\widetilde{K}$ as the space of all functions from T to $\overline{\mathbb{R}}$, where $\overline{\mathbb{R}} = \mathbb{R} \cup \{\infty\}$ is the compactification of $\mathbb{R}$, with the natural topology. Then $\widetilde{K}$ is a compact set. Furthermore, $\widetilde{\Omega} \subset \widetilde{K}$, and every set which is open in $\widetilde{\Omega}$ is also open in $\widetilde{K}$.

The set B' is compact in $\widetilde{K}$, since it is a closed subset of a compact set. Since every covering of B' with sets which are open in $\widetilde{\Omega}$ can be viewed as an open covering in $\widetilde{K}$, it admits a finite subcovering. Therefore, B' is compact in $\widetilde{\Omega}$. By extracting a finite subcovering from the sequence $B'_1, B'_2, \ldots$, we obtain $B' \subseteq \cup_{i=1}^{n} B'_i$ with $m(B') > \sum_{i=1}^{n} m(B'_i)$ for some n. This contradicts the finite additivity of m. We have thus proved that m is σ-additive on the algebra of cylindrical sets, and can be extended to the measure $\widetilde{\mathrm{P}}$ on the σ-algebra $\mathcal{B}$.

The uniqueness part of the theorem follows from the uniqueness of the extension in the Caratheodory Theorem. $\square$

Remark 12.10. We did not impose any requirements on the set of parameters T. The Kolmogorov Consistency Theorem applies, therefore, to families of finite-dimensional distributions indexed by elements of an arbitrary set. Furthermore, after making trivial modifications in the proof, we can claim the same result for processes whose state space is $\mathbb{R}^d$, $\mathbb{C}$, or any metric space with a finite or countable number of elements.

Unfortunately, the σ-algebra $\mathcal{B}$ is not rich enough for certain properties of a process to be described in terms of a measure on $(\widetilde{\Omega}, \mathcal{B})$. For example, the set $\{\widetilde{\omega} : |\widetilde{\omega}(t)| < C$ for all $t \in T\}$ does not belong to $\mathcal{B}$ if $T = \mathbb{R}^+$ or $\mathbb{R}$ (see Problem 2). Similarly, the set of all continuous functions does not belong to $\mathcal{B}$. At the same time, it is often important to consider random processes, whose typical realizations are bounded (or continuous, differentiable, etc.). The Kolmogorov Theorem alone is not sufficient in order to establish the existence of a process with properties beyond the prescribed finite-dimensional distributions.

We shall now consider several examples, where the existence of a random process is guaranteed by the Kolmogorov Theorem.

1. Homogeneous Sequences of Independent Random Trails. Let the parameter t be discrete. Given a probability measure P on $\mathbb{R}$, define the finite-dimensional measures $P_{t_1,\dots,t_k}$ as product measures:

$$P_{t_1,\dots,t_k}((\widetilde{\omega}(t_1),\dots,\widetilde{\omega}(t_k)) \in A_1 \times \dots \times A_k) = \prod_{i=1}^{k} P(A_i).$$

This family of measures clearly satisfies the consistency conditions.

2. Markov Chains. Assume that $t \in \mathbb{Z}^+$. Let $P(x, C)$ be a Markov transition function and μ_0 a probability measure on $\mathbb{R}$. We shall specify all the finite-dimensional measures $P_{t_0,\dots,t_k}$, where $t_0 = 0, \dots, t_k = k$,

$$P_{t_0,\dots,t_k}((\widetilde{\omega}(t_0), \widetilde{\omega}(t_1), \dots, \widetilde{\omega}(t_k)) \in A_0 \times A_1 \times \dots \times A_k) =$$

$$\int_{A_0} d\mu_0(x_0) \int_{A_1} P(x_0, dx_1) \int_{A_2} P(x_1, dx_2) \dots \int_{A_k} P(x_{k-1}, dx_k).$$

Again, it can be seen that this family of measures satisfies the consistency conditions.

3. Gaussian Processes. A random process X_t is called Gaussian if for any $t_1, t_2, \dots, t_k$ the joint probability distribution of $X_{t_1}, X_{t_2}, \dots, X_{t_k}$ is Gaussian. As shown in Sect. 9.3, such distributions are determined by the moments of the first two orders, that is by the expectation vector $m(t_i) = EX_{t_i}$ and the covariance matrix $B(t_i, t_j) = E(X_{t_i} - m_i)(X_{t_j} - m_j)$. Thus the finite-dimensional distributions of any Gaussian process are determined by a function of one variable $m(t)$ and a symmetric function of two variables $B(t, s)$.

Conversely, given a function $m(t)$ and a symmetric function of two variables $B(t, s)$ such that the $k \times k$ matrix $B(t_i, t_j)$ is non-negative definite for any $t_1, \dots, t_k$, we can define the finite-dimensional measure $P_{t_1,\dots,t_k}$ by

$$P_{t_1,\dots,t_k}(\{\widetilde{\omega} : (\widetilde{\omega}(t_1), \dots, \widetilde{\omega}(t_k)) \in A\}) = P'_{t_1,\dots,t_k}(A), \qquad (12.1)$$

where $P'_{t_1,\dots,t_k}$ is a Gaussian measure with the expectation vector $m(t_i)$ and the covariance matrix $B(t_i, t_j)$. The family of such measures satisfies the consistency conditions (see Problem 6).

12.3 Poisson Process

Let $\lambda > 0$. A process X_t is called a Poisson process with parameter λ if it has the following properties:

1. $X_0 = 0$ almost surely.
2. X_t is a process with independent increments, that is for $0 \leq t_1 \leq \ldots \leq t_k$ the variables $X_{t_1}, X_{t_2} - X_{t_1}, \ldots, X_{t_k} - X_{t_{k-1}}$ are independent.
3. For any $0 \leq s < t < \infty$ the random variable $X_t - X_s$ has Poisson distribution with parameter $\lambda(t - s)$.

Let us use the Kolmogorov Consistency Theorem to demonstrate the existence of a Poisson process with parameter λ.

For $0 \leq t_1 \leq \ldots \leq t_k$, let $\eta_1, \eta_2, \ldots, \eta_k$ be independent Poisson random variables with parameters $\lambda t_1, \lambda(t_2 - t_1), \ldots, \lambda(t_k - t_{k-1})$, respectively. Define $P'_{t_1, \ldots, t_k}$ to be the measure on $\mathbb{R}^k$ induced by the random vector

$$\eta = (\eta_1, \eta_1 + \eta_2, \ldots, \eta_1 + \eta_2 + \ldots + \eta_k).$$

Now we can define the family of finite-dimensional measures $P_{t_1, \ldots, t_k}$ by

$$P_{t_1, \ldots, t_k}(\{\widetilde{\omega} : (\widetilde{\omega}(t_1), \ldots, \widetilde{\omega}(t_k)) \in A\}) = P'_{t_1, \ldots, t_k}(A).$$

It can be easily seen that this family of measures satisfies the consistency conditions. Thus, by the Kolmogorov Theorem, there exists a process X_t with such finite-dimensional distributions. For $0 \leq t_1 \leq \ldots \leq t_k$ the random vector $(X_{t_1}, \ldots, X_{t_k})$ has the same distribution as η. Therefore, the random vector $(X_{t_1}, X_{t_2} - X_{t_1}, \ldots, X_{t_k} - X_{t_{k-1}})$ has the same distribution as $(\eta_1, \ldots, \eta_k)$, which shows that X_t is a Poisson process with parameter λ.

A Poisson process can be constructed explicitly as follows. Let $\xi_1, \xi_2, \ldots$ be a sequence of independent identically distributed random variables. The distribution of each ξ_i is assumed to be exponential with parameter λ, that is ξ_i have the density

$$p(u) = \begin{cases} \lambda e^{-\lambda u} & u \geq 0, \\ 0 & u < 0. \end{cases}$$

Define the process X_t, $t \geq 0$, as follows

$$X_t(\omega) = \sup\{n : \sum_{i \leq n} \xi_i(\omega) \leq t\}. \tag{12.2}$$

Here, a sum over an empty set of indices is assumed to be equal to zero. The process defined by (12.2) is a Poisson process (see Problem 8). In Sect. 14.3 we shall prove a similar statement for Markov processes with a finite number of states.

Now we shall discuss an everyday situation where a Poisson process appears naturally. Let us model the times between the arrival of consecutive customers to a store by random variables ξ_i. Thus, ξ_1 is the time between the opening of the store and the arrival of the first customer, ξ_2 is the time between the arrival of the first customer and the arrival of the second one, etc. It is reasonable to assume that ξ_i are independent identically distributed random variables.

It is also reasonable to assume that if no customers showed up by time t, then the distribution of the time remaining till the next customer shows up is the same as the distribution of each of ξ_i. More rigorously,

$$P(\xi_i - t \in A | \xi_i > t) = P(\xi_i \in A). \tag{12.3}$$

for any Borel set $A \subseteq \mathbb{R}$. If an unbounded random variable satisfies (12.3), then it has exponential distribution (see Problem 2 of Chap. 4). Therefore, the process X_t defined by (12.2) models the number of customers that have arrived to the store by time t.

12.4 Problems

1. Let $\widetilde{\Omega}_{\mathbb{R}}$ be the set of all functions $\widetilde{\omega} : \mathbb{R} \to \mathbb{R}$ and $\mathcal{B}_{\mathbb{R}}$ be the minimal σ-algebra containing all the cylindrical subsets of $\widetilde{\Omega}_{\mathbb{R}}$. Let $\widetilde{\Omega}_{\mathbb{Z}^+}$ be the set of all functions from $\mathbb{Z}^+$ to $\mathbb{R}$, and $\mathcal{B}_{\mathbb{Z}^+}$ be the minimal σ-algebra containing all the cylindrical subsets of $\widetilde{\Omega}_{\mathbb{Z}^+}$.
   Show that a set $S \subseteq \widetilde{\Omega}_{\mathbb{R}}$ belongs to $\mathcal{B}_{\mathbb{R}}$ if and only if one can find a set $B \in \mathcal{B}_{\mathbb{Z}^+}$ and an infinite sequence of real numbers $t_1, t_2, \ldots$ such that

$$S = \{\widetilde{\omega} : (\widetilde{\omega}(t_1), \widetilde{\omega}(t_2), \ldots) \in B\}.$$

2. Let $\widetilde{\Omega}$ be the set of all functions $\widetilde{\omega} : \mathbb{R} \to \mathbb{R}$ and $\mathcal{B}$ the minimal σ-algebra containing all the cylindrical subsets of $\widetilde{\Omega}$. Prove that the sets $\{\widetilde{\omega} \in \widetilde{\Omega} : |\widetilde{\omega}(t)| < C \text{ for all } t \in \mathbb{R}\}$ and $\{\widetilde{\omega} \in \widetilde{\Omega} : \widetilde{\omega} \text{ is continuous for all } t \in \mathbb{R}\}$ do not belong to $\mathcal{B}$. (Hint: use Problem 1.)

3. Let $\widetilde{\Omega}$ be the space of all functions from $\mathbb{R}$ to $\mathbb{R}$ and $\mathcal{B}$ the σ-algebra generated by cylindrical sets. Prove that the mapping $(\widetilde{\omega}, t) \to \widetilde{\omega}(t)$ from the product space $\widetilde{\Omega} \times \mathbb{R}$ to $\mathbb{R}$ is not measurable. (Hint: use Problem 2.)

4. Prove that two processes with a countable parameter set are indistinguishable if and only if they are modifications of one another. Give an example of two processes defined on an uncountable parameter set which are modifications of one another, but are not indistinguishable.

5. Assume that the random variables X_t, $t \in \mathbb{R}$, are independent and identically distributed with the distribution which is absolutely continuous with respect to the Lebesgue measure. Prove that the realizations of the process X_t, $t \in \mathbb{R}$, are discontinuous almost surely.

6. Prove that the family of measures defined by (12.1) satisfies the consistency conditions.

7. Let X_t^1 and X_t^2 be two independent Poisson processes with parameters λ_1 and λ_2 respectively. Prove that $X_t^1 + X_t^2$ is a Poisson process with parameter $\lambda_1 + \lambda_2$.

8. Prove that the process X_t defined by (12.2) is a Poisson process.

9. Let $X_t^1, \ldots, X_t^n$ be independent Poisson processes with parameters $\lambda_1, \ldots, \lambda_n$. Let

$$X_t = c_1 X_t^1 + \ldots + c_n X_t^n,$$

where $c_1, \ldots, c_n$ are positive constants. Find the probability distribution of the number of discontinuities of X_t on the segment $[0, 1]$.

10. Assume that the time intervals between the arrival of consecutive customers to a store are independent identically distributed random variables with exponential distribution with parameter λ. Let τ_n be the time of the arrival of the n-th customer. Find the distribution of τ_n.

If customers arrive at the rate of 3 a minute, what is the probability that the number of customers arriving in the first 2 min is equal to 3.

Conditional Expectations and Martingales

13.1 Conditional Expectations

For two events $A, B \in \mathcal{F}$ in a probability space $(\Omega, \mathcal{F}, P)$, we previously defined the conditional probability of A given B as

$$P(A|B) = \frac{P(A \bigcap B)}{P(B)}.$$

Similarly, we can define the conditional expectation of a random variable f given B as

$$E(f|B) = \frac{\int_B f(\omega)dP(\omega)}{P(B)},$$

provided that the integral on the right-hand side is finite and the denominator is different from zero.

We now introduce an important generalization of this notion by defining the conditional expectation of a random variable given a σ-subalgebra $\mathcal{G} \subseteq \mathcal{F}$.

Definition 13.1. *Let $(\Omega, \mathcal{F}, P)$ be a probability space, $\mathcal{G}$ a σ-subalgebra of $\mathcal{F}$, and $f \in L^1(\Omega, \mathcal{F}, P)$. The conditional expectation of f given $\mathcal{G}$, denoted by $E(f|\mathcal{G})$, is the random variable $g \in L^1(\Omega, \mathcal{G}, P)$ such that for any $A \in \mathcal{G}$*

$$\int_A f dP = \int_A g dP. \tag{13.1}$$

Note that for fixed f, the left-hand side of (13.1) is a σ-additive function defined on the σ-algebra $\mathcal{G}$. Therefore, the existence and uniqueness (up to a set of measure zero) of the function g are guaranteed by the Radon-Nikodym Theorem. Here are several simple examples.

If f is measurable with respect to $\mathcal{G}$, then clearly $E(f|\mathcal{G}) = f$. If f is independent of the σ-algebra $\mathcal{G}$, then $E(f|\mathcal{G}) = Ef$, since $\int_A f dP = P(A)Ef$ in this case. Thus the conditional expectation is reduced to ordinary expectation

L. Koralov and Y.G. Sinai, *Theory of Probability and Random Processes,*
Universitext, DOI 10.1007/978-3-540-68829-7_13,
© Springer-Verlag Berlin Heidelberg 2012

if f is independent of $\mathcal{G}$. This is the case, in particular, when $\mathcal{G}$ is the trivial σ-algebra, $\mathcal{G} = \{\emptyset, \Omega\}$.

If $\mathcal{G} = \{B, \Omega \backslash B, \emptyset, \Omega\}$, where $0 < P(B) < 1$, then

$$E(f|\mathcal{G}) = E(f|B)\chi_B + E(f|(\Omega \backslash B))\chi_{\Omega \backslash B}.$$

Thus, the conditional expectation of f with respect to the smallest σ-algebra containing B is equal to the constant $E(f|B)$ on the set B.

Concerning the notations, we shall often write $E(f|g)$ instead of $E(f|\sigma(g))$, if f and g are random variables on $(\Omega, \mathcal{F}, P)$. Likewise, we shall often write $P(A|\mathcal{G})$ instead of $E(\chi_A|\mathcal{G})$ to denote the conditional expectation of the indicator function of a set $A \in \mathcal{F}$. The function $P(A|\mathcal{G})$ will be referred to as the conditional probability of A given the σ-algebra $\mathcal{G}$.

13.2 Properties of Conditional Expectations

Let us list several important properties of conditional expectations. Note that since the conditional expectation is defined up to a set of measure zero, all the equalities and inequalities below hold almost surely.

1. If $f_1, f_2 \in L^1(\Omega, \mathcal{F}, P)$ and a, b are constants, then

$$E(af_1 + bf_2|\mathcal{G}) = aE(f_1|\mathcal{G}) + bE(f_2|\mathcal{G}).$$

2. If $f \in L^1(\Omega, \mathcal{F}, P)$, and $\mathcal{G}_1$ and $\mathcal{G}_2$ are σ-subalgebras of $\mathcal{F}$ such that $\mathcal{G}_2 \subseteq \mathcal{G}_1 \subseteq \mathcal{F}$, then
$$E(f|\mathcal{G}_2) = E(E(f|\mathcal{G}_1)|\mathcal{G}_2).$$

3. If $f_1, f_2 \in L^1(\Omega, \mathcal{F}, P)$ and $f_1 \leq f_2$, then $E(f_1|\mathcal{G}) \leq E(f_2|\mathcal{G})$.
4. $E(E(f|\mathcal{G})) = Ef$.
5. (Conditional Dominated Convergence Theorem) If a sequence of measurable functions f_n converges to a measurable function f almost surely, and

$$|f_n| \leq \varphi,$$

where φ is integrable on Ω, then $\lim_{n \to \infty} E(f_n|\mathcal{G}) = E(f|\mathcal{G})$ almost surely.
6. If $g, fg \in L^1(\Omega, \mathcal{F}, P)$, and f is measurable with respect to $\mathcal{G}$, then $E(fg|\mathcal{G}) = fE(g|\mathcal{G})$.

Properties 1–3 are clear. To prove property 4, it suffices to take $A = \Omega$ in the equality $\int_A f dP = \int_A E(f|\mathcal{G}) dP$ defining the conditional expectation.

To prove the Conditional Dominated Convergence Theorem, let us first assume that f_n is a monotonic sequence. Without loss of generality we may assume that f_n is monotonically non-decreasing (the case of a non-increasing sequence is treated similarly). Thus the sequence of functions $E(f_n|\mathcal{G})$ satisfies the assumptions of the Levi Convergence Theorem (see Sect. 3.5).

Let $g = \lim_{n\to\infty} \mathrm{E}(f_n|\mathcal{G})$. Then g is $\mathcal{G}$-measurable and $\int_A g d\mathrm{P} = \int_A f d\mathrm{P}$ for any $A \in \mathcal{G}$, again by the Levi Theorem.

If the sequence f_n is not necessarily monotonic, we can consider the auxiliary sequences $\overline{f}_n = \inf_{m\geq n} f_m$ and $\overline{\overline{f}}_n = \sup_{m\geq n} f_m$. These sequences are already monotonic and satisfy the assumptions placed on the sequence f_n. Therefore,

$$\lim_{n\to\infty} \mathrm{E}(\overline{f}_n|\mathcal{G}) = \lim_{n\to\infty} \mathrm{E}(\overline{\overline{f}}_n|\mathcal{G}) = \mathrm{E}(f|\mathcal{G}).$$

Since $\overline{f}_n \leq f_n \leq \overline{\overline{f}}_n$, the Dominated Convergence Theorem follows from the monotonicity of the conditional expectation (property 3).

To prove the last property, first we consider the case when f is the indicator function of a set $B \in \mathcal{G}$. Then for any $A \in \mathcal{G}$

$$\int_A \chi_B \mathrm{E}(g|\mathcal{G}) d\mathrm{P} = \int_{A\cap B} \mathrm{E}(g|\mathcal{G}) d\mathrm{P} = \int_{A\cap B} g d\mathrm{P} = \int_A \chi_B g d\mathrm{P},$$

which proves the statement for $f = \chi_B$. By linearity, the statement is also true for simple functions taking a finite number of values. Next, without loss of generality, we may assume that $f, g \geq 0$. Then we can find a non-decreasing sequence of simple functions f_n, each taking a finite number of values such that $\lim_{n\to\infty} f_n = f$ almost surely. We have $f_n g \to fg$ almost surely, and the Dominated Convergence Theorem for conditional expectations can be applied to the sequence $f_n g$ to conclude that

$$\mathrm{E}(fg|\mathcal{G}) = \lim_{n\to\infty} \mathrm{E}(f_n g|\mathcal{G}) = \lim_{n\to\infty} f_n \mathrm{E}(g|\mathcal{G}) = f\mathrm{E}(g|\mathcal{G}).$$

We now state Jensen's Inequality and the Conditional Jensen's Inequality, essential to our discussion of conditional expectations and martingales. The proofs of these statements can be found in many other textbooks, and we shall not provide them here (see "Real Analysis and Probability" by R. M. Dudley).

We shall consider a random variable f with values in $\mathbb{R}^d$ defined on a probability space $(\Omega, \mathcal{F}, \mathrm{P})$. Recall that a function $g : \mathbb{R}^d \to \mathbb{R}$ is called convex if $g(cx + (1-c)y) \leq cg(x) + (1-c)g(y)$ for all $x, y \in \mathbb{R}^d$, $0 \leq c \leq 1$.

Theorem 13.2 (Jensen's Inequality). *Let g be a convex (and consequently continuous) function on $\mathbb{R}^d$ and f a random variable with values in $\mathbb{R}^d$ such that $\mathrm{E}|f| < \infty$. Then, either $\mathrm{E}g(f) = +\infty$, or*

$$g(\mathrm{E}f) \leq \mathrm{E}g(f) < \infty.$$

Theorem 13.3 (Conditional Jensen's Inequality). *Let g be a convex function on $\mathbb{R}^d$ and f a random variable with values in $\mathbb{R}^d$ such that*

$$\mathrm{E}|f|, \mathrm{E}|g(f)| < \infty.$$

Let $\mathcal{G}$ be a σ-subalgebra of $\mathcal{F}$. Then almost surely

$$g(\mathrm{E}(f|\mathcal{G})) \leq \mathrm{E}(g(f)|\mathcal{G}).$$

Let $\mathcal{G}$ be a σ-subalgebra of $\mathcal{F}$. Let $H = L^2(\Omega, \mathcal{G}, \mathrm{P})$ be the closed linear subspace of the Hilbert space $L^2(\Omega, \mathcal{F}, \mathrm{P})$. Let us illustrate the use of the Conditional Jensen's Inequality by proving that for a random variable $f \in L^2(\Omega, \mathcal{F}, \mathrm{P})$, taking the conditional expectation $\mathrm{E}(f|\mathcal{G})$ is the same as taking the projection on H.

Lemma 13.4. *Let $f \in L^2(\Omega, \mathcal{F}, \mathrm{P})$ and P_H be the projection operator on the space H. Then*
$$\mathrm{E}(f|\mathcal{G}) = P_H f.$$

Proof. The function $\mathrm{E}(f|\mathcal{G})$ is square-integrable by the Conditional Jensen's Inequality applied to $g(x) = x^2$. Thus, $\mathrm{E}(f|\mathcal{G}) \in H$. It remains to show that $f - \mathrm{E}(f|\mathcal{G})$ is orthogonal to any $h \in H$. Since h is $\mathcal{G}$-measurable,
$$\mathrm{E}((f - \mathrm{E}(f|\mathcal{G}))\overline{h}) = \mathrm{E}\mathrm{E}((f - \mathrm{E}(f|\mathcal{G}))\overline{h}|\mathcal{G}) = \mathrm{E}(\overline{h}\mathrm{E}((f - \mathrm{E}(f|\mathcal{G}))|\mathcal{G})) = 0.$$

$\square$

13.3 Regular Conditional Probabilities

Let f and g be random variables on a probability space $(\Omega, \mathcal{F}, \mathrm{P})$. If g takes a finite or countable number of values $y_1, y_2, \ldots$, and the probabilities of the events $\{\omega : g(\omega) = y_i\}$ are positive, we can write, similarly to (4.1), the formula of full expectation
$$\mathrm{E}f = \sum_i \mathrm{E}(f|g = y_i)\mathrm{P}(g = y_i).$$

Let us derive an analogue to this formula, which will work when the number of values of g is not necessarily finite or countable. The sets $\Omega_y = \{\omega : g(\omega) = y\}$, where $y \in \mathbb{R}$, still form a partition of the probability space Ω, but the probability of each Ω_y may be equal to zero. Thus, we need to attribute meaning to the expression $\mathrm{E}(f|\Omega_y)$ (also denoted by $\mathrm{E}(f|g = y)$). One way to do this is with the help of the concept of a regular conditional probability, which we introduce below.

Let $(\Omega, \mathcal{F}, \mathrm{P})$ be a probability space and $\mathcal{G} \subseteq \mathcal{F}$ a σ-subalgebra. Let h be a measurable function from $(\Omega, \mathcal{F})$ to a measurable space $(X, \mathcal{B})$. To motivate the formal definition of a regular conditional probability, let us first assume that $\mathcal{G}$ is generated by a finite or countable partition $A_1, A_2, \ldots$ such that $\mathrm{P}(A_i) > 0$ for all i. In this case, for a fixed $B \in \mathcal{B}$, the conditional probability $\mathrm{P}(h \in B|\mathcal{G})$ is constant on each A_i equal to $\mathrm{P}(h \in B|A_i)$, as follows from the definition of the conditional probability. As a function of B, this expression is a probability measure on $(X, \mathcal{B})$. The concept of a regular conditional probability allows us to view $\mathrm{P}(h \in B|\mathcal{G})(\omega)$, for fixed ω, as a probability measure, even without the assumption that $\mathcal{G}$ is generated by a finite or countable partition.

Definition 13.5. *A function* $Q : \mathcal{B} \times \Omega \to [0,1]$ *is called a regular conditional probability of* h *given* $\mathcal{G}$ *if:*

1. *For each* $\omega \in \Omega$, *the function* $Q(\cdot, \omega) : \mathcal{B} \to [0,1]$ *is a probability measure on* $(X, \mathcal{B})$.
2. *For each* $B \in \mathcal{B}$, *the function* $Q(B, \cdot) : \Omega \to [0,1]$ *is* $\mathcal{G}$-*measurable.*
3. *For each* $B \in \mathcal{B}$, *the equality* $P(h \in B|\mathcal{G})(\omega) = Q(B, \omega)$ *holds almost surely.*

We have the following theorem, which guarantees the existence and uniqueness of a regular conditional probability when X is a complete separable metric space. (The proof of this theorem can be found in "Real Analysis and Probability" by R. M. Dudley.)

Theorem 13.6. *Let* $(\Omega, \mathcal{F}, P)$ *be a probability space and* $\mathcal{G} \subseteq \mathcal{F}$ *a* σ-*subalgebra. Let* X *be a complete separable metric space and* $\mathcal{B}$ *the* σ-*algebra of Borel sets of* X. *Take a measurable function* h *from* $(\Omega, \mathcal{F})$ *to* $(X, \mathcal{B})$. *Then there exists a regular conditional probability of* h *given* $\mathcal{G}$. *It is unique in the sense that if* Q *and* Q' *are regular conditional probabilities, then the measures* $Q(\cdot, \omega)$ *and* $Q'(\cdot, \omega)$ *coincide for almost all* ω.

The next lemma states that when the regular conditional probability exists, the conditional expectation can be written as an integral with respect to the measure $Q(\cdot, \omega)$.

Lemma 13.7. *Let the assumptions of Theorem 13.6 hold, and* $f : X \to \mathbb{R}$ *be a measurable function such that* $E(f(h(\omega)))$ *is finite. Then, for almost all* ω, *the function* f *is integrable with respect to* $Q(\cdot, \omega)$, *and*

$$E(f(h)|\mathcal{G})(\omega) = \int_X f(x)Q(dx, \omega) \quad \text{for almost all } \omega. \qquad (13.2)$$

Proof. First, let f be an indicator function of a measurable set, that is $f = \chi_B$ for $B \in \mathcal{B}$. In this case, the statement of the lemma is reduced to

$$P(h \in B|\mathcal{G})(\omega) = Q(B, \omega),$$

which follows from the definition of the regular conditional probability.

Since both sides of (13.2) are linear in f, the lemma also holds when f is a simple function with a finite number of values. Now, let f be a non-negative measurable function such that $E(f(h(\omega)))$ is finite. One can find a sequence of non-negative simple functions f_n, each taking a finite number of values, such that $f_n \to f$ monotonically from below. Thus, $E(f_n(h)|\mathcal{G})(\omega) \to E(f(h)|\mathcal{G})(\omega)$ almost surely by the Conditional Dominated Convergence Theorem. Therefore, the sequence $\int_X f_n(x)Q(dx, \omega)$ is bounded almost surely, and $\int_X f_n(x)Q(dx, \omega) \to \int_X f(x)Q(dx, \omega)$ for almost all ω by the Levi Monotonic Convergence Theorem. This justifies (13.2) for non-negative f.

Finally, if f is not necessarily non-negative, it can be represented as a difference of two non-negative functions. $\qquad \square$

Example. Assume that Ω is a complete separable metric space, $\mathcal{F}$ is the σ-algebra of its Borel sets, and $(X, \mathcal{B}) = (\Omega, \mathcal{F})$. Let P be a probability measure on $(\Omega, \mathcal{F})$, and f and g be random variables on $(\Omega, \mathcal{F}, \mathrm{P})$. Let h be the identity mapping from Ω to itself, and let $\mathcal{G} = \sigma(g)$. In this case, (13.2) takes the form

$$\mathrm{E}(f|g)(\omega) = \int_\Omega f(\widetilde{\omega})Q(d\widetilde{\omega}, \omega) \quad \text{for almost all } \omega. \tag{13.3}$$

Let P_g be the measure on $\mathbb{R}$ induced by the mapping $g : \Omega \to \mathbb{R}$. For any $B \in \mathcal{B}$, the function $Q(B, \cdot)$ is constant on each level set of g, since it is measurable with respect to $\sigma(g)$. Therefore, for almost all y (with respect to the measure P_g), we can define measures $Q_y(\cdot)$ on $(\Omega, \mathcal{F})$ by putting $Q_{g(\omega)}(B) = Q(B, \omega)$.

The function $\mathrm{E}(f|g)$ is constant on each level set of g. Therefore, we can define $\mathrm{E}(f|g = y) = \mathrm{E}(f|g)(\omega)$, where ω is such that $g(\omega) = y$. This function is defined up to a set of measure zero (with respect to the measure P_g). In order to calculate the expectation of f, we can write

$$\mathrm{E}f = \mathrm{E}(\mathrm{E}(f|g)) = \int_\mathbb{R} \mathrm{E}(f|g = y)d\mathrm{P}_g(y) = \int_\mathbb{R}(\int_\Omega f(\widetilde{\omega})dQ_y(\widetilde{\omega}))d\mathrm{P}_g(y),$$

where the second equality follows from the change of variable formula in the Lebesgue integral. It is possible to show that the measure Q_y is supported on the event $\Omega_y = \{\omega : g(\omega) = y\}$ for P_g−almost all y (we do not prove this statement here). Therefore, we can write the expectation as a double integral

$$\mathrm{E}f = \int_\mathbb{R}(\int_{\Omega_y} f(\widetilde{\omega})dQ_y(\widetilde{\omega}))d\mathrm{P}_g(y).$$

This is the formula of the full mathematical expectation.

Example. Let h be a random variable with values in $\mathbb{R}$, f the identity mapping on $\mathbb{R}$, and $\mathcal{G} = \sigma(g)$. Then Lemma 13.7 states that

$$\mathrm{E}(h|g)(\omega) = \int_\mathbb{R} xQ(dx, \omega) \quad \text{for almost all } \omega,$$

where Q is the regular conditional probability of h given $\sigma(g)$. Assume that h and g have a joint probability density $p(x, y)$, which is a continuous function satisfying $0 < \int_\mathbb{R} p(x, y)dx < \infty$ for all y. It is easy to check that

$$Q(B, \omega) = \int_B p(x, g(\omega))dx(\int_\mathbb{R} p(x, g(\omega))dx)^{-1}$$

has the properties required of the regular conditional probability. Therefore,

$$\mathrm{E}(h|g)(\omega) = \int_\mathbb{R} xp(x, g(\omega))dx(\int_\mathbb{R} p(x, g(\omega))dx)^{-1} \quad \text{for almost all } \omega,$$

and

$$\mathrm{E}(h|g = y) = \int_\mathbb{R} xp(x, y)dx(\int_\mathbb{R} p(x, y)dx)^{-1} \quad \text{for } \mathrm{P}_g\text{−almost all } y.$$

13.4 Filtrations, Stopping Times, and Martingales

Let $(\Omega, \mathcal{F})$ be a measurable space and T a subset of $\mathbb{R}$ or $\mathbb{Z}$.

Definition 13.8. *A collection of σ-subalgebras $\mathcal{F}_t \subseteq \mathcal{F}$, $t \in T$, is called a filtration if $\mathcal{F}_s \subseteq \mathcal{F}_t$ for all $s \leq t$.*

Definition 13.9. *A random variable τ with values in the parameter set T is a stopping time of the filtration $\mathcal{F}_t$ if $\{\tau \leq t\} \in \mathcal{F}_t$ for each $t \in T$.*

Remark 13.10. Sometimes it will be convenient to allow τ to take values in $T \cup \{\infty\}$. In this case, τ is still called a stopping time if $\{\tau \leq t\} \in \mathcal{F}_t$ for each $t \in T$.

Example. Let $T = \mathbb{N}$ and Ω be the space of all functions $\omega : \mathbb{N} \to \{-1, 1\}$. (In other words, Ω is the space of infinite sequences made of -1's and 1's.) Let $\mathcal{F}_n$ be the smallest σ-algebra which contains all the sets of the form

$$\{\omega : \omega(1) = a_1, \ldots, \omega(n) = a_n\},$$

where $a_1, \ldots, a_n \in \{-1, 1\}$. Let $\mathcal{F}$ be the smallest σ-algebra containing all $\mathcal{F}_n$, $n \geq 1$. The space $(\Omega, \mathcal{F})$ can be used to model an infinite sequence of games, where the outcome of each game is either a loss or a gain of one dollar. Let

$$\tau(\omega) = \min\{n : \sum_{i=1}^{n} \omega(i) = 3\}.$$

Thus, τ is the first time when a gambler playing the game accumulates three dollars in winnings. (Note that $\tau(\omega) = \infty$ for some ω.) It is easy to demonstrate that τ is a stopping time. Let

$$\sigma(\omega) = \min\{n : \omega(n+1) = -1\}.$$

Thus, a gambler stops at time σ if the next game will result in a loss. Following such a strategy involves looking at the outcome of a future game before deciding whether to play it. Indeed, it is easy to check that σ does not satisfy the definition of a stopping time.

Remark 13.11. Recall the following notation: if x and y are real numbers, then $x \wedge y = \min(x, y)$ and $x \vee y = \max(x, y)$.

Lemma 13.12. *If σ and τ are stopping times of a filtration $\mathcal{F}_t$, then $\sigma \wedge \tau$ is also a stopping time.*

Proof. We need to show that $\{\sigma \wedge \tau \leq t\} \in \mathcal{F}_t$ for any $t \in T$, which immediately follows from

$$\{\sigma \wedge \tau \leq t\} = \{\sigma \leq t\} \bigcup \{\tau \leq t\} \in \mathcal{F}_t.$$

$\square$

In fact, if σ and τ are stopping times, then $\sigma \vee \tau$ is also a stopping time. If, in addition, $\sigma, \tau \geq 0$, then $\sigma + \tau$ is also a stopping time (see Problem 7).

Definition 13.13. *Let τ be a stopping time of the filtration $\mathcal{F}_t$. The σ-algebra of events determined prior to the stopping time τ, denoted by $\mathcal{F}_\tau$, is the collection of events $A \in \mathcal{F}$ for which $A \bigcap \{\tau \leq t\} \in \mathcal{F}_t$ for each $t \in T$.*

Clearly, $\mathcal{F}_\tau$ is a σ-algebra. Moreover, τ is $\mathcal{F}_\tau$-measurable since

$$\{\tau \leq c\} \bigcap \{\tau \leq t\} = \{\tau \leq c \wedge t\} \in \mathcal{F}_t,$$

and therefore $\{\tau \leq c\} \in \mathcal{F}_\tau$ for each c. If σ and τ are two stopping times such that $\sigma \leq \tau$, then $\mathcal{F}_\sigma \subseteq \mathcal{F}_\tau$. Indeed, if $A \in \mathcal{F}_\sigma$, then

$$A \bigcap \{\tau \leq t\} = (A \bigcap \{\sigma \leq t\}) \bigcap \{\tau \leq t\} \in \mathcal{F}_t.$$

Now let us consider a process X_t together with a filtration $\mathcal{F}_t$ defined on a common probability space.

Definition 13.14. *A random process X_t is called adapted to a filtration $\mathcal{F}_t$ if X_t is $\mathcal{F}_t$-measurable for each $t \in T$.*

An example of a stopping time is provided by the first time when a continuous process hits a closed set.

Lemma 13.15. *Let X_t be a continuous $\mathbb{R}^d$-valued process adapted to a filtration $\mathcal{F}_t$, where $t \in \mathbb{R}^+$. Let K be a closed set in $\mathbb{R}^d$ and $s \geq 0$. Let*

$$\tau^s(\omega) = \inf\{t \geq s, X_t(\omega) \in K\}$$

be the first time, following s, when the process hits K. Then τ^s is a stopping time.

Proof. For an open set U, define

$$\tau_U^s(\omega) = \inf\{t \geq s, X_t(\omega) \in U\},$$

where the infimum of the empty set is $+\infty$. First, we show that the set $\{\omega : \tau_U^s(\omega) < t\}$ belongs to $\mathcal{F}_t$ for any $t \in \mathbb{R}^+$. Indeed, from the continuity of the process it easily follows that

$$\{\tau_U^s < t\} = \bigcup_{u \in \mathbb{Q}, s < u < t} \{X_u \in U\},$$

and the right-hand side of this equality belongs to $\mathcal{F}_t$. Now, for the set K, we define the open sets $U_n = \{x \in \mathbb{R}^d : \text{dist}(x, K) < 1/n\}$. We claim that for $t > s$,

$$\{\tau^s \leq t\} = \bigcap_{n=1}^{\infty} \{\tau_{U_n}^s < t\}. \tag{13.4}$$

Indeed, if $\tau^s(\omega) \leq t$, then for each n the trajectory $X_u(\omega)$ enters the open set U_n for some u, $s < u < t$, due to the continuity of the process. Thus ω belongs to the event on the right-hand side of (13.4).

Conversely, if ω belongs to the event on the right-hand side of (13.4), then there is a non-decreasing sequence of times u_n such that $s < u_n < t$ and $X_{u_n}(\omega) \in U_n$. Taking $u = \lim_{n \to \infty} u_n$, we see that $u \leq t$ and $X_u(\omega) \in K$, again due to the continuity of the process. This means that $\tau^s(\omega) \leq t$, which justifies (13.4).

Since the event on the right-hand side of (13.4) belongs to $\mathcal{F}_t$, we see that $\{\tau^s \leq t\}$ belongs to $\mathcal{F}_t$ for $t > s$. Furthermore, $\{\tau^s \leq s\} = \{X_s \in K\} \in \mathcal{F}_s$. We have thus proved that τ^s is a stopping time. $\qquad\square$

For a given random process, a simple example of a filtration is that generated by the process itself:

$$\mathcal{F}_t^X = \sigma(X_s, s \leq t).$$

Clearly, X_t is adapted to the filtration $\mathcal{F}_t^X$.

Definition 13.16. *A family $(X_t, \mathcal{F}_t)_{t \in T}$ is called a martingale if the process X_t is adapted to the filtration $\mathcal{F}_t$, $X_t \in L^1(\Omega, \mathcal{F}, P)$ for all t, and*

$$X_s = \mathrm{E}(X_t | \mathcal{F}_s) \qquad \text{for} \quad s \leq t.$$

If the equal sign is replaced by $\leq$ or $\geq$, then $(X_t, \mathcal{F}_t)_{t \in T}$ is called a submartingale or supermartingale respectively.

We shall often say that X_t is a martingale, without specifying a filtration, if it is clear from the context what the parameter set and the filtration are.

If one thinks of X_t as the fortune of a gambler at time t, then a martingale is a model of a fair game (any information available by time s does not affect the fact that the expected increment in the fortune over the time period from s to t is equal to zero). More precisely, $\mathrm{E}(X_t - X_s | \mathcal{F}_s) = 0$.

If $(X_t, \mathcal{F}_t)_{t \in T}$ is a martingale and f is a convex function such that $f(X_t)$ is integrable for all t, then $(f(X_t), \mathcal{F}_t)_{t \in T}$ is a submartingale. Indeed, by the Conditional Jensen's Inequality,

$$f(X_s) = f(\mathrm{E}(X_t | \mathcal{F}_s)) \leq \mathrm{E}(f(X_t) | \mathcal{F}_s).$$

For example, if $(X_t, \mathcal{F}_t)_{t \in T}$ is a martingale, then $(|X_t|, \mathcal{F}_t)_{t \in T}$ is a submartingale, If, in addition, X_t is square-integrable, then $(X_t^2, \mathcal{F}_t)_{t \in T}$ is a submartingale.

13.5 Martingales with Discrete Time

In this section we study martingales with discrete time ($T = \mathbb{N}$). In the next section we shall state the corresponding results for continuous time martingales, which will lead us to the notion of an integral of a random process with respect to a continuous martingale.

Our first theorem states that any submartingale can be decomposed, in a unique way, into a sum of a martingale and a non-decreasing process adapted to the filtration $(\mathcal{F}_{n-1})_{n \geq 2}$.

Theorem 13.17 (Doob Decomposition). *If $(X_n, \mathcal{F}_n)_{n \in \mathbb{N}}$ is a submartingale, then there exist two random processes, M_n and A_n, with the following properties:*

1. $X_n = M_n + A_n$ *for $n \geq 1$.*
2. $(M_n, \mathcal{F}_n)_{n \in \mathbb{N}}$ *is a martingale.*
3. $A_1 = 0$, A_n *is $\mathcal{F}_{n-1}$-measurable for $n \geq 2$.*
4. A_n *is non-decreasing, that is*

$$A_n(\omega) \leq A_{n+1}(\omega)$$

almost surely for all $n \geq 1$.

If another pair of processes $\overline{M}_n, \overline{A}_n$ has the same properties, then $M_n = \overline{M}_n$, $A_n = \overline{A}_n$ almost surely.

Proof. Assuming that the processes M_n and A_n with the required properties exist, we can write for $n \geq 2$

$$X_{n-1} = M_{n-1} + A_{n-1},$$

$$X_n = M_n + A_n.$$

Taking the difference and then the conditional expectation with respect to $\mathcal{F}_{n-1}$, we obtain

$$\mathrm{E}(X_n | \mathcal{F}_{n-1}) - X_{n-1} = A_n - A_{n-1}.$$

This shows that A_n is uniquely defined by the process X_n and the random variable A_{n-1}. The random variable M_n is also uniquely defined, since $M_n = X_n - A_n$. Since $M_1 = X_1$ and $A_1 = 0$, we see, by induction on n, that the pair of processes M_n, A_n with the required properties is unique.

Furthermore, given a submartingale X_n, we can use the relations

$$M_1 = X_1, \quad A_1 = 0,$$

$$A_n = \mathrm{E}(X_n | \mathcal{F}_{n-1}) - X_{n-1} + A_{n-1}, \quad M_n = X_n - A_n, \quad n \geq 2,$$

to define inductively the processes M_n and A_n. Clearly, they have properties 1, 3 and 4. In order to verify property 2, we write

$$\mathrm{E}(M_n | \mathcal{F}_{n-1}) = \mathrm{E}(X_n - A_n | \mathcal{F}_{n-1}) = \mathrm{E}(X_n | \mathcal{F}_{n-1}) - A_n$$

$$= X_{n-1} - A_{n-1} = M_{n-1}, \quad n \geq 2,$$

which proves that $(M_n, \mathcal{F}_n)_{n \in \mathbb{N}}$ is a martingale. $\square$

If $(X_n, \mathcal{F}_n)$ is an adapted process and τ is a stopping time, then $X_{\tau(\omega)}(\omega)$ is a random variable measurable with respect to the σ-algebra $\mathcal{F}_\tau$. Indeed, one needs to check that $\{X_\tau \in B\} \cap \{\tau \leq n\} \in \mathcal{F}_n$ for any Borel set B of the real line and each n. This is true since τ takes only integer values and $\{X_m \in B\} \in \mathcal{F}_n$ for each $m \leq n$.

In order to develop an intuitive understanding of the next theorem, one can again think of a martingale as a model of a fair game. In a fair game, a gambler cannot increase or decrease the expectation of his fortune by entering the game at a point of time $\sigma(\omega)$, and then quitting the game at $\tau(\omega)$, provided that he decides to enter and leave the game based only on the information available by the time of the decision (that is, without looking into the future).

Theorem 13.18 (Optional Sampling Theorem). *If $(X_n, \mathcal{F}_n)_{n \in \mathbb{N}}$ is a submartingale and σ and τ are two stopping times such that $\sigma \leq \tau \leq k$ for some $k \in \mathbb{N}$, then*

$$X_\sigma \leq \mathrm{E}(X_\tau | \mathcal{F}_\sigma).$$

If $(X_n, \mathcal{F}_n)_{n \in \mathbb{N}}$ is a martingale or a supermartingale, then the same statement holds with the $\leq$ sign replaced by $=$ or $\geq$ respectively.

Proof. The case of $(X_n, \mathcal{F}_n)_{n \in \mathbb{N}}$ being a supermartingale is equivalent to considering the submartingale $(-X_n, \mathcal{F}_n)_{n \in \mathbb{N}}$. Thus, without loss of generality, we may assume that $(X_n, \mathcal{F}_n)_{n \in \mathbb{N}}$ is a submartingale.

Let $A \in \mathcal{F}_\sigma$. For $1 \leq m \leq n$ we define

$$A_m = A \cap \{\sigma = m\}, \quad A_{m,n} = A_m \cap \{\tau = n\},$$

$$B_{m,n} = A_m \cap \{\tau > n\}, \quad C_{m,n} = A_m \cap \{\tau \geq n\}.$$

Note that $B_{m,n} \in \mathcal{F}_n$, since $\{\tau > n\} = \Omega \setminus \{\tau \leq n\} \in \mathcal{F}_n$. Therefore, by definition of a submartingale,

$$\int_{B_{m,n}} X_n d\mathrm{P} \leq \int_{B_{m,n}} X_{n+1} d\mathrm{P}.$$

Since $C_{m,n} = A_{m,n} \bigcup B_{m,n}$,

$$\int_{C_{m,n}} X_n d\mathrm{P} \leq \int_{A_{m,n}} X_n d\mathrm{P} + \int_{B_{m,n}} X_{n+1} d\mathrm{P},$$

and thus, since $B_{m,n} = C_{m,n+1}$,

$$\int_{C_{m,n}} X_n d\mathrm{P} - \int_{C_{m,n+1}} X_{n+1} d\mathrm{P} \leq \int_{A_{m,n}} X_n d\mathrm{P}.$$

By taking the sum from $n = m$ to k, and noting that we have a telescopic sum on the left-hand side, we obtain

$$\int_{A_m} X_m dP \le \int_{A_m} X_\tau dP,$$

were we used that $A_m = C_{m,m}$. By taking the sum from $m = 1$ to k, we obtain

$$\int_A X_\sigma dP \le \int_A X_\tau dP.$$

Since $A \in \mathcal{F}_\sigma$ was arbitrary, this completes the proof of the theorem.   □

Definition 13.19. *A set of random variables $\{f_s\}_{s \in S}$ is said to be uniformly integrable if*

$$\lim_{\lambda \to \infty} \sup_{s \in S} \int_{\{|f_s| > \lambda\}} |f_s| dP = 0.$$

Remark 13.20. The Optional Sampling Theorem is, in general, not true for unbounded stopping times σ and τ. If, however, we assume that the random variables $X_n, n \in \mathbb{N}$, are uniformly integrable, then the theorem remains valid even for unbounded σ and τ.

Remark 13.21. There is an equivalent way to define uniform integrability (see Problem 9). Namely, a set of random variables $\{f_s\}_{s \in S}$ is uniformly integrable if

(1) There is a constant K such that $\int_\Omega |f_s| dP \le K$ for all $s \in S$, and
(2) For any $\varepsilon > 0$ one can find $\delta > 0$ such that $\int_A |f_s(\omega)| dP(\omega) \le \varepsilon$ for all $s \in S$, provided that $P(A) \le \delta$.

For a random process X_n and a constant $\lambda > 0$, we define the event $A(\lambda, n) = \{\omega : \max_{1 \le i \le n} X_i(\omega) \ge \lambda\}$. From the Chebyshev Inequality it follows that $\lambda P(\{X_n \ge \lambda\}) \le \mathrm{E}\max(X_n, 0)$. If $(X_n, \mathcal{F}_n)$ is a submartingale, we can make a stronger statement. Namely, we shall now use the Optional Sampling Theorem to show that the event $\{X_n \ge \lambda\}$ on the left-hand side can be replaced by $A(\lambda, n)$.

Theorem 13.22 (Doob Inequality). *If $(X_n, \mathcal{F}_n)$ is a submartingale, then for any $n \in \mathbb{N}$ and any $\lambda > 0$,*

$$\lambda P(A(\lambda, n)) \le \int_{A(\lambda, n)} X_n dP \le \mathrm{E}\max(X_n, 0).$$

Proof. We define the stopping time σ to be the first moment when $X_i \ge \lambda$ if $\max_{i \le n} X_i \ge \lambda$ and put $\sigma = n$ if $\max_{i \le n} X_i < \lambda$. The stopping time τ is defined simply as $\tau = n$. Since $\sigma \le \tau$, the Optional Sampling Theorem can be applied to the pair of stopping times σ and τ. Note that $A(\lambda, n) \in \mathcal{F}_\sigma$ since

$$A(\lambda, n) \bigcap \{\sigma \le m\} = \{\max_{i \le m} X_i \ge \lambda\} \in \mathcal{F}_m.$$

Therefore, since $X_\sigma \geq \lambda$ on $A(\lambda, n)$,

$$\lambda P(A(\lambda, n)) \leq \int_{A(\lambda,n)} X_\sigma dP \leq \int_{A(\lambda,n)} X_n dP \leq E \max(X_n, 0),$$

where the second inequality follows from the Optional Sampling Theorem. $\square$

Remark 13.23. Suppose that $\xi_1, \xi_2, \ldots$ is a sequence of independent random variables with finite mathematical expectations and variances, $m_i = E\xi_i$, $V_i = \text{Var}\xi_i$. One can obtain the Kolmogorov Inequality of Sect. 7.1 by applying Doob's Inequality to the submartingale $\zeta_n = (\xi_1 + \ldots + \xi_n - m_1 - \ldots - m_n)^2$.

13.6 Martingales with Continuous Time

In this section we shall formulate the statements of the Doob Decomposition, the Optional Sampling Theorem, and the Doob Inequality for continuous time martingales. The proofs of these results rely primarily on the corresponding statements for the case of martingales with discrete time. We shall not provide additional technical details, but interested readers may refer to "Brownian Motion and Stochastic Calculus" by I. Karatzas and S. Shreve for the complete proofs.

Before formulating the results, we introduce some new notations and definitions.

Given a filtration $(\mathcal{F}_t)_{t \in \mathbb{R}^+}$ on a probability space $(\Omega, \mathcal{F}, P)$, we define the filtration $(\mathcal{F}_{t+})_{t \in \mathbb{R}^+}$ as follows: $A \in \mathcal{F}_{t+}$ if and only if $A \in \mathcal{F}_{t+\delta}$ for any $\delta > 0$. We shall say that $(\mathcal{F}_t)_{t \in \mathbb{R}^+}$ is right-continuous if $\mathcal{F}_t = \mathcal{F}_{t+}$ for all $t \in \mathbb{R}^+$.

Recall that a set $A \subseteq \Omega$ is said to be P-negligible if there is an event $B \in \mathcal{F}$ such that $A \subseteq B$ and $P(B) = 0$.

We shall often impose the following technical assumption on our filtration.

Definition 13.24. *A filtration $(\mathcal{F}_t)_{t \in \mathbb{R}^+}$ is said to satisfy the usual conditions if it is right-continuous and all the P-negligible events from $\mathcal{F}$ belong to $\mathcal{F}_0$.*

We shall primarily be interested in processes whose every realization is right-continuous (right-continuous processes), or every realization is continuous (continuous processes). It will be clear that in the results stated below the assumption that a process is right-continuous (continuous) can be replaced by the assumption that the process is indistinguishable from a right-continuous (continuous) process.

Later we shall need the following lemma, which we state now without a proof. (A proof can be found in "Brownian Motion and Stochastic Calculus" by I. Karatzas and S. Shreve.)

Lemma 13.25. *Let $(X_t, \mathcal{F}_t)_{t \in \mathbb{R}^+}$ be a submartingale with filtration which satisfies the usual conditions. If the function $f : t \to EX_t$ from $\mathbb{R}^+$ to $\mathbb{R}$ is right-continuous, then there exists a right-continuous modification of the process X_t*

which is also adapted to the filtration $\mathcal{F}_t$ (and therefore is also a submartingale).

We formulate the theorem on the decomposition of continuous submartingales.

Theorem 13.26 (Doob-Meyer Decomposition). *Let $(X_t, \mathcal{F}_t)_{t \in \mathbb{R}^+}$ be a continuous submartingale with filtration which satisfies the usual conditions. Let S_a be the set of all stopping times bounded by a. Assume that for every $a > 0$ the set of random variables $\{X_\tau\}_{\tau \in S_a}$ is uniformly integrable. Then there exist two continuous random processes M_t and A_t such that:*

1. *$X_t = M_t + A_t$ for all $t \geq 0$ almost surely.*
2. *$(M_t, \mathcal{F}_t)_{t \in \mathbb{R}^+}$ is a martingale.*
3. *$A_0 = 0$, A_t is adapted to the filtration $\mathcal{F}_t$.*
4. *A_t is non-decreasing, that is $A_s(\omega) \leq A_t(\omega)$ if $s \leq t$ for every ω.*

If another pair of processes $\overline{M}_t, \overline{A}_t$ has the same properties, then M_t is indistinguishable from $\overline{M}_t$ and A_t is indistinguishable from $\overline{A}_t$.

We can also formulate the Optional Sampling Theorem for continuous time submartingales. If τ is a stopping time of a filtration $\mathcal{F}_t$, and the process X_t is adapted to the filtration $\mathcal{F}_t$ and right-continuous, then it is not difficult to show that X_τ is $\mathcal{F}_\tau$-measurable (see Problems 1 and 2 in Chap. 19).

Theorem 13.27 (Optional Sampling Theorem). *If $(X_t, \mathcal{F}_t)_{t \in \mathbb{R}^+}$ is a right-continuous submartingale, and σ and τ are two stopping times such that $\sigma \leq \tau \leq r$ for some $r \in \mathbb{R}^+$, then*

$$X_\sigma \leq \mathrm{E}(X_\tau | \mathcal{F}_\sigma).$$

If $(X_t, \mathcal{F}_t)_{t \in \mathbb{R}^+}$ is a either martingale or a supermartingale, then the same statement holds with the $\leq$ sign replaced by $=$ or $\geq$ respectively.

Remark 13.28. As in the case of discrete time, the Optional Sampling Theorem remains valid even for unbounded σ and τ if the random variables $X_t, t \in \mathbb{R}^+$, are uniformly integrable.

The proof of the following lemma relies on a simple application of the Optional Sampling Theorem.

Lemma 13.29. *If $(X_t, \mathcal{F}_t)_{t \in \mathbb{R}^+}$ is a right-continuous (continuous) martingale, τ is a stopping time of the filtration $\mathcal{F}_t$, and $Y_t = X_{t \wedge \tau}$, then $(Y_t, \mathcal{F}_t)_{t \in \mathbb{R}^+}$ is also a right-continuous (continuous) martingale.*

Proof. Let us show that $\mathrm{E}(Y_t - Y_s | \mathcal{F}_s) = 0$ for $s \leq t$. We have

$$\mathrm{E}(Y_t - Y_s | \mathcal{F}_s) = \mathrm{E}(X_{t \wedge \tau} - X_{s \wedge \tau} | \mathcal{F}_s) = \mathrm{E}((X_{(t \wedge \tau) \vee s} - X_s) | \mathcal{F}_s).$$

The expression on the right-hand side of this equality is equal to zero by the Optional Sampling Theorem. Since $t \wedge \tau$ is a continuous function of t, the right-continuity (continuity) of Y_t follows from the right-continuity (continuity) of X_t. □

Finally, we formulate the Doob Inequality for continuous time submartingales.

Theorem 13.30 (Doob Inequality). *If $(X_t, \mathcal{F}_t)$ is a right-continuous submartingale, then for any $t \in \mathbb{R}^+$ and any $\lambda > 0$*

$$\lambda P(A(\lambda, t)) \leq \int_{A(\lambda, t)} X_t dP \leq \operatorname{E} \max(X_t, 0),$$

where $A(\lambda, t) = \{\omega : \sup_{0 \leq s \leq t} X_s(\omega) \geq \lambda\}$.

13.7 Convergence of Martingales

We first discuss convergence of martingales with discrete time.

Definition 13.31. *A martingale $(X_n, \mathcal{F}_n)_{n \in \mathbb{N}}$ is said to be right-closable if there is a random variable $X_\infty \in L^1(\Omega, \mathcal{F}, P)$ such that $\operatorname{E}(X_\infty | \mathcal{F}_n) = X_n$ for all $n \in \mathbb{N}$.*

The random variable X_∞ is sometimes referred to as the last element of the martingale.

We can define $\mathcal{F}_\infty$ as the minimal σ-algebra containing $\mathcal{F}_n$ for all n. For a right-closable martingale we can define $X'_\infty = \operatorname{E}(X_\infty | \mathcal{F}_\infty)$. Then X'_∞ also serves as the last element since

$$\operatorname{E}(X'_\infty | \mathcal{F}_n) = \operatorname{E}(\operatorname{E}(X_\infty | \mathcal{F}_\infty) | \mathcal{F}_n) = \operatorname{E}(X_\infty | \mathcal{F}_n) = X_n.$$

Therefore, without loss of generality, we shall assume from now on that, for a right-closable martingale, the last element X_∞ is $\mathcal{F}_\infty$-measurable.

Theorem 13.32. *A martingale is right-closable if and only if it is uniformly integrable (that is the sequence of random variables $X_n, n \in \mathbb{N}$, is uniformly integrable).*

We shall only prove that a right-closable martingale is uniformly integrable. The proof of the converse statement is slightly more complicated, and we omit it here. Interested readers may find it in "Real Analysis and Probability" by R. M. Dudley.

Proof. We need to show that

$$\lim_{\lambda \to \infty} \sup_{n \in \mathbb{N}} \int_{\{|X_n| > \lambda\}} |X_n| dP = 0.$$

Since $|\cdot|$ is a convex function,

$$|X_n| = |\operatorname{E}(X_\infty | \mathcal{F}_n)| \leq \operatorname{E}(|X_\infty| | \mathcal{F}_n)$$

by the Conditional Jensen's Inequality. Therefore,

$$\int_{\{|X_n|>\lambda\}} |X_n|d\mathrm{P} \le \int_{\{|X_n|>\lambda\}} |X_\infty|d\mathrm{P}.$$

Since $|X_\infty|$ is integrable and the integral is absolutely continuous with respect to the measure P, it is sufficient to prove that

$$\lim_{\lambda\to\infty} \sup_{n\in\mathbb{N}} \mathrm{P}\{|X_n| > \lambda\} = 0.$$

By the Chebyshev Inequality,

$$\lim_{\lambda\to\infty} \sup_{n\in\mathbb{N}} \mathrm{P}\{|X_n| > \lambda\} \le \lim_{\lambda\to\infty} \sup_{n\in\mathbb{N}} \mathrm{E}|X_n|/\lambda \le \lim_{\lambda\to\infty} \mathrm{E}|X_\infty|/\lambda = 0,$$

which proves that a right-closable martingale is uniformly integrable.   □

The fact that a martingale is right-closable is sufficient to establish convergence in probability and in L^1.

Theorem 13.33 (Doob). *Let $(X_n, \mathcal{F}_n)_{n\in\mathbb{N}}$ be a right-closable martingale. Then*

$$\lim_{n\to\infty} X_n = X_\infty$$

almost surely and in $L^1(\Omega, \mathcal{F}, \mathrm{P})$.

Proof. (Due to C.W. Lamb.) Let $\mathcal{K} = \bigcup_{n\in\mathbb{N}} \mathcal{F}_n$. Let $\mathcal{G}$ be the collection of sets which can be approximated by sets from $\mathcal{K}$. Namely, $A \in \mathcal{G}$ if for any $\varepsilon > 0$ there is $B \in \mathcal{K}$ such that $\mathrm{P}(A \Delta B) < \varepsilon$. It is clear that $\mathcal{K}$ is a π-system, and that $\mathcal{G}$ is a Dynkin system. Therefore, $\mathcal{F}_\infty = \sigma(\mathcal{K}) \subseteq \mathcal{G}$ by Lemma 4.13.

Let F be the set of functions which are in $L^1(\Omega, \mathcal{F}, \mathrm{P})$ and are measurable with respect to $\mathcal{F}_n$ for some $n < \infty$. We claim that F is dense in $L^1(\Omega, \mathcal{F}_\infty, \mathrm{P})$. Indeed, any indicator function of a set from $\mathcal{F}_\infty$ can be approximated by elements of F, as we just demonstrated. Therefore, the same is true for finite linear combinations of indicator functions which, in turn, are dense in $L^1(\Omega, \mathcal{F}_\infty, \mathrm{P})$.

Since X_∞ is $\mathcal{F}_\infty$-measurable, for any $\varepsilon > 0$ we can find $Y_\infty \in F$ such that $\mathrm{E}|X_\infty - Y_\infty| \le \varepsilon^2$. Let $Y_n = \mathrm{E}(Y_\infty | \mathcal{F}_n)$. Then $(X_n - Y_n, \mathcal{F}_n)_{n\in\mathbb{N}}$ is a martingale. Therefore, $(|X_n - Y_n|, \mathcal{F}_n)_{n\in\mathbb{N}}$ is a submartingale, as shown in Sect. 13.4, and $\mathrm{E}|X_n - Y_n| \le \mathrm{E}|X_\infty - Y_\infty|$. By Doob's Inequality (Theorem 13.22),

$$\mathrm{P}(\sup_{n\in\mathbb{N}} |X_n - Y_n| > \varepsilon) \le \sup_{n\in\mathbb{N}} \mathrm{E}|X_n - Y_n|/\varepsilon \le \mathrm{E}|X_\infty - Y_\infty|/\varepsilon \le \varepsilon.$$

Note that $Y_n = Y_\infty$ for large enough n, since Y_∞ is $\mathcal{F}_n$ measurable for some finite n. Therefore,

$$\mathrm{P}(\limsup_{n\to\infty} X_n - Y_\infty > \varepsilon) \le \varepsilon \quad \text{and} \quad \mathrm{P}(\liminf_{n\to\infty} X_n - Y_\infty < -\varepsilon) \le \varepsilon.$$

Also, by the Chebyshev Inequality, $P(|X_\infty - Y_\infty| > \varepsilon) \leq \varepsilon$. Therefore,

$$P(\limsup_{n\to\infty} X_n - X_\infty > 2\varepsilon) \leq 2\varepsilon \quad \text{and} \quad P(\liminf_{n\to\infty} X_n - X_\infty < -2\varepsilon) \leq 2\varepsilon.$$

Since $\varepsilon > 0$ was arbitrary, this implies that $\lim_{n\to\infty} X_n = X_\infty$ almost surely.

As shown above, for each $\varepsilon > 0$ we have the inequalities $E|X_\infty - Y_\infty| \leq \varepsilon^2$, $E|X_n - Y_n| \leq E|X_\infty - Y_\infty|$, while $Y_n = Y_\infty$ for all sufficiently large n. This implies the convergence of X_n to X_∞ in $L^1(\Omega, \mathcal{F}, P)$. □

Example (Polya Urn Scheme). Consider an urn containing one black and one white ball. At time step n we take a ball randomly out of the urn and replace it with two balls of the same color.

More precisely, consider two processes A_n (number of black balls) and B_n (number of white balls). Then $A_0 = B_0 = 1$, and A_n, B_n, $n \geq 1$, are defined inductively as follows: $A_n = A_{n-1} + \xi_n$, $B_n = B_{n-1} + (1 - \xi_n)$, where ξ_n is a random variable such that

$$P(\xi_n = 1|\mathcal{F}_{n-1}) = \frac{A_{n-1}}{A_{n-1} + B_{n-1}}, \quad \text{and} \quad P(\xi_n = 0|\mathcal{F}_{n-1}) = \frac{B_{n-1}}{A_{n-1} + B_{n-1}},$$

and $\mathcal{F}_{n-1}$ is the σ-algebra generated by all A_k, B_k with $k \leq n - 1$. Let $X_n = A_n/(A_n + B_n)$ be the proportion of black balls. Let us show that $(X_n, \mathcal{F}_n)_{n \geq 0}$ is a martingale. Indeed,

$$E(X_n - X_{n-1}|\mathcal{F}_{n-1}) = E(\frac{A_n}{A_n + B_n} - \frac{A_{n-1}}{A_{n-1} + B_{n-1}}|\mathcal{F}_{n-1}) =$$

$$E(\frac{(A_{n-1} + B_{n-1})\xi_n - A_{n-1}}{(A_n + B_n)(A_{n-1} + B_{n-1})}|\mathcal{F}_{n-1}) =$$

$$\frac{1}{A_n + B_n}E(\xi_n - \frac{A_{n-1}}{A_{n-1} + B_{n-1}})|\mathcal{F}_{n-1}) = 0,$$

as is required of a martingale. Here we used that $A_n + B_n = A_{n-1} + B_{n-1} + 1$, and is therefore $\mathcal{F}_{n-1}$-measurable. The martingale $(X_n, \mathcal{F}_n)_{n \geq 0}$ is uniformly integrable, simply because X_n are bounded by one. Therefore, by Theorem 13.33, there is a random variable X_∞ such that $\lim_{n\to\infty} X_n = X_\infty$ almost surely.

We can actually write the distribution of X_∞ explicitly. The variable A_n can take integer values between 1 and $n + 1$. We claim that $P(A_n = k) = 1/(n + 1)$ for all $1 \leq k \leq n + 1$. Indeed, the statement is obvious for $n = 0$. For $n \geq 1$, by induction,

$$P(A_n = k) = P(A_{n-1} = k - 1; \xi_n = 1) + P(A_{n-1} = k; \xi_n = 0) =$$

$$\frac{1}{n} \cdot \frac{k-1}{n+1} + \frac{1}{n} \cdot \frac{n-k+1}{n+1} = \frac{1}{n+1}.$$

This means that $P(X_n = k/(n+2)) = 1/(n+1)$ for $1 \leq k \leq n+1$. Since the sequence X_n converges to X_∞ almost surely, it also converges in distribution. Therefore, the distribution of X_∞ is uniform on the interval $[0, 1]$.

If $(X_n, \mathcal{F}_n)_{n \in \mathbb{N}}$ is bounded in $L^1(\Omega, \mathcal{F}, \mathrm{P})$ (that is $\mathrm{E}|X_n| \leq c$ for some constant c and all n), we cannot claim that it is right-closable. Yet, the L^1-boundedness still guarantees almost sure convergence, although not necessarily to the last element of the martingale (which does not exist unless the martingale is uniformly integrable). We state the following theorem without a proof.

Theorem 13.34 (Doob). *Let $(X_n, \mathcal{F}_n)_{n \in \mathbb{N}}$ be a $L^1(\Omega, \mathcal{F}, \mathrm{P})$-bounded martingale. Then*

$$\lim_{n \to \infty} X_n = Y$$

almost surely, where Y is some random variable from $L^1(\Omega, \mathcal{F}, \mathrm{P})$.

Remark 13.35. Although the random variable Y belongs to $L^1(\Omega, \mathcal{F}, \mathrm{P})$, the sequence X_n need not converge to Y in $L^1(\Omega, \mathcal{F}, \mathrm{P})$.

Let us briefly examine the convergence of submartingales. Let $(X_n, \mathcal{F}_n)_{n \in \mathbb{N}}$ be an $L^1(\Omega, \mathcal{F}, \mathrm{P})$-bounded submartingale, and let $X_n = M_n + A_n$ be its Doob Decomposition. Then $\mathrm{E}A_n = \mathrm{E}(X_n - M_n) = \mathrm{E}(X_n - M_1)$. Thus, A_n is a monotonically non-decreasing sequence of random variables which is bounded in $L^1(\Omega, \mathcal{F}, \mathrm{P})$. By the Levi Monotonic Convergence Theorem, there exists the almost sure limit $A = \lim_{n \to \infty} A_n \in L^1(\Omega, \mathcal{F}, \mathrm{P})$.

Since A_n are bounded in $L^1(\Omega, \mathcal{F}, \mathrm{P})$, so too are M_n. Since A_n are non-negative random variables bounded from above by A, they are uniformly integrable. Therefore, if $(X_n, \mathcal{F}_n)_{n \in \mathbb{N}}$ is a uniformly integrable submartingale, then $(M_n, \mathcal{F}_n)_{n \in \mathbb{N}}$ is a uniformly integrable martingale. Upon gathering the above arguments, and applying Theorems 13.33 and 13.34, we obtain the following lemma.

Lemma 13.36. *Let a submartingale $(X_n, \mathcal{F}_n)_{n \in \mathbb{N}}$ be bounded in $L^1(\Omega, \mathcal{F}, \mathrm{P})$. Then*

$$\lim_{n \to \infty} X_n = Y$$

almost surely, where Y is some random variable from $L^1(\Omega, \mathcal{F}, \mathrm{P})$. If X_n are uniformly integrable, then the convergence is also in $L^1(\Omega, \mathcal{F}, \mathrm{P})$.

Although our discussion of martingale convergence has been focused so far on martingales with discrete time, the same results carry over to the case of right-continuous martingales with continuous time. In Definition 13.31 and Theorems 13.33 and 13.34 we only need to replace the parameter $n \in \mathbb{N}$ by $t \in \mathbb{R}^+$. Since the proof of Lemma 13.36 in the continuous time case relies on the Doob-Meyer Decomposition, in order to make it valid in the continuous time case, we must additionally assume that the filtration satisfies the usual conditions and that the submartingale is continuous.

13.8 Problems

1. Let $g : \mathbb{R} \to \mathbb{R}$ be a measurable function which is not convex. Show that there is a random variable f on some probability space such that $E|f| < \infty$ and $-\infty < Eg(f) < g(Ef) < \infty$.
2. Let ξ and η be two random variables with finite expectations such that $E(\xi|\eta) \geq \eta$ and $E(\eta|\xi) \geq \xi$. Prove that $\xi = \eta$ almost surely.
3. Let $(X_n, \mathcal{F}_n)_{n \in \mathbb{N}}$ be a square-integrable martingale with $EX_1 = 0$. Show that for each $c > 0$

$$P(\max_{1 \leq i \leq n} X_i \geq c) \leq \frac{\mathrm{Var}(X_n)}{\mathrm{Var}(X_n) + c^2}.$$

4. Let $(\xi_1, \ldots, \xi_n)$ be a Gaussian vector with zero mean and covariance matrix B. Find the distribution of the random variable $E(\xi_1|\xi_2, \ldots, \xi_n)$.
5. Let $A = \{(x, y) \in \mathbb{R}^2 : |x - y| < a, |x + y| < b\}$, where $a, b > 0$. Assume that the random vector (ξ_1, ξ_2) is uniformly distributed on A. Find the distribution of $E(\xi_1|\xi_2)$.
6. Let ξ_1, ξ_2, ξ_3 be independent identically distributed bounded random variables with density $p(x)$. Find the distribution of

$$E(\max(\xi_1, \xi_2, \xi_3)| \min(\xi_1, \xi_2, \xi_3))$$

in terms of the density p.

7. Prove that if σ and τ are stopping times of a filtration $\mathcal{F}_t$, then so is $\sigma \vee \tau$. If, in addition, $\sigma, \tau \geq 0$, then $\sigma + \tau$ is a stopping time.
8. Let $\xi_1, \xi_2, \ldots$ be independent $N(0, 1)$ distributed random variables. Let $S_n = \xi_1 + \ldots + \xi_n$ and $X_n = e^{S_n - n/2}$. Let $\mathcal{F}_n^X$ be the σ-algebra generated by $X_1, \ldots, X_n$. Prove that $(X_n, \mathcal{F}_n^X)_{n \in \mathbb{N}}$ is a martingale.
9. Prove that the definition of uniform integrability given in Remark 13.21 is equivalent to Definition 13.19.
10. A man tossing a coin wins one point for heads and five points for tails. The game stops when the man accumulates at least 1,000 points. Estimate with an accuracy ± 2 the expectation of the length of the game.
11. Let X_n be a process adapted to a filtration $\mathcal{F}_n$, $n \in \mathbb{N}$. Let $M > 0$ and $\tau(\omega) = \min(n : |X_n(\omega)| \geq M)$ (where $\tau(\omega) = \infty$ if $|X_n(\omega)| < M$ for all n). Prove that τ is a stopping time of the filtration $\mathcal{F}_n$.
12. Let a martingale $(X_n, \mathcal{F}_n)_{n \in \mathbb{N}}$ be uniformly integrable. Let the stopping time τ be defined as in the previous problem. Prove that $(X_{n \wedge \tau}, \mathcal{F}_{n \wedge \tau})_{n \in \mathbb{N}}$ is a uniformly integrable martingale.
13. Let N_n, $n \geq 1$, be the size of a population of bacteria at time step n. At each time step each bacteria produces a number of offspring and dies. The number of offspring is independent for each bacteria and is distributed according to the Poisson law with parameter $\lambda = 2$. Assuming that $N_1 = a > 0$, find the probability that the population will eventually die, that is find $P(N_n = 0 \text{ for some } n \geq 1)$. (Hint: find c such that $\exp(-cN_n)$ is a martingale.)

14. Ann and Bob are gambling at a casino. In each game the probability of winning a dollar is 48%, and the probability of loosing a dollar is 52%. Ann decided to play 20 games, but will stop after 2 games if she wins them both. Bob decided to play 20 games, but will stop after 10 games if he wins at least 9 out of the first 10. What is larger: the amount of money Ann is expected to loose, or the amount of money Bob is expected to loose?

15. Let $(X_t, \mathcal{F}_t)_{t \in \mathbb{R}}$ be a martingale with continuous realizations. For $0 \leq s \leq t$, find $\mathrm{E}(\int_0^t X_u du | \mathcal{F}_s)$.

16. Consider an urn containing A_0 black balls and B_0 white balls. At time step n we take a ball randomly out of the urn and replace it with two balls of the same color. Let X_n denote the proportion of the black balls. Prove that X_n converges almost surely, and find the distribution of the limit.

14

Markov Processes with a Finite State Space

14.1 Definition of a Markov Process

In this section we define a homogeneous Markov process with values in a finite state space. We can assume that the state space X is the set of the first r positive integers, that is $X = \{1, \ldots, r\}$.

Let $P(t)$ be a family of $r \times r$ stochastic matrices indexed by the parameter $t \in [0, \infty)$. The elements of $P(t)$ will be denoted by $P_{ij}(t)$, $1 \leq i, j \leq r$. We assume that the family $P(t)$ forms a semi-group, that is $P(s)P(t) = P(s+t)$ for any $s, t \geq 0$. Since $P(t)$ are stochastic matrices, the semi-group property implies that $P(0)$ is the identity matrix. Let μ be a distribution on X.

Let $\widetilde{\Omega}$ be the set of all functions $\widetilde{\omega} : \mathbb{R}^+ \to X$ and $\mathcal{B}$ be the σ-algebra generated by all the cylindrical sets. Define a family of finite-dimensional distributions $P_{t_0, \ldots, t_k}$, where $0 = t_0 \leq t_1 \leq \ldots \leq t_k$, as follows

$$P_{t_0, \ldots, t_k}(\widetilde{\omega}(t_0) = i_0, \widetilde{\omega}(t_1) = i_1, \ldots, \widetilde{\omega}(t_k) = i_k)$$

$$= \mu_{i_0} P_{i_0 i_1}(t_1) P_{i_1 i_2}(t_2 - t_1) \ldots P_{i_{k-1} i_k}(t_k - t_{k-1}).$$

It can be easily seen that this family of finite-dimensional distributions satisfies the consistency conditions. By the Kolmogorov Consistency Theorem, there is a process X_t with values in X with these finite-dimensional distributions. Any such process will be called a homogeneous Markov process with the family of transition matrices $P(t)$ and the initial distribution μ. (Since we do not consider non-homogeneous Markov processes in this section, we shall refer to X_t simply as a Markov process).

Lemma 14.1. *Let X_t be a Markov process with the family of transition matrices $P(t)$. Then, for $0 \leq s_1 \leq \ldots \leq s_k$, $t \geq 0$, and $i_1, \ldots, i_k, j \in X$, we have*

$$P(X_{s_k+t} = j | X_{s_1} = i_1, \ldots, X_{s_k} = i_k) = P(X_{s_k+t} = j | X_{s_k} = i_k) = P_{i_k j}(t) \tag{14.1}$$

if the conditional probability on the left-hand side is defined.

L. Koralov and Y.G. Sinai, *Theory of Probability and Random Processes*,   201
Universitext, DOI 10.1007/978-3-540-68829-7_14,
© Springer-Verlag Berlin Heidelberg 2012

The proof of this lemma is similar to the arguments in Sect. 5.2, and thus will not be provided here. As in Sect. 5.2, it is easy to see that for a Markov process with the family of transition matrices $P(t)$ and the initial distribution μ the distribution of X_t is $\mu P(t)$.

Definition 14.2. *A distribution π is said to be stationary for a semi-group of Markov transition matrices $P(t)$ if $\pi P(t) = \pi$ for all $t \geq 0$.*

As in the case of discrete time we have the Ergodic Theorem.

Theorem 14.3. *Let $P(t)$ be a semi-group of Markov transition matrices such that for some t all the matrix entries of $P(t)$ are positive. Then there is a unique stationary distribution π for the semi-group of transition matrices. Moreover, $\sup_{i,j \in X} |P_{ij}(t) - \pi_j|$ converges to zero exponentially fast as $t \to \infty$.*

This theorem can be proved similarly to the Ergodic Theorem for Markov chains (Theorem 5.9). We leave the details as an exercise for the reader.

14.2 Infinitesimal Matrix

In this section we consider semi-groups of Markov transition matrices which are differentiable at zero. Namely, assume that there exist the following limits

$$Q_{ij} = \lim_{t \downarrow 0} \frac{P_{ij}(t) - I_{ij}}{t}, \quad 1 \leq i, j \leq r, \tag{14.2}$$

where I is the identity matrix.

Definition 14.4. *If the limits in (14.2) exist for all $1 \leq i, j \leq r$, then the matrix Q is called the infinitesimal matrix of the semigroup $P(t)$.*

Since $P_{ij}(t) \geq 0$ and $I_{ij} = 0$ for $i \neq j$, the off-diagonal elements of Q are non-negative. Moreover,

$$\sum_{j=1}^{r} Q_{ij} = \sum_{j=1}^{r} \lim_{t \downarrow 0} \frac{P_{ij}(t) - I_{ij}}{t} = \lim_{t \downarrow 0} \frac{\sum_{j=1}^{r} P_{ij}(t) - 1}{t} = 0,$$

or, equivalently,

$$Q_{ii} = -\sum_{j \neq i} Q_{ij}.$$

Lemma 14.5. *If the limits in (14.2) exist, then the transition matrices are differentiable for all $t \in \mathbb{R}^+$ and satisfy the following systems of ordinary differential equations.*

$$\frac{dP(t)}{dt} = P(t)Q \quad (forward \ system).$$

$$\frac{dP(t)}{dt} = QP(t) \quad (backward \ system).$$

The derivatives at $t = 0$ should be understood as one-sided derivatives.

Proof. Due to the semi-group property of $P(t)$,

$$\lim_{h \downarrow 0} \frac{P(t+h) - P(t)}{h} = P(t) \lim_{h \downarrow 0} \frac{P(h) - I}{h} = P(t)Q. \tag{14.3}$$

This shows, in particular, that $P(t)$ is right-differentiable. Let us prove that $P(t)$ is left-continuous. For $t > 0$ and $0 \leq h \leq t$,

$$P(t) - P(t-h) = P(t-h)(P(h) - I).$$

All the elements of $P(t-h)$ are bounded, while all the elements of $(P(h) - I)$ tend to zero as $h \downarrow 0$. This establishes the continuity of $P(t)$.

For $t > 0$,

$$\lim_{h \downarrow 0} \frac{P(t) - P(t-h)}{h} = \lim_{h \downarrow 0} P(t-h) \lim_{h \downarrow 0} \frac{P(h) - I}{h} = P(t)Q. \tag{14.4}$$

Combining (14.3) and (14.4), we obtain the forward system of equations.

Due to the semi-group property of $P(t)$, for $t \geq 0$,

$$\lim_{h \downarrow 0} \frac{P(t+h) - P(t)}{h} = \lim_{h \downarrow 0} \frac{P(h) - I}{h} P(t) = QP(t),$$

and similarly, for $t > 0$,

$$\lim_{h \downarrow 0} \frac{P(t) - P(t-h)}{h} = \lim_{h \downarrow 0} \frac{P(h) - I}{h} \lim_{h \downarrow 0} P(t-h) = QP(t).$$

This justifies the backward system of equations.   □

The system $dP(t)/dt = P(t)Q$ with the initial condition $P_0 = I$ has the unique solution $P(t) = \exp(tQ)$. Thus, the transition matrices can be uniquely expressed in terms of the infinitesimal matrix.

Let us note another property of the infinitesimal matrix. If π is a stationary distribution for the semi-group of transition matrices $P(t)$, then

$$\pi Q = \lim_{t \downarrow 0} \frac{\pi P(t) - \pi}{t} = 0.$$

Conversely, if $\pi Q = 0$ for some distribution π, then

$$\pi P(t) = \pi \exp(tQ) = \pi(I + tQ + \frac{t^2 Q^2}{2!} + \frac{t^3 Q^3}{3!} + \cdots) = \pi.$$

Thus, π is a stationary distribution for the family $P(t)$.

14.3 A Construction of a Markov Process

Let μ be a probability distribution on X and $P(t)$ be a differentiable semi-group of transition matrices with the infinitesimal matrix Q. Assume that $Q_{ii} < 0$ for all i.

On an intuitive level, a Markov process with the family of transition matrices $P(t)$ and initial distribution μ can be described as follows. At time $t = 0$ the process is distributed according to μ. If at time t the process is in a state i, then it will remain in the same state for time τ, where τ is a random variable with exponential distribution. The parameter of the distribution depends on i, but does not depend on t. After time τ the process goes to another state, where it remains for exponential time, and so on. The transition probabilities depend on i, but not on the moment of time t.

Now let us justify the above description and relate the transition times and transition probabilities to the infinitesimal matrix. Let Q be an $r \times r$ matrix with $Q_{ii} < 0$ for all i. Assume that there are random variables ξ, τ_i^n, $1 \leq i \leq r$, $n \in \mathbb{N}$, and η_i^n, $1 \leq i \leq r$, $n \in \mathbb{N}$, defined on a common probability space, with the following properties.

1. The random variable ξ takes values in X and has distribution μ.
2. For any $1 \leq i \leq r$, the random variables τ_i^n, $n \in \mathbb{N}$, are identically distributed according to the exponential distribution with parameter $r_i = -Q_{ii}$.
3. For any $1 \leq i \leq r$, the random variables η_i^n, $n \in \mathbb{N}$, take values in $X \setminus \{i\}$ and are identically distributed with $\mathrm{P}(\eta_i^n = j) = -Q_{ij}/Q_{ii}$ for $j \neq i$.
4. The random variables ξ, τ_i^n, η_i^n, $1 \leq i \leq r$, $n \in \mathbb{N}$, are independent.

We inductively define two sequences of random variables: σ^n, $n \geq 0$, with values in $\mathbb{R}^+$, and ξ^n, $n \geq 0$, with values in X. Let $\sigma^0 = 0$ and $\xi^0 = \xi$. Assume that σ^m and ξ^m have been defined for all $m < n$, where $n \geq 1$, and set

$$\sigma^n = \sigma^{n-1} + \tau_{\xi^{n-1}}^n.$$

$$\xi^n = \eta_{\xi^{n-1}}^n.$$

We shall treat σ^n as the time till the n-th transition takes place, and ξ^n as the n-th state visited by the process. Thus, define

$$X_t = \xi^n \quad \text{for} \quad \sigma^n \leq t < \sigma^{n+1}. \tag{14.5}$$

Lemma 14.6. *Assume that the random variables ξ, τ_i^n, $1 \leq i \leq r$, $n \in \mathbb{N}$, and η_i^n, $1 \leq i \leq r$, $n \in \mathbb{N}$, are defined on a common probability space and satisfy assumptions 1–4 above. Then the process X_t defined by (14.5) is a Markov process with the family of transition matrices $P(t) = \exp(tQ)$ and initial distribution μ.*

Sketch of the Proof. It is clear from (14.5) that the initial distribution of X_t is μ. Using the properties of τ_i^n and η_i^n it is possible to show that, for $k \neq j$,

$$P(X_0 = i, X_t = k, X_{t+h} = j)$$

$$= P(X_0 = i, X_t = k)(P(\tau_k^1 < h)P(\xi_k^1 = j) + o(h))$$

$$= P(X_0 = i, X_t = k)(Q_{kj}h + o(h)) \quad \text{as} \quad h \downarrow 0.$$

In other words, the main contribution to the probability on the left-hand side comes from the event that there is exactly one transition between the states k and j during the time interval $[t, t+h)$.

Similarly,

$$P(X_0 = i, X_t = j, X_{t+h} = j)$$

$$= P(X_0 = i, X_t = j)(P(\tau_j^1 \geq h) + o(h))$$

$$= P(X_0 = i, X_t = j)(1 + Q_{jj}h + o(h)) \quad \text{as} \quad h \downarrow 0,$$

that is, the main contribution to the probability on the left-hand side comes from the event that there are no transitions during the time interval $[t, t+h]$.

Therefore,

$$\sum_{k=1}^{r} P(X_0 = i, X_t = k, X_{t+h} = j)$$

$$= P(X_0 = i, X_t = j) + h \sum_{k=1}^{r} P(X_0 = i, X_t = k)Q_{kj} + o(h).$$

Let $R_{ij}(t) = P(X_0 = i, X_t = j)$. The last equality can be written as

$$R_{ij}(t+h) = R_{ij}(t) + h \sum_{k=1}^{r} R_{ik}(t)Q_{kj} + o(h).$$

Using matrix notation,

$$\lim_{h \downarrow 0} \frac{R(t+h) - R(t)}{h} = R(t)Q.$$

The existence of the left derivative is justified similarly. Therefore,

$$\frac{dR(t)}{dt} = R(t)Q \quad \text{for} \quad t \geq 0.$$

Note that $R_{ij}(0) = \mu_i$ for $i = j$, and $R_{ij}(0) = 0$ for $i \neq j$. These are the same equation and initial condition that are satisfied by the matrix-valued function $\mu_i P_{ij}(t)$. Therefore,

$$R_{ij}(t) = P(X_0 = i, X_t = j) = \mu_i P_{ij}(t). \tag{14.6}$$

In order to prove that X_t is a Markov process with the family of transition matrices $P(t)$, it is sufficient to demonstrate that

$$P(X_{t_0} = i_0, X_{t_1} = i_1, \ldots, X_{t_k} = i_k)$$

$$= \mu_{i_0} P_{i_0 i_1}(t_1) P_{i_1 i_2}(t_2 - t_1) \ldots P_{i_{k-1} i_k}(t_k - t_{k-1}).$$

for $0 = t_0 \leq t_1 \leq \ldots \leq t_k$. The case $k = 1$ has been covered by (14.6). The proof for $k > 1$ is similar and is based on induction on k. $\square$

14.4 A Problem in Queuing Theory

Markov processes with a finite or countable state space are used in the Queuing Theory. In this section we consider one basic example.

Assume that there are r identical devices designed to handle incoming requests. The times between consecutive requests are assumed to be independent exponentially distributed random variables with parameter λ. At a given time, each device may be either free or busy servicing one request. An incoming request is serviced by any of the free devices and, if all the devices are busy, the request is rejected. The times to service each request are assumed to be independent exponentially distributed random variables with parameter μ. They are also assumed to be independent of the arrival times of the requests.

Let us model the above system by a process with the state space $X = \{0, 1, \ldots, r\}$. A state of the process corresponds to the number of devices busy servicing requests. If there are no requests in the system, the time till the first one arrives is exponential with parameter λ. If there are r requests in the system, the time till the first one of them is serviced is an exponential random variable with parameter $r\mu$. If there are $1 \leq i \leq r - 1$ requests in the system, the time till either one of them is serviced, or a new request arrives, is an exponential random variable with parameter $\lambda + i\mu$. Therefore, the process remains in a state i for a time which is exponentially distributed with parameter

$$\gamma(i) = \begin{cases} \lambda & \text{if } i = 0, \\ \lambda + i\mu & \text{if } 1 \leq i \leq r - 1, \\ i\mu & \text{if } i = r. \end{cases}$$

If the process is in the state $i = 0$, it can only make a transition to the state $i = 1$, which corresponds to an arrival of a request. From a state $1 \leq i \leq r - 1$ the process can make a transition either to state $i-1$ or to state $i+1$. The former corresponds to completion of one of i requests being serviced before the arrival of a new request. Therefore, the probability of transition from i to $i - 1$ is equal to the probability that the smallest of i exponential random variables with parameter μ is less than an exponential random variable with parameter λ (all the random variables are independent). This probability is equal to $i\mu/(i\mu + \lambda)$. Consequently, the transition probability from i to $i+1$ is equal to $\lambda/(i\mu+\lambda)$. Finally, if the process is in the state r, it can only make a transition to the state $r-1$.

Let the initial state of the process X_t be independent of the arrival times of the requests and the times it takes to service the requests. Then the process X_t satisfies the assumptions of Lemma 14.6 (see the discussion before Lemma 14.6). The matrix Q is the $(r+1) \times (r+1)$ tridiagonal matrix with the vectors $\gamma(i)$, $0 \leq i \leq r$, on the diagonal, $u(i) \equiv \lambda$, $1 \leq i \leq r$, above the diagonal, and $l(i) = i\mu$, $1 \leq i \leq r$, below the diagonal. By Lemma 14.6, the process X_t is Markov with the family of transition matrices $P(t) = \exp(tQ)$.

It is not difficult to prove that all the entries of $\exp(tQ)$ are positive for some t, and therefore the Ergodic Theorem is applicable. Let us find the stationary distribution for the family of transition matrices $P(t)$. As noted in Sect. 14.2, a distribution π is stationary for $P(t)$ if and only if $\pi Q = 0$. It is easy to verify that the solution of this linear system, subject to the conditions $\pi(i) \geq 0$, $0 \leq i \leq r$, and $\sum_{i=0}^{r} \pi(i) = 1$, is

$$\pi(i) = \frac{(\lambda/\mu)^i/i!}{\sum_{j=0}^{r}(\lambda/\mu)^j/j!}, \quad 0 \leq i \leq r.$$

14.5 Problems

1. Let $P(t)$ be a differentiable semi-group of Markov transition matrices with the infinitesimal matrix Q. Assume that $Q_{ij} \neq 0$ for $1 \leq i,j \leq r$. Prove that for every $t > 0$ all the matrix entries of $P(t)$ are positive. Prove that there is a unique stationary distribution π for the semi-group of transition matrices. (Hint: represent Q as $(Q + cI) - cI$ with a constant c sufficiently large so that to make all the elements of the matrix $Q + cI$ non-negative.)

2. Let $P(t)$ be a differentiable semi-group of transition matrices. Prove that if all the elements of $P(t)$ are positive for some t, then all the elements of $P(t)$ are positive for all $t > 0$.

3. Let $P(t)$ be a differentiable semi-group of Markov transition matrices with the infinitesimal matrix Q. Assuming that Q is self-adjoint, find a stationary distribution for the semi-group $P(t)$.

4. Let X_t be a Markov process with a differentiable semi-group of transition matrices and initial distribution μ such that $\mu(i) > 0$ for $1 \leq i \leq r$. Prove that $\mathrm{P}(X_t = i) > 0$ for all i.

5. Consider a taxi station where taxis and customers arrive according to Poisson processes. The taxis arrive at the rate of 1 per min, and the customers at the rate of 2 per min. A taxi will wait only if there are no other taxis waiting already. A customer will wait no matter how many other customers are in line. Find the probability that there is a taxi waiting at a given moment and the average number of customers waiting in line.

6. A company gets an average of five calls an hour from prospective clients. It takes a company representative an average of 20 min to handle one call (the distribution of time to handle one call is exponential). A prospective client who cannot immediately talk to a representative never calls again. For each prospective client that talks to a representative the company makes $1,000. How many representatives should the company maintain if each is paid $10 an hour?

Wide-Sense Stationary Random Processes

15.1 Hilbert Space Generated by a Stationary Process

Let $(\Omega, \mathcal{F}, \mathrm{P})$ be a probability space. Consider a complex-valued random process X_t on this probability space, and assume that $\mathrm{E}|X_t|^2 < \infty$ for all $t \in T$ ($T = \mathbb{R}$ or $\mathbb{Z}$).

Definition 15.1. *A random process X_t is called wide-sense stationary if there exist a constant m and a function $b(t)$, $t \in T$, called the expectation and the covariance of the random process, respectively, such that $\mathrm{E}X_t = m$ and $\mathrm{E}(X_t \overline{X}_s) = b(t-s)$ for all $t, s \in T$.*

This means that the expectation of random variables X_t is constant, and the covariance depends only on the distance between the points on the time axis. In the remaining part of this section we shall assume that $\mathrm{E}X_t \equiv 0$, the general case requiring only trivial modifications.

Let $\widehat{H}$ be the subspace of $L^2(\Omega, \mathcal{F}, \mathrm{P})$ consisting of functions which can be represented as finite linear combinations of the form $\xi = \sum_{s \in S} c_s X_s$ with complex coefficients c_s. Here S is an arbitrary finite subset of T, and the equality is understood in the sense of $L^2(\Omega, \mathcal{F}, \mathrm{P})$. Thus $\widehat{H}$ is a vector space over the field of complex numbers. The inner product on $\widehat{H}$ is induced from $L^2(\Omega, \mathcal{F}, \mathrm{P})$. Namely, for $\xi = \sum_{s \in S_1} c_s X_s$ and $\eta = \sum_{s \in S_2} d_s X_s$,

$$(\xi, \eta) = \mathrm{E}(\xi \overline{\eta}) = \sum_{s_1 \in S_1} \sum_{s_2 \in S_2} c_{s_1} \overline{d}_{s_2} \mathrm{E}(X_{s_1} \overline{X}_{s_2}).$$

In particular, $(X_s, X_s) = \mathrm{E}|X_s|^2$. Let H denote the closure of $\widehat{H}$ with respect to this inner product. Thus $\xi \in H$ if one can find a Cauchy sequence $\xi_n \in \widehat{H}$ such that $\mathrm{E}|\xi - \xi_n|^2 \to 0$ as $n \to \infty$. In particular, the sum of an infinite series $\sum c_s X_s$ (if the series converges) is contained in H. In general, however, not every $\xi \in H$ can be represented as an infinite sum of this form. Note that $\mathrm{E}\xi = 0$ for each $\xi \in H$ since the same is true for all elements of $\widehat{H}$.

L. Koralov and Y.G. Sinai, *Theory of Probability and Random Processes*, Universitext, DOI 10.1007/978-3-540-68829-7_15, © Springer-Verlag Berlin Heidelberg 2012

Definition 15.2. *The space H is called the Hilbert space generated by the random process X_t.*

We shall now define a family of operators U^t on the Hilbert space generated by a wide-sense stationary random process. The operator U^t is first defined on the elements of $\widehat{H}$ as follows:

$$U^t \sum_{s \in S} c_s X_s = \sum_{s \in S} c_s X_{s+t}. \tag{15.1}$$

This definition will make sense if we show that $\sum_{s \in S_1} c_s X_s = \sum_{s \in S_2} d_s X_s$ implies that $\sum_{s \in S_1} c_s X_{s+t} = \sum_{s \in S_2} d_s X_{s+t}$.

Lemma 15.3. *The operators U^t are correctly defined and preserve the inner product, that is $(U^t \xi, U^t \eta) = (\xi, \eta)$ for $\xi, \eta \in \widehat{H}$.*

Proof. Since the random process is stationary,

$$\left(\sum_{s \in S_1} c_s X_{s+t}, \sum_{s \in S_2} d_s X_{s+t} \right) = \sum_{s_1 \in S_1, s_2 \in S_2} c_{s_1} \overline{d}_{s_2} \mathrm{E}(X_{s_1+t} \overline{X}_{s_2+t})$$

$$= \sum_{s_1 \in S_1, s_2 \in S_2} c_{s_1} \overline{d}_{s_2} \mathrm{E}(X_{s_1} \overline{X}_{s_2}) = \left(\sum_{s \in S_1} c_s X_s, \sum_{s \in S_2} d_s X_s \right). \tag{15.2}$$

If $\sum_{s \in S_1} c_s X_s = \sum_{s \in S_2} d_s X_s$, then

$$\left(\sum_{s \in S_1} c_s X_{s+t} - \sum_{s \in S_2} d_s X_{s+t}, \sum_{s \in S_1} c_s X_{s+t} - \sum_{s \in S_2} d_s X_{s+t} \right)$$

$$= \left(\sum_{s \in S_1} c_s X_s - \sum_{s \in S_2} d_s X_s, \sum_{s \in S_1} c_s X_s - \sum_{s \in S_2} d_s X_s \right) = 0,$$

that is, the operator U^t is well-defined. Furthermore, for $\xi = \sum_{s \in S_1} c_s X_s$ and $\eta = \sum_{s \in S_2} d_s X_s$, the equality (15.2) implies $(U^t \xi, U^t \eta) = (\xi, \eta)$, that is U^t preserves the inner product. $\square$

Recall the following definition.

Definition 15.4. *Let H be a Hilbert space. A linear operator $U : H \to H$ is called unitary if it is a bijection that preserves the inner product, i.e., $(U\xi, U\eta) = (\xi, \eta)$ for all $\xi, \eta \in H$.*

The inverse operator U^{-1} is then also unitary and $U^* = U^{-1}$, where U^* is the adjoint operator.

In our case, both the domain and the range of U^t are dense in H for any $t \in T$. Since U^t preserves the inner product, it can be extended by continuity from $\widehat{H}$ to H, and the extension, also denoted by U^t, is a unitary operator.

By (15.1), the operators U^t on $\widehat{H}$ form a group, that is U^0 is the identity operator and $U^t U^s = U^{t+s}$. By continuity, the same is true for operators U^t on H.

15.2 Law of Large Numbers for Stationary Random Processes

Let X_n be a wide-sense stationary random process with discrete time. As before, we assume that $EX_n \equiv 0$. Consider the time averages

$$(X_k + X_{k+1} + \ldots + X_{k+n-1})/n,$$

which clearly belong to $\widehat{H}$. In the case of discrete time we shall use the notation $U = U^1$.

Theorem 15.5 (Law of Large Numbers). *There exists $\eta \in H$ such that*

$$\lim_{n \to \infty} \frac{X_k + \ldots + X_{k+n-1}}{n} = \eta \quad (in \ H)$$

for all k. The limit η does not depend on k and is invariant under U, that is $U\eta = \eta$.

We shall derive Theorem 15.5 from the so-called von Neumann Ergodic Theorem for unitary operators.

Theorem 15.6 (von Neumann Ergodic Theorem). *Let U be a unitary operator in a Hilbert space H. Let P be the orthogonal projection onto the subspace $H_0 = \{\varphi : \varphi \in H, U\varphi = \varphi\}$. Then for any $\xi \in H$,*

$$\lim_{n \to \infty} \frac{\xi + \ldots + U^{n-1}\xi}{n} = P\xi. \tag{15.3}$$

Proof. If $\xi \in H_0$, then (15.3) is obvious with $P\xi = \xi$. If ξ is of the form $\xi = U\xi_1 - \xi_1$ with some $\xi_1 \in H$, then

$$\lim_{n \to \infty} \frac{\xi + \ldots + U^{n-1}\xi}{n} = \lim_{n \to \infty} \frac{U^n \xi_1 - \xi_1}{n} = 0.$$

Furthermore, $P\xi = 0$. Indeed, take any $\alpha \in H_0$. Since $\alpha = U\alpha$, we have

$$(\xi, \alpha) = (U\xi_1 - \xi_1, \alpha) = (U\xi_1, \alpha) - (\xi_1, \alpha) = (U\xi_1, U\alpha) - (\xi_1, \alpha) = 0.$$

Therefore, the statement of the theorem holds for all ξ of the form $\xi = U\xi_1 - \xi_1$.

The next step is to show that, if $\xi^{(r)} \to \xi$ and the statement of the theorem is valid for each $\xi^{(r)}$, then it is valid for ξ.

Indeed, let $\eta^{(r)} = P\xi^{(r)}$. Take any $\varepsilon > 0$ and find r such that $||\xi^{(r)} - \xi|| \leq \varepsilon/3$. Then $||\eta^{(r)} - P\xi|| = ||P(\xi^{(r)} - \xi)|| \leq ||\xi^{(r)} - \xi|| \leq \varepsilon/3$. Therefore,

$$||\frac{\xi + \ldots + U^{n-1}\xi}{n} - P\xi|| \leq ||\frac{\xi^{(r)} + \ldots + U^{n-1}\xi^{(r)}}{n} - \eta^{(r)}||$$

$$+ \frac{1}{n}\left(||\xi^{(r)} - \xi|| + ||U\xi^{(r)} - U\xi|| + \ldots + ||U^{n-1}\xi^{(r)} - U^{n-1}\xi|| \right) + ||\eta^{(r)} - P\xi||,$$

which can be made smaller than ε by selecting sufficiently large n.

Now let us finish the proof of the theorem. Take an arbitrary $\xi \in H$ and write $\xi = P\xi + \xi_1$, where $\xi_1 \in H_0^{\perp}$. We must show that

$$\lim_{n \to \infty} \frac{\xi_1 + \ldots + U^{n-1}\xi_1}{n} = 0.$$

In order to prove this statement, it is sufficient to show that the set of all vectors of the form $U\xi - \xi$, $\xi \in H$, is dense in $H_0^{\perp}$.

Assume the contrary. Then one can find $\alpha \in H$, $\alpha \neq 0$, such that $\alpha \perp H_0$ and α is orthogonal to any vector of the form $U\xi - \xi$. If this is the case, then

$$(U\alpha - \alpha, U\alpha - \alpha) = (U\alpha, U\alpha - \alpha) = (\alpha, \alpha - U^{-1}\alpha) = (\alpha, U(U^{-1}\alpha) - U^{-1}\alpha) = 0,$$

that is $U\alpha - \alpha = 0$. Thus $\alpha \in H_0$, which is a contradiction. $\quad\square$

Proof of Theorem 15.5. We have $X_k = U^k X_0$. If $X_0 = \eta + \eta_0$, where $\eta \in H_0$, $\eta_0 \perp H_0$, then

$$X_k = U^k X_0 = U^k(\eta + \eta_0) = \eta + U^k\eta_0.$$

Since $U^k\eta_0 \perp H_0$, we have $PX_k = \eta$, which does not depend on k. Thus Theorem 15.5 follows from the von Neumann Ergodic Theorem.

$\quad\square$

Let us show that, in our case, the space H_0 is at most one-dimensional. Indeed, write $X_0 = \eta + \eta_0$, where $\eta \in H_0$, $\eta_0 \perp H_0$. We have already seen that $X_k = \eta + U^k\eta_0$ and $U^k\eta_0 \perp H_0$.

Assume that there exists $\overline{\eta} \in H_0$, $\overline{\eta} \perp \eta$. Then $\overline{\eta} \perp X_k$ for any k. Therefore, $\overline{\eta} \perp \sum c_k X_k$ for any finite linear combination $\sum c_k X_k$, and thus $\overline{\eta} = 0$. $\quad\square$

Now we can improve Theorem 15.5 in the following way.

Theorem 15.7. *Either, for every $\xi \in H$,*

$$\lim_{n \to \infty} \frac{\xi + \ldots + U^{n-1}\xi}{n} = 0,$$

or there exists a vector $\eta \in H$, $||\eta|| = 1$, such that

$$\lim_{n \to \infty} \frac{\xi + \ldots + U^{n-1}\xi}{n} = (\xi, \eta) \cdot \eta.$$

15.3 Bochner Theorem and Other Useful Facts

In this section we shall state, without proof, the Bochner Theorem and some facts from measure theory to be used later in this chapter.

Recall that a function $f(x)$ defined on $\mathbb{R}$ or $\mathbb{N}$ is called non-negative definite if the inequality

$$\sum_{i,j=1}^{n} f(x_i - x_j)c_i\bar{c}_j \geq 0$$

holds for any $n > 0$, any $x_1, \ldots, x_n \in \mathbb{R}$ (or $\mathbb{N}$), and any complex numbers $c_1, \ldots, c_n$.

Theorem 15.8 (Bochner Theorem). *There is a one-to-one correspondence between the set of continuous non-negative definite functions and the set of finite measures on the Borel σ-algebra of $\mathbb{R}$. Namely, if ρ is a finite measure, then*

$$f(x) = \int_{\mathbb{R}} e^{i\lambda x} d\rho(\lambda) \tag{15.4}$$

is non-negative definite. Conversely, any continuous non-negative definite function can be represented in this form.

Similarly, there is a one-to-one correspondence between the set of non-negative definite functions on $\mathbb{N}$ and the set of finite measures on $[0,1)$, which is given by

$$f(n) = \int_{[0,1)} e^{2\pi i \lambda n} d\rho(\lambda).$$

We shall only prove that the expression on the right-hand side of (15.4) defines a non-negative definite function. For the converse statement we refer the reader to "Generalized Functions", Volume 4, by I.M. Gelfand and N.Y. Vilenkin.

Let $x_1, \ldots, x_n \in \mathbb{R}$ and $c_1, \ldots, c_n \in \mathbb{Z}$. Then

$$\sum_{i,j=1}^{n} f(x_i - x_j)c_i\bar{c}_j = \sum_{i,j=1}^{n} c_i\bar{c}_j \int_{\mathbb{R}} e^{i\lambda(x_i - x_j)} d\rho(\lambda)$$

$$= \int_{\mathbb{R}} \left(\sum_{i=1}^{n} c_i e^{i\lambda x_i}\right)\overline{\left(\sum_{j=1}^{n} c_j e^{i\lambda x_j}\right)} d\rho(\lambda) = \int_{\mathbb{R}} \left|\sum_{i=1}^{n} c_i e^{i\lambda x_i}\right|^2 d\rho(\lambda) \geq 0.$$

This proves that f is non-negative definite.

Theorem 15.9. *Let ρ be a finite measure on $\mathcal{B}([0,1))$. The set of trigonometric polynomials $p(\lambda) = \sum_{n=1}^{k} c_n e^{2\pi i n \lambda}$ is dense in the Hilbert space $L^2([0,1), \mathcal{B}([0,1)), \rho)$.*

Sketch of the Proof. From the definition of the integral, it is easy to show that the set of simple functions $g = \sum_{i=1}^{k} a_i \chi_{A_i}$ is dense in $L^2([0,1), \mathcal{B}([0,1)), \rho)$. Using the construction described in Sect. 3.4, one can show that for any set $A \in \mathcal{B}([0,1))$, the indicator function χ_A can be approximated by step functions of the form $f = \sum_{i=1}^{k} b_i \chi_{I_i}$, where I_i are disjoint subintervals of $[0,1)$. A step function can be approximated by continuous functions, while any continuous function can be uniformly approximated by trigonometric polynomials. $\square$

Let C be the set of functions which are continuous on $[0,1)$ and have the left limit $\lim_{\lambda \uparrow 1} f(\lambda) = f(0)$. This is a Banach space with the norm $\|f\|_C = \sup_{\lambda \in [0,1)} f(\lambda)$. The following theorem is a particular case of the Riesz Representation Theorem.

Theorem 15.10. *Let $\mathcal{B}$ be the σ-algebra of Borel subsets of $[0,1)$. For any linear continuous functional ψ on C there is a unique signed measure μ on $\mathcal{B}$ such that*

$$\psi(\varphi) = \int_{[0,1)} \varphi d\mu \tag{15.5}$$

for all $\varphi \in C$.

Remark 15.11. Clearly, the right-hand side of (15.5) defines a linear continuous functional for any signed measure μ. If μ is such that $\int_{[0,1)} \varphi d\mu = 0$ for all $\varphi \in C$, then, by the uniqueness part of Theorem 15.10, μ is identically zero.

15.4 Spectral Representation of Stationary Random Processes

In this section we again consider processes with discrete time and zero expectation. Let us start with a simple example. Define

$$X_n = \sum_{k=1}^{K} \alpha_k e^{2\pi i \lambda_k n} \tag{15.6}$$

Here, λ_k are real numbers and α_k are random variables. Assume that α_k are such that $E\alpha_k = 0$ and $E\alpha_{k_1} \overline{\alpha}_{k_2} = \beta_{k_1} \cdot \delta(k_1 - k_2)$, where $\delta(0) = 1$ and $\delta(k) = 0$ for $k \neq 0$.

Let us check that X_n is a wide-sense stationary random process. We have

$$E(X_{n_1} \overline{X}_{n_2}) = E\Big(\sum_{k_1=1}^{K} \alpha_{k_1} e^{2\pi i \lambda_{k_1} n_1} \cdot \sum_{k_2=1}^{K} \overline{\alpha}_{k_2} e^{-2\pi i \lambda_{k_2} n_2} \Big)$$

$$= \sum_{k=1}^{K} \beta_k e^{2\pi i \lambda_k (n_1 - n_2)} = b(n_1 - n_2),$$

which shows that X_n is stationary. We shall prove that any stationary process with zero mean can be represented in a form similar to (15.6), with the sum replaced by an integral with respect to an orthogonal random measure, which will be defined below.

Consider the covariance function $b(n)$ for a stationary process X_n, which is given by $b(n_1 - n_2) = \mathrm{E}(X_{n_1}\overline{X}_{n_2})$. The function b is non-negative definite, since for any finite set $n_1, \ldots, n_k$, and any complex numbers $c_1, \ldots, c_k$, we have

$$\sum_{i,j=1}^{k} b(n_i - n_j)c_{n_i}\bar{c}_{n_j} = \sum_{i,j=1}^{k} \mathrm{E}(X_{n_i}\overline{X}_{n_j})c_{n_i}\bar{c}_{n_j} = \mathrm{E}|\sum_{i=1}^{k} c_{n_i}X_{n_i}|^2 \geq 0.$$

Now we can use the Bochner Theorem to represent the numbers $b(n)$ as Fourier coefficients of a finite measure on the unit circle,

$$b(n) = \int_{[0,1)} e^{2\pi i \lambda n} d\rho(\lambda).$$

Definition 15.12. *The measure ρ is called the spectral measure of the process X_n.*

Theorem 15.13. *Consider the space $L^2 = L^2([0,1), \mathcal{B}([0,1)), \rho)$ of square-integrable functions on $[0,1)$ (with respect to the measure ρ). There exists an isomorphism $\psi : H \to L^2$ of the spaces H and L^2 such that $\psi(U\xi) = e^{2\pi i \lambda}\psi(\xi)$.*

Proof. Denote by $\hat{L}^2$ the space of all finite trigonometric polynomials on the interval $[0,1)$, that is functions of the form $p(\lambda) = \sum c_n e^{2\pi i n \lambda}$. This space is dense in L^2 by Theorem 15.9.

Take $\xi = \sum c_n X_n \in \hat{H}$ and put $\psi(\xi) = \sum c_n e^{2\pi i n \lambda}$. It is clear that ψ maps $\hat{H}$ linearly onto $\hat{L}^2$. Also, for $\xi_1 = \sum c_n X_n$ and $\xi_2 = \sum d_n X_n$,

$$(\xi_1, \xi_2) = \sum_{n_1, n_2} c_{n_1}\bar{d}_{n_2} \mathrm{E}(X_{n_1}\overline{X}_{n_2}) = \sum_{n_1, n_2} c_{n_1}\bar{d}_{n_2} b(n_1 - n_2)$$

$$= \sum_{n_1, n_2} c_{n_1}\bar{d}_{n_2} \int_{[0,1)} e^{2\pi i \lambda(n_1 - n_2)} d\rho(\lambda)$$

$$= \int_{[0,1)} \left(\sum_{n_1} c_{n_1} e^{2\pi i \lambda n_1} \right) \overline{\left(\sum_{n_2} d_{n_2} e^{2\pi i \lambda n_2} \right)} d\rho(\lambda) = \int_{[0,1)} \psi(\xi_1)\overline{\psi(\xi_2)} d\rho(\lambda).$$

Thus, ψ is an isometry between $\hat{H}$ and $\hat{L}^2$. Therefore it can be extended by continuity to an isometry of H and L^2.

For $\xi = \sum c_n X_n$, we have $U\xi = \sum c_n X_{n+1}$, $\psi(\xi) = \sum c_n e^{2\pi i \lambda n}$, and $\psi(U\xi) = \sum c_n e^{2\pi i \lambda(n+1)} = e^{2\pi i \lambda}\psi(\xi)$. The equality $\psi(U\xi) = e^{2\pi i \lambda}\psi(\xi)$ remains true for all ξ by continuity. $\quad\square$

Corollary 15.14. *If $\rho(\{0\}) = 0$, then $H_0 = 0$ and the time averages in the Law of Large Numbers converge to zero.*

Proof. The space H_0 consists of U-invariant vectors of H. Take $\eta \in H_0$ and let $f(\lambda) = \psi(\eta) \in L^2$. Then,

$$f(\lambda) = \psi(\eta) = \psi(U\eta) = e^{2\pi i \lambda} \psi(\eta) = e^{2\pi i \lambda} f(\lambda),$$

or

$$(e^{2\pi i \lambda} - 1) f(\lambda) = 0,$$

where the equality is understood in the sense of $L^2([0,1), \mathcal{B}([0,1)), \rho)$. Since $\rho(\{0\}) = 0$, and the function $e^{2\pi i \lambda} - 1$ is different from zero on $(0,1)$, the function $f(\lambda)$ must be equal to zero almost surely with respect to ρ, and thus the norm of f is zero.   □

The arguments for the following corollary are the same as above.

Corollary 15.15. *If $\rho(\{0\}) > 0$, then $\psi(H_0)$ is the one-dimensional space of functions concentrated at zero.*

15.5 Orthogonal Random Measures

Take a Borel subset $\Delta \subseteq [0,1)$ and set

$$Z(\Delta) = \psi^{-1}(\chi_\Delta) \in H \subseteq L^2(\Omega, \mathcal{F}, \mathrm{P}). \tag{15.7}$$

Here, χ_Δ is the indicator of Δ. Let us now study the properties of $Z(\Delta)$ as a function of Δ.

Lemma 15.16. *For any Borel sets Δ, we have*

$$EZ(\Delta) = 0. \tag{15.8}$$

For any Borel sets Δ_1 and $\Delta_2 \subseteq [0,1)$, we have

$$EZ(\Delta_1)\overline{Z(\Delta_2)} = \rho(\Delta_1 \cap \Delta_2). \tag{15.9}$$

If $\Delta = \bigcup_{k=1}^{\infty} \Delta_k$ and $\Delta_{k_1} \cap \Delta_{k_2} = \emptyset$ for $k_1 \neq k_2$, then

$$Z(\Delta) = \sum_{k=1}^{\infty} Z(\Delta_k), \tag{15.10}$$

where the sum is understood as a limit of partial sums in the space $L^2(\Omega, \mathcal{F}, \mathrm{P})$.

Proof. The first statement is true since $E\xi = 0$ for all $\xi \in H$. The second statement of the lemma follows from

$$EZ(\Delta_1)\overline{Z(\Delta_2)} = E\psi^{-1}(\chi_{\Delta_1})\overline{\psi^{-1}(\chi_{\Delta_2})}$$

$$= \int_{[0,1)} \chi_{\Delta_1}(\lambda)\chi_{\Delta_2}(\lambda)d\rho(\lambda) = \rho(\Delta_1 \bigcap \Delta_2).$$

The third statement holds since $\chi_\Delta = \sum_{k=1}^{\infty} \chi_{\Delta_k}$ in $L^2([0,1), \mathcal{B}([0,1)), \rho)$. □

A function Z with values in $L^2(\Omega, \mathcal{F}, P)$ defined on a σ-algebra is called an orthogonal random measure if it satisfies (15.8), (15.9), and (15.10). In particular, if $Z(\Delta)$ is given by (15.7), it is called the (random) spectral measure of the process X_n.

The non-random measure ρ, which in this case is a finite measure on $[0,1)$, may in general be a σ-finite measure on $\mathbb{R}$ (or $\mathbb{R}^n$, as in the context of random fields, for example).

Now we shall introduce the notion of an integral with respect to an orthogonal random measure. The integral shares many properties with the usual Lebesgue integral, but differs in some respects.

For each $f \in L^2([0,1), \mathcal{B}([0,1)), \rho)$, one can define a random variable $I(f) \in L^2(\Omega, \mathcal{F}, P)$ such that:

(a) $I(c_1 f_1 + c_2 f_2) = c_1 I(f_1) + c_2 I(f_2)$.
(b) $E|I(f)|^2 = \int_{[0,1)} |f(\lambda)|^2 d\rho(\lambda)$.

The precise definition is as follows. For a finite linear combination of indicator functions

$$f = \sum c_k \chi_{\Delta_k},$$

set

$$I(f) = \sum c_k Z(\Delta_k).$$

Thus, the correspondence $f \to I(f)$ is a linear map which preserves the inner product. Therefore, it can be extended by continuity to $L^2([0,1), \mathcal{B}([0,1)), \rho)$.

We shall write $I(f) = \int f(\lambda)dZ(\lambda)$ and call it the integral with respect to the orthogonal random measure $Z(\Delta)$.

Note that when Z is a spectral measure, the maps $f \to I(f)$ and $f \to \psi^{-1}(f)$ are equal, since they are both isomorphisms of $L^2([0,1), \mathcal{B}([0,1)), \rho)$ onto H and coincide on all the indicator functions. Therefore, we can recover the process X_n, given its random spectral measure:

$$X_n = \int_0^1 e^{2\pi i \lambda n} dZ(\lambda). \qquad (15.11)$$

This formula is referred to as the spectral decomposition of the wide-sense stationary random process.

Given any orthogonal random measure $Z(\Delta)$, this formula defines a wide-sense stationary random process. Thus, we have established a one-to-one correspondence between wide-sense stationary random processes with zero mean and random measures on $[0, 1)$.

Given a stationary process X_n, its spectral measure $Z(\Delta)$, and an arbitrary function $f(\lambda) \in L^2([0,1), \mathcal{B}([0,1)), \rho)$, we can define a random process

$$Y_n = \int_{[0,1)} e^{2\pi i n \lambda} f(\lambda) dZ(\lambda).$$

Since

$$\mathrm{E}(Y_{n_1} \overline{Y}_{n_2}) = \int_{[0,1)} e^{2\pi i (n_1 - n_2)\lambda} |f(\lambda)|^2 d\rho(\lambda),$$

the process Y_n is wide-sense stationary with spectral measure equal to $d\rho_Y(\lambda) = |f(\lambda)|^2 d\rho(\lambda)$. Since $Y_n \in H$ for each n, the linear space H_Y generated by the process Y_n is a subspace of H. The question of whether the opposite inclusion holds is answered by the following lemma.

Lemma 15.17. *Let $f(\lambda) \in L^2([0,1), \mathcal{B}([0,1)), \rho)$, and the processes X_n and Y_n be as above. Then $H_Y = H$ if and only if $f(\lambda) > 0$ almost everywhere with respect to the measure ρ.*

Proof. In the spectral representation, the space H_Y consists of those elements of $L^2 = L^2([0,1), \mathcal{B}([0,1)), \rho)$ which can be approximated in the L^2 norm by finite sums of the form $\sum c_n e^{2\pi i \lambda n} f(\lambda)$. If $f(\lambda) = 0$ on a set A with $\rho(A) > 0$, then the indicator function $\chi_A(\lambda)$ cannot be approximated by such sums.

Conversely, assume that $f(\lambda) > 0$ almost surely with respect to the measure ρ, and that the sums of the form $\sum c_n e^{2\pi i \lambda n} f(\lambda)$ are not dense in L^2. Then there exists $g(\lambda) \in L^2$, $g \neq 0$, such that

$$\int_{[0,1)} P(\lambda) f(\lambda) g(\lambda) d\rho(\lambda) = 0$$

for any finite trigonometric polynomial $P(\lambda)$. Note that the signed measure $d\mu(\lambda) = f(\lambda) g(\lambda) d\rho(\lambda)$ is not identically zero, and $\int_{[0,1)} P(\lambda) d\mu(\lambda) = 0$ for any $P(\lambda)$. Since trigonometric polynomials are dense in $C([0,1])$, we obtain $\int_{[0,1)} \varphi(\lambda) d\mu(\lambda) = 0$ for any continuous function φ. By Remark 15.11, $\mu = 0$, which is a contradiction. $\square$

15.6 Linear Prediction of Stationary Random Processes

In this section we consider stationary random processes with discrete time, and assume that $\mathrm{E}X_n \equiv 0$. For each k_1 and k_2 such that $-\infty \leq k_1 \leq k_2 \leq \infty$, we define the subspace $H_{k_1}^{k_2}$ of the space H as the closure of the space of all finite sums $\sum c_n X_n$ for which $k_1 \le n \le k_2$. The operator of projection on the space $H_{k_1}^{k_2}$ will be denoted by $P_{k_1}^{k_2}$.

We shall be interested in the following problem: given m and k_0 such that $m < k_0$, we wish to find the best approximation of X_{k_0} by elements of $H_{-\infty}^m$. Due to stationarity, it is sufficient to consider $k_0 = 0$.

More precisely, let

$$h_{-m} = \inf \|X_0 - \sum_{n \le -m} c_n X_n\|,$$

where the infimum is taken over all finite sums, and $\| \ \|$ is the $L^2(\Omega, \mathcal{F}, P)$ norm. These quantities have a natural geometric interpretation. Namely, we can write

$$X_0 = P_{-\infty}^{-m} X_0 + \xi_{-m}^0, \quad P_{-\infty}^{-m} X_0 \in H_{-\infty}^{-m}, \quad \xi_{-m}^0 \perp H_{-\infty}^{-m}.$$

Then $h_{-m} = \|\xi_{-m}^0\|$.

Definition 15.18. *A random process X_n is called linearly non-deterministic if $h_{-1} > 0$.*

Definition 15.19. *A random process X_n is called linearly regular if $P_{-\infty}^{-m} X_0 \to 0$ as $m \to \infty$.*

Thus, a process X_n is linearly non-deterministic if it is impossible to approximate X_0 with arbitrary accuracy by the linear combinations $\sum_{n \le -1} c_n X_n$. A process is linearly regular if the best approximation of X_0 by the linear combinations $\sum_{n \le -m} c_n X_n$ tends to zero as $m \to \infty$.

The main problems in the theory of linear prediction are finding the conditions under which the process X_n is linearly regular, and finding the value of h_{-1}.

Theorem 15.20. *A process X_n is linearly regular if and only if it is linearly non-deterministic and ρ is absolutely continuous with respect to the Lebesgue measure.*

In order to prove this theorem, we shall need the following fact about the geometry of Hilbert spaces, which we state here as a lemma.

Lemma 15.21. *Assume that in a Hilbert space H there is a decreasing sequence of subspaces $\{L_m\}$, that is $L_{m+1} \subseteq L_m$. Let $L_\infty = \bigcap_m L_m$. For every $h \in H$, let*

$$h = h_m' + h_m'', \quad h_m' \in L_m, \quad h_m'' \perp L_m,$$

and

$$h = h' + h'', \quad h' \in L_\infty, \quad h'' \perp L_\infty.$$

Then, $h' = \lim_{m\to\infty} h_m'$ and $h'' = \lim_{m\to\infty} h_m''$.

Proof. Let us show that h'_m is a Cauchy sequence of vectors. If we assume the contrary, then there is $\varepsilon > 0$ and a subsequence h'_{m_k} such that $|h'_{m_{k+1}} - h'_{m_k}| \geq \varepsilon$ for all $k \geq 0$. This implies that $|h'_{m_0} - h'_{m_k}|$ can be made arbitrarily large for sufficiently large k, since all the vectors $h'_{m_{k+1}} - h'_{m_k}$ are perpendicular to each other. This contradicts $|h'_m| \leq |h|$ for all m.

Let $\overline{h'} = \lim_{m \to \infty} h'_m$. Then $\overline{h'} \in L_m$ for all m, and thus $\overline{h'} \in L_\infty$. On the other hand, the projection of $\overline{h'}$ onto L_∞ is equal to h', since the projection of each of the vectors h'_m onto L_∞ is equal to h'. We conclude that $\overline{h'} = h'$. Since $h''_m = h - h'_m$, we have $h'' = \lim_{m \to \infty} h''_m$. $\qquad\square$

Proof of Theorem 15.20. Let us show that X_n is linearly regular if and only if $\bigcap_m H^m_{-\infty} = \{0\}$. Indeed, if $\bigcap_m H^m_{-\infty} = \{0\}$, then, applying Lemma 15.21 with $L_m = H^{-m}_{-\infty}$, we see that in the expansion

$$X_0 = P^{-m}_{-\infty} X_0 + \xi^0_{-m}, \quad \text{where} \quad P^{-m}_{-\infty} X_0 \in H^{-m}_{-\infty}, \ \xi^0_{-m} \perp H^{-m}_{-\infty},$$

the first term $P^{-m}_{-\infty} X_0$ converges to the projection of X_0 onto $\bigcap_m H^{-m}_{-\infty}$, which is zero.

Conversely, if $P^{-m}_{-\infty} X_0 \to 0$ as $m \to \infty$, then in the expansion

$$X_n = P^{-m}_{-\infty} X_n + \xi^n_{-m}, \quad \text{where} \quad P^{-m}_{-\infty} X_n \in H^{-m}_{-\infty}, \ \xi^n_{-m} \perp H^{-m}_{-\infty}, \quad (15.12)$$

the first term $P^{-m}_{-\infty} X_n$ tends to zero as $m \to \infty$ since, due to stationarity, $P^{-m}_{-\infty} X_n = U^n P^{-m-n}_{-\infty} X_0$. Therefore, for each finite linear combination $\eta = \sum c_n X_n$, the first term $P^{-m}_{-\infty} \eta$ in the expansion

$$\eta = P^{-m}_{-\infty} \eta + \xi_{-m}, \quad \text{where} \quad P^{-m}_{-\infty} \eta \in H^{-m}_{-\infty}, \ \xi_{-m} \perp H^{-m}_{-\infty},$$

tends to zero as $m \to \infty$. By continuity, the same is true for every $\eta \in H$. Thus, the projection of every η on the space $\bigcap_m H^{-m}_{-\infty}$ is equal to zero, which implies that $\bigcap_m H^{-m}_{-\infty} = \{0\}$.

Let ξ^n_{-1} be defined as in the expansion (15.12). Then $\xi^k_{-1} \in H^k_{-\infty}$, while $\xi^n_{-1} \perp H^k_{-\infty}$ if $k < n$. Therefore,

$$(\xi^k_{-1}, \xi^n_{-1}) = 0, \quad \text{if} \ k \neq n. \quad (15.13)$$

Let us show that X_n is linearly regular if and only if $\{\xi^n_{-1}\}$ is an orthogonal basis in H. Indeed, if $\{\xi^n_{-1}\}$ is a basis, then

$$X_0 = \sum_{n=-\infty}^{\infty} c_n \xi^n_{-1}$$

and $\sum_{n=-\infty}^{\infty} c_n^2 \|\xi^n_{-1}\|^2 < \infty$. Note that

$$\|P^{-m}_{-\infty} X_0\|^2 = \sum_{n=-\infty}^{-m} c_n^2 \|\xi^n_{-1}\|^2 \to 0,$$

and therefore the process is linearly regular.

In order to prove the converse implication, let us represent H as the following direct sum:

$$H = (\bigcap_m H^m_{-\infty}) \oplus (\bigoplus_m (H^m_{-\infty} \ominus H^{m-1}_{-\infty})).$$

If X_n is linearly regular, then $\bigcap_m H^m_{-\infty} = \{0\}$. On the other hand, $H^m_{-\infty} \ominus H^{m-1}_{-\infty}$ is the one-dimensional subspace generated by ξ^m_{-1}. Therefore, $\{\xi^m_{-1}\}$ is a basis in H.

Let f be the spectral representation of ξ^0_{-1}. Since $\xi^m_{-1} = U^m \xi^0_{-1}$ and due to (15.13), we have

$$\int_{[0,1)} |f(\lambda)|^2 e^{2\pi i\lambda m} d\rho(\lambda) = (\xi^m_{-1}, \xi^0_{-1}) = \delta(m)||\xi^0_{-1}||^2,$$

where $\delta(m) = 1$ if $m = 0$ and $\delta(m) = 0$ otherwise. Thus,

$$\int_{[0,1)} e^{2\pi i\lambda m} d\overline{\rho}(\lambda) = \delta(m)||\xi^0_{-1}||^2 \text{ for } m \in \mathbb{Z},$$

where $d\overline{\rho}(\lambda) = |f(\lambda)|^2 d\rho(\lambda)$. This shows that $d\overline{\rho}(\lambda) = ||\xi^0_{-1}||^2 d\lambda$, that is $\overline{\rho}$ is a constant multiple of the Lebesgue measure.

Assume that the process is linearly non-deterministic and ρ is absolutely continuous with respect to the Lebesgue measure. Then $||\xi^0_{-1}|| \neq 0$, since the process is linearly non-deterministic. Note that $f(\lambda) > 0$ almost everywhere with respect to the measure ρ, since $|f(\lambda)|^2 d\rho(\lambda) = ||\xi^0_{-1}||^2 d\lambda$ and ρ is absolutely continuous with respect to the Lebesgue measure. By Lemma 15.17, the space generated by $\{\xi^m_{-1}\}$ coincides with H, and therefore the process is linearly regular.

Conversely, if the process is linearly regular, then $||\xi^0_{-1}|| \neq 0$, and thus the process is non-deterministic. Since $f(\lambda) > 0$ almost everywhere with respect to the measure ρ (by Lemma 15.17), and

$$|f(\lambda)|^2 d\rho(\lambda) = ||\xi^0_{-1}||^2 d\lambda,$$

the measure ρ is absolutely continuous with respect to the Lebesgue measure. $\square$

Consider the spectral measure ρ and its decomposition $\rho = \rho_0 + \rho_1$, where ρ_0 which is absolutely continuous with respect to the Lebesgue measure and ρ_1 is singular with respect to the Lebesgue measure. Let $p_0(\lambda)$ be the Radon-Nikodym derivative of ρ_0 with respect to the Lebesgue measure, that is $p_0(\lambda)d\lambda = d\rho_0(\lambda)$. The following theorem was proved independently by Kolmogorov and Wiener.

Theorem 15.22 (Kolmogorov-Wiener). *For any wide-sense stationary process with zero mean we have*

$$h_{-1} = \exp(\frac{1}{2} \int_{[0,1)} \ln p_0(\lambda) d\lambda), \qquad (15.14)$$

where the right-hand side is set to be equal to zero if the integral in the exponent is equal to $-\infty$.

In particular, this theorem implies that if the process X_n is linearly non-deterministic, then $p_0(\lambda) > 0$ almost everywhere with respect to the Lebesgue measure.

The proof of the Kolmogorov-Wiener Theorem is rather complicated. We shall only sketch it for a particular case, namely when the spectral measure is absolutely continuous, and the density p_0 is a positive twice continuously differentiable periodic function. The latter means that there is a positive periodic function $\overline{p}_0$ with period one, which is twice continuously differentiable, such that $p_0(\lambda) = \overline{p}_0(\lambda)$ for all $\lambda \in [0, 1)$.

Sketch of the proof. Let us take the vectors $v_1^{(n)} = X_{-n+1}, \dots, v_n^{(n)} = X_0$ in the space H, and consider the $n \times n$ matrix $M^{(n)}$ with elements $M_{ij}^{(n)} = \mathrm{E}(v_i^{(n)} \overline{v_j^{(n)}}) = b(i-j)$ (called the Gram matrix of the vectors $v_1^{(n)}, \dots, v_n^{(n)}$). It is well known that the determinant of the Gram matrix is equal to the square of the volume (in the sense of the Hilbert space H) spanned by the vectors $v_1^{(n)}, \dots, v_n^{(n)}$. More precisely, let us write

$$v_j^{(n)} = w_j^{(n)} + h_j^{(n)}, \quad 1 \le j \le n,$$

where $w_1^{(n)} = 0$ and, for $j > 1$, $w_j^{(n)}$ is the orthogonal projection of $v_j^{(n)}$ on the space spanned by the vectors $v_i^{(n)}$ with $i < j$. Then

$$\det M^{(n)} = \prod_{j=1}^{n} ||h_j^{(n)}||^2.$$

After taking the logarithm on both sides and dividing by $2n$, we obtain

$$\frac{1}{2n} \ln(\det M^{(n)}) = \frac{1}{n} \sum_{j=1}^{n} \ln ||h_j^{(n)}||.$$

Due to the stationarity of the process, the norms $||h_j^{(n)}||$ depend only on j, but not on n. Moreover, $\lim_{j \to \infty} ||h_j^{(n)}|| = h_{-1}$. Therefore, the right-hand side of the last equality tends to $\ln h_{-1}$ when $n \to \infty$, which implies that

$$h_{-1} = \exp(\frac{1}{2} \lim_{n \to \infty} \frac{1}{n} \ln(\det M^{(n)})).$$

Since the matrix $M^{(n)}$ is Hermitian, it has n real eigenvalues, which will be denoted by $\gamma_1^{(n)}, \gamma_2^{(n)}, \ldots, \gamma_n^{(n)}$. Thus,

$$h_{-1} = \exp(\frac{1}{2} \lim_{n \to \infty} \frac{1}{n} \sum_{j=1}^{n} \ln \gamma_j^{(n)}). \qquad (15.15)$$

Let c_1 and c_2 be positive constants such that $c_1 \leq p_0(\lambda) \leq c_2$ for all $\lambda \in [0, 1)$. Let us show that $c_1 \leq \gamma_j^{(n)} \leq c_2$ for $j = 1, \ldots, n$. Indeed, if $z = (z_1, \ldots, z_n)$ is a vector of complex numbers, then from the spectral representation of the process it follows that

$$(M^{(n)}z, z) = E|\sum_{j=1}^{n} z_j X_{-n+j}|^2 = \int_{[0,1)} |\sum_{j=1}^{n} z_j e^{2\pi i \lambda j}|^2 p_0(\lambda) d\lambda.$$

At the same time,

$$c_1|z|^2 = c_1 \sum_{j=1}^{n} |z_j|^2 = c_1 \int_{[0,1)} |\sum_{j=1}^{n} z_j e^{2\pi i \lambda j}|^2 d\lambda$$

$$\leq \int_{[0,1)} |\sum_{j=1}^{n} z_j e^{2\pi i \lambda j}|^2 p_0(\lambda) d\lambda$$

$$\leq c_2 \int_{[0,1)} |\sum_{j=1}^{n} z_j e^{2\pi i \lambda j}|^2 d\lambda = c_2 \sum_{j=1}^{n} |z_j|^2 = c_2|z|^2.$$

Therefore, $c_1|z|^2 \leq (M^{(n)}z, z) \leq c_2|z|^2$, which gives the bound on the eigenvalues.

Let f be a continuous function on the interval $[c_1, c_2]$. We shall prove that

$$\lim_{n \to \infty} \frac{1}{n} \sum_{j=1}^{n} f(\gamma_j^{(n)}) = \int_{[0,1)} f(p_0(\lambda)) d\lambda, \qquad (15.16)$$

in order to apply it to the function $f(x) = \ln(x)$,

$$\lim_{n \to \infty} \frac{1}{n} \sum_{j=1}^{n} \ln \gamma_j^{(n)} = \int_{[0,1)} \ln p_0(\lambda) d\lambda.$$

The statement of the theorem will then follow from (15.15).

Since both sides of (15.16) are linear in f, and any continuous function can be uniformly approximated by polynomials, it is sufficient to prove (15.16) for the functions of the form $f(x) = x^r$, where r is a positive integer. Let r be fixed. The trace of the matrix $(M^{(n)})^r$ is equal to the sum of its eigenvalues, that is $\sum_{j=1}^{n} (\gamma_j^{(n)})^r = \mathrm{Tr}((M^{(n)})^r)$. Therefore, (15.16) becomes

$$\lim_{n\to\infty} \frac{1}{n} \mathrm{Tr}((M^{(n)})^r) = \int_{[0,1)} (p_0(\lambda))^r d\lambda. \qquad (15.17)$$

Let us discretize the spectral measure ρ in the following way. Divide the segment $[0,1)$ into n equal parts $\Delta_j^{(n)} = [\frac{j}{n}, \frac{j+1}{n})$, $j = 0, \ldots, n-1$, and consider the discrete measure $\rho^{(n)}$ concentrated at the points $\frac{j}{n}$ such that

$$\rho^{(n)}\left(\frac{j}{n}\right) = \rho(\Delta_j^{(n)}) = \int_{[\frac{j}{n}, \frac{j+1}{n})} p_0(\lambda) d\lambda.$$

Consider the $n \times n$ matrix $\widetilde{M}^{(n)}$ with the following elements:

$$\widetilde{M}_{ij}^{(n)} = \widetilde{b}^{(n)}(i-j), \quad \text{where} \quad \widetilde{b}^{(n)}(j) = \int_{[0,1)} e^{2\pi i\lambda j} d\rho^{(n)}(\lambda).$$

Recall that

$$M_{ij}^{(n)} = b^{(n)}(i-j), \quad \text{where} \quad b^{(n)}(j) = \int_{[0,1)} e^{2\pi i\lambda j} d\rho(\lambda).$$

Consider n vectors V_j, $j = 1, \ldots, n$, each of length n. The k-th element of V_j is defined as $\exp(2\pi i \frac{k(j-1)}{n})$. Clearly, these are the eigenvectors of the matrix $\widetilde{M}^{(n)}$ with eigenvalues $\widetilde{\gamma}_j^{(n)} = n\rho^{(n)}(\frac{j-1}{n})$. Therefore,

$$\lim_{n\to\infty} \frac{1}{n} \mathrm{Tr}((\widetilde{M}^{(n)})^r) = \lim_{n\to\infty} \frac{1}{n} \sum_{j=1}^{n} (\widetilde{\gamma}_j^{(n)})^r$$

$$= \lim_{n\to\infty} \frac{1}{n} \sum_{j=1}^{n} (n\rho^{(n)}(\frac{j-1}{n}))^r = \int_{[0,1)} (p_0(\lambda))^r d\lambda.$$

It remains to show that

$$\lim_{n\to\infty} \frac{1}{n}(\mathrm{Tr}((\widetilde{M}^{(n)})^r) - \mathrm{Tr}((M^{(n)})^r)) = 0.$$

The trace of the matrix $(M^{(n)})^r$ can be expressed in terms of the elements of the matrix $M^{(n)}$ as follows:

$$\mathrm{Tr}((M^{(n)})^r) = \sum_{j_1,\ldots,j_r=1}^{n} b^{(n)}(j_1 - j_2)b^{(n)}(j_2 - j_3)\ldots b^{(n)}(j_r - j_1). \qquad (15.18)$$

Similarly,

$$\mathrm{Tr}((\widetilde{M}^{(n)})^r) = \sum_{j_1,\ldots,j_r=1}^{n} \widetilde{b}^{(n)}(j_1 - j_2)\widetilde{b}^{(n)}(j_2 - j_3)\ldots\widetilde{b}^{(n)}(j_r - j_1). \quad (15.19)$$

Note that

$$b^{(n)}(j) = \int_{[0,1)} e^{2\pi i\lambda j} d\rho(\lambda) = \int_{[0,1)} e^{2\pi i\lambda j} p_0(\lambda) d\lambda$$

$$= -\frac{1}{(2\pi)^2 j^2} \int_{[0,1)} e^{2\pi i\lambda j} p_0''(\lambda) d\lambda,$$

where the last equality is due to integration by parts (twice) and the periodicity of the function $p_0(\lambda)$. Thus,

$$|b^{(n)}(j)| \leq \frac{k_1}{j^2} \quad (15.20)$$

for some constant k_1. A similar estimate can be obtained for $\widetilde{b}^{(n)}(j)$. Namely,

$$|\widetilde{b}^{(n)}(j)| \leq \frac{k_2}{(\mathrm{dist}(j, n\mathbb{Z}))^2}, \quad (15.21)$$

where k_2 is a constant and $\mathrm{dist}(j, n\mathbb{Z}) = \min_{p \in \mathbb{Z}} |j - np|$. In order to obtain this estimate, we can write

$$\widetilde{b}^{(n)}(j) = \int_{[0,1)} e^{2\pi i\lambda j} d\rho^{(n)}(\lambda) = \sum_{k=0}^{n-1} e^{2\pi i\frac{k}{n}j} \rho^{(n)}(\frac{k}{n}),$$

and then apply Abel transform (a discrete analogue of integration by parts) to the sum on the right-hand side of this equality. We leave the details of the argument leading to (15.21) to the reader.

Using the estimate (15.20), we can modify the sum on the right-hand side of (15.18) as follows:

$$\mathrm{Tr}((M^{(n)})^r) = \widehat{\sum_{j_1,\ldots,j_r=1}^{n}} b^{(n)}(j_1 - j_2)b^{(n)}(j_2 - j_3)\ldots b^{(n)}(j_r - j_1) + \delta_1(t, n),$$

where $\widehat{\sum}$ means that the sum is taken over those $j_1,\ldots,j_r$ which satisfy $\mathrm{dist}(j_k - j_{k+1}, n\mathbb{Z}) \leq t$ for $k = 1,\ldots,r-1$, and $\mathrm{dist}(j_r - j_1, n\mathbb{Z}) \leq t$. The remainder can be estimated as follows:

$$|\delta_1(t, n)| \leq n\varepsilon_1(t), \quad \text{where} \quad \lim_{t\to\infty} \varepsilon_1(t) = 0.$$

Similarly,

$$\mathrm{Tr}((\widetilde{M}^{(n)})^r) = \widehat{\sum_{j_1,\ldots,j_r=1}^{n}} \widetilde{b}^{(n)}(j_1 - j_2)\widetilde{b}^{(n)}(j_2 - j_3)\ldots\widetilde{b}^{(n)}(j_r - j_1) + \delta_2(t, n)$$

with

$$|\delta_2(t, n)| \leq n\varepsilon_2(t), \quad \text{where} \quad \lim_{t\to\infty} \varepsilon_2(t) = 0.$$

The difference $\frac{1}{n}|\delta_2(t, n) - \delta_1(t, n)|$ can be made arbitrarily small for all sufficiently large t. Therefore, it remains to demonstrate that for each fixed value of t we have

$$\lim_{n\to\infty} \frac{1}{n} \Big(\widehat{\sum}_{j_1,\dots,j_r=1}^n \widetilde{b}^{(n)}(j_1 - j_2)\widetilde{b}^{(n)}(j_2 - j_3)\dots \widetilde{b}^{(n)}(j_r - j_1)$$

$$- \widehat{\sum}_{j_1,\dots,j_r=1}^n b^{(n)}(j_1 - j_2)b^{(n)}(j_2 - j_3)\dots b^{(n)}(j_r - j_1)\Big) = 0.$$

From the definitions of $\widetilde{b}^{(n)}(j)$ and $b^{(n)}(j)$ it immediately follows that

$$\lim_{n\to\infty} \sup_{j:\mathrm{dist}(j,\{np\})\leq t} |\widetilde{b}^{(n)}(j) - b^{(n)}(j)| = 0.$$

It remains to note that the number of terms in each of the sums does not exceed $n(2t + 1)^{r-1}$. $\qquad\square$

15.7 Stationary Random Processes with Continuous Time

In this section we consider a stationary random processes X_t, $t \in \mathbb{R}$, and assume that $\mathrm{E}X_t \equiv 0$. In addition, we assume that the covariance function $b(t) = \mathrm{E}(X_t\overline{X}_0)$ is continuous.

Lemma 15.23. *If the covariance function $b(t)$ of a stationary random process is continuous at $t = 0$, then it is continuous for all t.*

Proof. Let t be fixed. Then

$$|b(t + h) - b(t)| = |\mathrm{E}((X_{t+h} - X_t)\overline{X}_0)|$$

$$\leq \sqrt{\mathrm{E}|X_0|^2 \mathrm{E}|X_{t+h} - X_t|^2} = \sqrt{b(0)(2b(0) - b(h) - b(-h))},$$

which tends to zero as $h \to 0$, thus showing that $b(t)$ is continuous. $\qquad\square$

It is worth noting that the continuity of $b(t)$ is equivalent to the continuity of the process in the L^2 sense. Indeed,

$$\mathrm{E}|X_{t+h} - X_t|^2 = 2b(0) - b(h) - b(-h),$$

which tends to zero as $h \to 0$ if b is continuous. Conversely, if $\mathrm{E}|X_h - X_0|^2$ tends to zero, then $\lim_{h\to 0} \mathrm{Re}(b(h)) = b(0)$, since $b(h) = \overline{b(-h)}$. We also have $\lim_{h\to 0} \mathrm{Im}(b(h)) = 0$, since $|b(h)| \leq b(0)$ for all h.

We shall now state how the results proved above for the random processes with discrete time carry over to the continuous case.

Recall the definition of the operators U^t from Sect. 15.1. If the covariance function is continuous, then the group of unitary operators U^t is strongly continuous, that is $\lim_{t \to 0} U^t \eta = \eta$ for any $\eta \in H$. The von Neumann Ergodic Theorem now takes the following form.

Theorem 15.24. *Let U^t be a strongly continuous group of unitary operators in a Hilbert space H. Let P be the orthogonal projection onto $H_0 = \{\varphi : \varphi \in H, U^t \varphi = \varphi$ for all $t\}$. Then for any $\xi \in H$,*

$$\lim_{T \to \infty} \frac{1}{T} \int_0^T U^t \xi \, dt = \lim_{T \to \infty} \frac{1}{T} \int_0^T U^{-t} \xi \, dt = P\xi.$$

As in the case of processes with discrete time, the von Neumann Ergodic Theorem implies the following Law of Large Numbers.

Theorem 15.25. *Let X_t be a wide-sense stationary process with continuous covariance. There exists $\eta \in H$ such that*

$$\lim_{T \to \infty} \frac{1}{T} \int_{T_0}^{T_0+T} X_t \, dt = \eta$$

for any T_0. The limit η is invariant for the operators U^t.

The covariance function $b(t)$ is now a continuous non-negative definite function defined for $t \in \mathbb{R}$. The Bochner Theorem states that there is a one-to-one correspondence between the set of such functions and the set of finite measures on the real line. Namely, $b(t)$ is the Fourier transform of some measure ρ, which is called the spectral measure of the process X_t,

$$b(t) = \int_{\mathbb{R}} e^{i\lambda t} d\rho(\lambda), \quad -\infty < t < \infty.$$

The theorem on the spectral isomorphism is now stated as follows.

Theorem 15.26. *There exists an isomorphism ψ of the Hilbert spaces H and $L^2(\mathbb{R}, \mathcal{B}([0,1)), \rho)$ such that $\psi(U^t \xi) = e^{i\lambda t} \psi(\xi)$.*

The random spectral measure $Z(\Delta)$ for the process X_t is now defined for $\Delta \subseteq \mathbb{R}$ by the same formula

$$Z(\Delta) = \psi^{-1}(\chi_\Delta).$$

Given the random spectral measure, we can recover the process X_t via

$$X_t = \int_{\mathbb{R}} e^{i\lambda t} dZ(\lambda).$$

As in the case of discrete time, the subspace $H_{-\infty}^t$ of the space H is defined as the closure of the space of all finite sums $\sum c_s X_s$ for which $s \leq t$.

Definition 15.27. *A random process X_t is called linearly regular if $P_{-\infty}^{-t} X_0 \to 0$ as $t \to \infty$.*

We state the following theorem without providing a proof.

Theorem 15.28 (Krein). *A wide-sense stationary process with continuous covariance function is linearly regular if and only if the spectral measure ρ is absolutely continuous with respect to the Lebesgue measure and*

$$\int_{-\infty}^{\infty} \frac{\ln p_0(\lambda)}{1 + \lambda^2} d\lambda > -\infty$$

for the spectral density $p_0(\lambda) = d\rho/d\lambda$.

15.8 Problems

1. Let X_t, $t \in \mathbb{R}$, be a bounded wide-sense stationary process. Assume that $X_t(\omega)$ is continuous for almost all ω. Prove that the process Y_n, $n \in \mathbb{Z}$, defined by $Y_n(\omega) = \int_n^{n+1} X_t(\omega) dt$ is wide-sense stationary, and express its covariance function in terms of the covariance of X_t.

2. Let $b(n)$, $n \in \mathbb{Z}$, be the covariance of a zero-mean stationary random process. Prove that if $b(n) \leq 0$ for all $n \neq 0$, then $\sum_{n \in \mathbb{Z}} |b(n)| < \infty$.

3. Let X_t, $t \in \mathbb{R}$, be a stationary Gaussian process, and H the Hilbert space generated by the process. Prove that every element of H is a Gaussian random variable.

4. Give an example of a wide-sense stationary random process X_n such that the time averages

$$(X_0 + X_1 + \ldots + X_{n-1})/n$$

converge to a limit which is not a constant.

5. Let X_t, $t \in \mathbb{R}$, be a wide-sense stationary process with covariance function b and spectral measure ρ. Find the covariance function and the spectral measure of the process $Y_t = X_{2t}$.
   Assume that $X_t(\omega)$ is differentiable and $X_t(\omega), X_t'(\omega)$ are bounded by a constant c for almost all ω. Find the covariance function and the spectral measure of the process $Z_t = X_t'$.

6. Let X_n, $n \in \mathbb{Z}$, be a wide-sense stationary process with spectral measure ρ. Under what conditions on ρ does there exist a wide-sense stationary process Y_n such that

$$X_n = 2Y_n - Y_{n-1} - Y_{n+1}, \quad n \in \mathbb{Z}.$$

7. Let X_n, $n \in \mathbb{Z}$, be a wide-sense stationary process with zero mean and spectral measure ρ. Assume that the spectral measure is absolutely

continuous, and the density p is a twice continuously differentiable periodic function. Prove that there exists the limit

$$\lim_{n \to \infty} \frac{E(X_0 + \ldots + X_{n-1})^2}{n},$$

and find its value.

8. Prove that a homogeneous ergodic Markov chain with the state space $\{1, \ldots, r\}$ is a wide-sense stationary process and its spectral measure has a continuous density.

9. Let X_n, $n \in \mathbb{Z}$, be a wide-sense stationary process. Assume that the spectral measure of X_n has a density p which satisfies $c_1 \leq p(\lambda) \leq c_2$ for some positive c_1, c_2 and all λ. Find the spectral representation of the projection of X_0 onto the space spanned by $\{X_n\}$, $n \neq 0$.

10. Let X_n, $n \in \mathbb{Z}$, be a wide-sense stationary process and $f : \mathbb{Z} \to \mathbb{C}$ a complex-valued function. Prove that for any $K > 0$ the process $Y_n = \sum_{|k| \leq K} f(k) X_{n+k}$ is wide-sense stationary.
Express the spectral measure of the process Y_n in terms of the spectral measure of the process X_n. Prove that if X_n is a linearly regular process, then so is Y_n.

11. Let ξ_1, ξ_2, ... be a sequence of bounded independent identically distributed random variables. Is the process $X_n = \xi_n - \xi_{n-1}$ linearly regular?

12. Let ξ_1, ξ_2, ... be a sequence of bounded independent identically distributed random variables. Consider the process $X_n = \xi_n - c\xi_{n-1}$, where $c \in \mathbb{R}$. Find the best linear prediction of X_0 provided that $X_{-1}, X_{-2}, \ldots$ are known (i.e., find the projection of X_0 on $H_{-\infty}^{-1}$).

13. Let X_n, $n \in \mathbb{Z}$, be a wide-sense stationary process. Assume that the projection of X_0 on $H_{-\infty}^{-1}$ is equal to cX_{-1}, where $0 < c < 1$. Find the spectral measure of the process.

Strictly Stationary Random Processes

16.1 Stationary Processes and Measure Preserving Transformations

Again, we start with a process X_t over a probability space $(\Omega, \mathcal{F}, \mathrm{P})$ with discrete time, that is $T = \mathbb{Z}$ or $T = \mathbb{Z}^+$. We assume that X_t takes values in a measurable space $(S, \mathcal{G})$. In most cases, $(S, \mathcal{G}) = (\mathbb{R}, \mathcal{B}(\mathbb{R}))$. Let $t_1, \ldots t_k$ be arbitrary moments of time and $A_1, \ldots, A_k \in \mathcal{G}$.

Definition 16.1. *A random process X_t is called strictly stationary if for any $t_1, \ldots, t_k \in T$ and $A_1, \ldots, A_k$ the probabilities*

$$\mathrm{P}(X_{t_1+t} \in A_1, \ldots, X_{t_k+t} \subset A_k)$$

do not depend on t, where $t \in T$.

By induction, if the above probability for $t = 1$ is equal to that for $t = 0$ for any $t_1, \ldots, t_k$ and $A_1, \ldots, A_k$, then the process is strictly stationary.

In this chapter the word "stationary" will always mean strictly stationary. Let us give several simple examples of stationary random processes.

Example. A sequence of independent identically distributed random variables is a stationary process (see Problem 1).

Example. Let $X = \{1, \ldots, r\}$ and P be an $r \times r$ stochastic matrix with elements p_{ij}, $1 \leq i, j \leq r$. In Chap. 5 we defined the Markov chain generated by an initial distribution π and the stochastic matrix P as a certain measure on the space of sequences $\omega : \mathbb{Z}^+ \to X$. Assuming that $\pi P = \pi$, let us modify the definition to the case of sequences $\omega : \mathbb{Z} \to X$.

Let $\widetilde{\Omega}$ be the set of all functions $\omega : \mathbb{Z} \to X$. Let $\mathcal{B}$ be the σ-algebra generated by the cylindrical subsets of $\widetilde{\Omega}$, and $\widetilde{\mathrm{P}}$ the measure on $(\widetilde{\Omega}, \mathcal{B})$ for which

$$\widetilde{\mathrm{P}}(\omega_k = i_0, \ldots, \omega_{k+n} = i_n) = \pi_{i_0} \cdot p_{i_0 i_1} \cdot \ldots \cdot p_{i_{n-1} i_n}$$

L. Koralov and Y.G. Sinai, *Theory of Probability and Random Processes*, Universitext, DOI 10.1007/978-3-540-68829-7_16, © Springer-Verlag Berlin Heidelberg 2012

for each $i_0, \ldots, i_n \in X$, $k \in \mathbb{Z}$, and $n \in \mathbb{Z}^+$. By the Kolmogorov Consistency Theorem, such $\tilde{P}$ exists and is unique. The term "Markov chain with the stationary distribution π and the transition matrix P" can be applied to the measure $\tilde{P}$ as well as to any process with values in X and time $T = \mathbb{Z}$ which induces the measure $\tilde{P}$ on $(\tilde{\Omega}, \mathcal{B})$.

It is not difficult to show that a Markov chain with the stationary distribution π is a stationary process (see Problem 2).

Example. A Gaussian process which is wide-sense stationary is also strictly stationary (see Problem 3).

Let us now discuss measure preserving transformations, which are closely related to stationary processes. Various properties of groups of measure preserving transformations are studied in the branch of mathematics called Ergodic Theory.

By a measure preserving transformation on a probability space $(\Omega, \mathcal{F}, P)$ we mean a measurable mapping $T : \Omega \to \Omega$ such that

$$P(T^{-1}C) = P(C) \quad \text{whenever} \quad C \in \mathcal{F}.$$

By the change of variables formula in the Lebesgue integral, the preservation of measure implies

$$\int_\Omega f(T\omega)dP(\omega) = \int_\Omega f(\omega)dP(\omega)$$

for any $f \in L^1(\Omega, \mathcal{F}, P)$. Conversely, this property implies that T is measure preserving (it is sufficient to consider f equal to the indicator function of the set C).

Let us assume now that we have a measure preserving transformation T and an arbitrary measurable function $f : \Omega \to \mathbb{R}$. We can define a random process X_t, $t \in \mathbb{Z}^+$, by

$$X_t(\omega) = f(T^t\omega).$$

Note that if T is one-to-one and T^{-1} is measurable, then the same formula defines a random process with time $\mathbb{Z}$. Let us demonstrate that the process X_t defined in this way is stationary. Let $t_1, \ldots, t_k$ and $A_1, \ldots, A_k$ be fixed, and $C = \{\omega : f(T^{t_1}\omega) \in A_1, \ldots, f(T^{t_k}\omega) \in A_k\}$. Then,

$$P(X_{t_1+1} \in A_1, \ldots, X_{t_k+1} \in A_k)$$

$$= P(\{\omega : f(T^{t_1+1}\omega) \in A_1, \ldots, f(T^{t_k+1}\omega) \in A_k\}) = P(\{\omega : T\omega \in C\})$$

$$= P(C) = P(X_{t_1} \in A_1, \ldots, X_{t_k} \in A_k),$$

which means that X_t is stationary.

Conversely, let us start with a stationary random process X_t. Let $\tilde{\Omega}$ now be the space of functions defined on the parameter set of the process, $\mathcal{B}$ the minimal σ-algebra containing all the cylindrical sets, and $\widetilde{P}$ the measure on $(\widetilde{\Omega}, \mathcal{B})$ induced by the process X_t (as in Sect. 12.1). We can define the shift transformation $T : \widetilde{\Omega} \to \widetilde{\Omega}$ by

$$T\widetilde{\omega}(t) = \widetilde{\omega}(t+1).$$

From the stationarity of the process it follows that the transformation T preserves the measure $\widetilde{P}$. Indeed, if C is an elementary cylindrical set of the form

$$C = \{\widetilde{\omega} : \widetilde{\omega}(t_1) \in A_1, \ldots, \widetilde{\omega}(t_k) \in A_k\}, \qquad (16.1)$$

then

$$\widetilde{P}(T^{-1}C) = P(X_{t_1+1} \in A_1, \ldots X_{t_k+1} \in A_k)$$
$$= P(X_{t_1} \in A_1, \ldots, X_{t_k} \in A_k) = \widetilde{P}(C).$$

Since all the sets of the form (16.1) form a π-system, from Lemma 4.13 it follows that $\widetilde{P}(T^{-1}C) = \widetilde{P}(C)$ for all $C \in \mathcal{B}$.

Let us define the function $f : \widetilde{\Omega} \to \mathbb{R}$ by $f(\widetilde{\omega}) = \widetilde{\omega}(0)$. Then the process $Y_t = f(T^t\widetilde{\omega}) = \widetilde{\omega}(t)$ defined on $(\widetilde{\Omega}, \mathcal{B}, \widetilde{P})$ clearly has the same finite-dimensional distributions as the original process X_t.

We have thus seen that measure preserving transformations can be used to generate stationary processes, and that any stationary process is equal, in distribution, to a process given by a measure preserving transformation.

16.2 Birkhoff Ergodic Theorem

One of the most important statements of the theory of stationary processes and ergodic theory is the Birkhoff Ergodic Theorem. We shall prove it in a rather general setting.

Let $(\Omega, \mathcal{F}, P)$ be a probability space and $T : \Omega \to \Omega$ a transformation preserving P. For $f \in L^1(\Omega, \mathcal{F}, P)$, we define the function Uf by the formula $Uf(\omega) = f(T\omega)$.

We shall be interested in the behavior of the time averages

$$A_n f = \frac{1}{n}(f + Uf + \cdots + U^{n-1}f).$$

The von Neumann Ergodic Theorem states that for $f \in L^2(\Omega, \mathcal{F}, P)$, the sequence $A_n f$ converges in $L^2(\Omega, \mathcal{F}, P)$ to a U-invariant function. It is natural to ask whether the almost sure convergence takes place.

Theorem 16.2 (Birkhoff Ergodic Theorem). *Let $(\Omega, \mathcal{F}, P)$ be a probability space, and $T : \Omega \to \Omega$ a transformation preserving the measure P. Then for any $f \in L^1(\Omega, \mathcal{F}, P)$ there exists $\overline{f} \in L^1(\Omega, \mathcal{F}, P)$ such that*

1. $A_n f \to \overline{f}$ both P-*almost surely and in* $L^1(\Omega, \mathcal{F}, P)$ *as* $n \to \infty$.
2. $U\overline{f} = \overline{f}$ *almost surely.*
3. $\int_{\Omega} \overline{f} dP = \int_{\Omega} f dP$.

Proof. Let us show that the convergence almost surely of the time averages implies all the other statements of the theorem. We begin by deriving the L^1-convergence from the convergence almost surely. By the definition of the time averages, for any $f \in L^1(\Omega, \mathcal{F}, P)$, we have

$$|A_n f|_{L^1} \leq |f|_{L^1}.$$

If f is a bounded function, then, by the Lebesgue Dominated Convergence Theorem, the convergence almost surely for $A_n f$ implies the L^1-convergence. Now take an arbitrary $f \in L^1$, and assume that $A_n f$ converges to $\overline{f}$ almost surely. For any $\varepsilon > 0$, we can write $f = f_1 + f_2$, where f_1 is bounded and $|f_2|_{L^1} < \varepsilon/3$. Since $A_n f_1$ converge in L^1 as $n \to \infty$, there exists N such that $|A_n f_1 - A_m f_1|_{L^1} < \varepsilon/3$ for $n, m > N$. Then

$$|A_n f - A_m f|_{L^1} \leq |A_n f_1 - A_m f_1|_{L^1} + |A_n f_2|_{L^1} + |A_m f_2|_{L^1} < \varepsilon.$$

Thus, the sequence $A_n f$ is fundamental in L^1 and, therefore, converges. Clearly, the limit in L^1 is equal to $\overline{f}$. Since

$$\int_{\Omega} A_n f dP = \int_{\Omega} f dP,$$

the L^1-convergence of the time averages implies the third statement of the theorem.

To establish the U-invariance of the limit function $\overline{f}$, we note that

$$U A_n f - A_n f = \frac{1}{n}(U^n f - f),$$

and, therefore,

$$|U A_n f - A_n f|_{L^1} \leq \frac{2|f|_{L^1}}{n}.$$

Therefore, by the L^1-convergence of the time averages $A_n f$, we have $U\overline{f} = \overline{f}$.

Now we prove the almost sure convergence of the time averages. The main step in the proof of the Ergodic Theorem that we present here is an estimate called the Maximal Ergodic Theorem. A weaker, but similar, estimate was a key step in Birkhoff's original paper.

For $f \in L^1(\Omega, \mathcal{F}, P)$, we define the functions f^* and f_* by the formulas

$$f^* = \sup_n A_n f \quad \text{and} \quad f_* = \inf_n A_n f.$$

In order to establish the almost sure convergence the of time averages, we shall obtain bounds on the measures of the sets

$$A(\alpha, f) = \{\omega \in \Omega : f^*(\omega) > \alpha\} \quad \text{and} \quad B(\alpha, f) = \{\omega \in \Omega : f_*(\omega) < \alpha\}.$$

Theorem 16.3 (Maximal Ergodic Theorem). *For any $\alpha \in \mathbb{R}$, we have*

$$\alpha P(A(\alpha, f)) \leq \int_{A(\alpha,f)} f \, dP, \quad \alpha P(B(\alpha, f)) \geq \int_{B(\alpha,f)} f \, dP.$$

Proof. The following proof is due to Adriano Garsia (1973). First, note that the second inequality follows from the first one by applying it to $-f$ and $-\alpha$. Next, observe that it suffices to prove the inequality

$$\int_{A(0,f)} f \geq 0, \tag{16.2}$$

since the general case follows by considering $f' = f - \alpha$. To prove (16.2), set $f_0 = 0$ and, for $n \geq 1$,

$$f_n = \max(f, f + Uf, \ldots, f + \cdots + U^{n-1}f).$$

To establish (16.2), it suffices to prove that

$$\int_{\{f_{n+1}>0\}} f \geq 0 \tag{16.3}$$

holds for all $n \geq 0$. For a function g, denote $g^+ = \max(g, 0)$ and observe that $U(g^+) = (Ug)^+$. Note that

$$f_n \leq f_{n+1} \leq f + Uf_n^+, \tag{16.4}$$

and therefore

$$\int_{\{f_{n+1}>0\}} f \geq \int_{\{f_{n+1}>0\}} f_{n+1} - \int_{\{f_{n+1}>0\}} Uf_n^+ \geq \int_{\Omega} f_{n+1}^+ - \int_{\Omega} Uf_n^+,$$

where the first inequality is due the second inequality in (16.4). Since, on the one hand, $\int_{\Omega} Uf_n^+ = \int_{\Omega} f_n^+$ and, on the other, $f_{n+1}^+ \geq f_n^+$, we conclude that $\int_{\Omega} f_{n+1}^+ - \int_{\Omega} Uf_n^+ \geq 0$. $\qquad \square$

Remark 16.4. In the proof of the Maximal Ergodic Theorem we did not use the fact that P is a probability measure. Therefore, the theorem is applicable to any measure space with a finite non-negative measure.

We now complete the proof of the Birkhoff Ergodic Theorem. For $\alpha, \beta \in \mathbb{R}$, $\alpha < \beta$, denote

$$E_{\alpha,\beta} = \{\omega \in \Omega : \liminf_{n \to \infty} A_n f(\omega) < \alpha < \beta < \limsup_{n \to \infty} A_n f(\omega)\}.$$

If the averages $A_n f$ do not converge P-almost surely, then there exist $\alpha, \beta \in \mathbb{R}$ such that $P(E_{\alpha,\beta}) > 0$. The set $E_{\alpha,\beta}$ is T-invariant. We may therefore apply the Maximal Ergodic Theorem to the transformation T, restricted to $E_{\alpha,\beta}$. We have

$$\{\omega \in E_{\alpha,\beta} : f^*(\omega) > \beta\} = E_{\alpha,\beta}.$$

Therefore, by Theorem 16.3,

$$\int_{E_{\alpha,\beta}} f \geq \beta P(E_{\alpha,\beta}). \qquad (16.5)$$

Similarly,

$$\{\omega \in E_{\alpha,\beta} : f_*(\omega) < \alpha\} = E_{\alpha,\beta},$$

and therefore

$$\int_{E_{\alpha,\beta}} f \leq \alpha P(E_{\alpha,\beta}). \qquad (16.6)$$

The Eqs. (16.5) and (16.6) imply $\alpha P(E_{\alpha,\beta}) \geq \beta P(E_{\alpha,\beta})$. This is a contradiction, which finally establishes the almost sure convergence of the time averages and completes the proof of the Birkhoff Ergodic Theorem. $\qquad\square$

16.3 Ergodicity, Mixing, and Regularity

Let $(\Omega, \mathcal{F}, P)$ be a probability space and $T : \Omega \to \Omega$ a transformation preserving P. We shall consider the stationary random process $X_t = f(T^t\omega)$, where $f \in L^1(\Omega, \mathcal{F}, P)$.

The main conclusion from the Birkhoff Ergodic Theorem is the Strong Law of Large Numbers. Namely, for any stationary random process there exists the almost sure limit

$$\lim_{n\to\infty} \frac{1}{n} \sum_{t=0}^{n-1} U^t f = \lim_{n\to\infty} \frac{1}{n} \sum_{t=0}^{n-1} X_t = \overline{f}(\omega).$$

In the laws of large numbers for sums of independent random variables studied in Chap. 7, the limit $\overline{f}(\omega)$ was a constant: $\overline{f}(\omega) = EX_t$. For a general stationary process this may not be the case. In order to study this question in detail, we introduce the following definitions.

Definition 16.5. *Let $(\Omega, \mathcal{F}, P)$ be a probability space, and $T : \Omega \to \Omega$ a transformation preserving the measure P. A random variable f is called T-invariant (mod 0) if $f(T\omega) = f(\omega)$ almost surely. An event $A \in \mathcal{F}$ is called T-invariant (mod 0) if its indicator function $\chi_A(\omega)$ is T-invariant (mod 0).*

Definition 16.6. *A measure preserving transformation T is called ergodic if each invariant (mod 0) function is a constant almost surely.*

It is easily seen that a measure preserving transformation is ergodic if and only if every T-invariant (mod 0) event has measure one or zero (see Problem 5).

As stated in the Birkhoff Ergodic Theorem, the limit of the time averages $\overline{f}(\omega)$ is T-invariant (mod 0), and therefore $\overline{f}(\omega)$ is a constant almost surely in the case of ergodic T. Since $\int \overline{f}(\omega)dP(\omega) = \int f(\omega)dP(\omega)$, the limit of the time averages equals the mathematical expectation.

Note that the T-invariant (mod 0) events form a σ-algebra. Let us denote it by $\mathcal{G}$. If T is ergodic, then $\mathcal{G}$ contains only events of measure zero and one. Since $\overline{f}(\omega)$ is T-invariant (mod 0), it is measurable with respect to $\mathcal{G}$. If $A \in \mathcal{G}$, then

$$\int_A \overline{f}(\omega)dP = \lim_{n\to\infty} \frac{1}{n}\sum_{t=0}^{n-1}\int_A f(T^t\omega)dP$$

$$= \lim_{n\to\infty}\frac{1}{n}\sum_{t=0}^{n-1}\int_\Omega f(T^t\omega)\chi_A(T^t\omega)dP = \int_\Omega f(\omega)\chi_A(\omega)dP = \int_A f(\omega)dP.$$

Therefore, by the definition of conditional expectation, $\overline{f} = \mathrm{E}(f|\mathcal{G})$.

Our next goal is to study conditions under which T is ergodic.

Definition 16.7. *A measure preserving transformation T is called mixing if for any events $B_1, B_2 \in \mathcal{F}$ we have*

$$\lim_{n\to\infty} \mathrm{P}(B_1 \cap T^{-n}B_2) = \mathrm{P}(B_1)\mathrm{P}(B_2). \tag{16.7}$$

The mixing property can be restated as follows. For any two bounded measurable functions f_1 and f_2

$$\lim_{n\to\infty}\int_\Omega f_1(\omega)f_2(T^n\omega)dP(\omega) = \int_\Omega f_1(\omega)dP(\omega)\int_\Omega f_2(\omega)dP(\omega)$$

(see Problem 6). The function $\rho(n) = \int f_1(\omega)f_2(T^n\omega)dP(\omega)$ is called the time-covariance function. Mixing means that all time-covariance functions tend to zero as $n \to \infty$, provided that at least one of the integrals $\int f_1 dP$ or $\int f_2 dP$ is equal to zero. Mixing implies ergodicity. Indeed, if B is an invariant (mod 0) event, then by (16.7)

$$\mathrm{P}(B) = \mathrm{P}(B \cap T^{-n}B) = \mathrm{P}^2(B),$$

that is $\mathrm{P}(B)$ is either one or zero.

We can formulate the corresponding definitions for stationary processes. Recall that there is a shift transformation on the space of functions defined on the parameter set of the process, with the measure associated to the process.

Definition 16.8. *A stationary process X_t is called ergodic if the corresponding shift transformation is ergodic. The process X_t is called mixing if the corresponding shift transformation is mixing.*

Let us stress the distinction between the ergodicity (mixing) of the underlying transformation T and the ergodicity (mixing) of the process $X_t = f(T^t \omega)$. If T is ergodic (mixing), then X_t is ergodic (mixing). However, this is not a necessary condition if f is fixed. The process X_t may be ergodic (mixing) according to Definition 16.8, even if the transformation T is not: for example, if f is a constant. The ergodicity (mixing) of the process X_t is the property which is determined by the distribution of the process, rather than by the underlying measure.

Now we shall introduce another important notion of the theory of stationary processes. Let the parameter set T be the set of all integers. For $-\infty \le k_1 \le k_2 \le \infty$, let $\mathcal{F}_{k_1}^{k_2} \subseteq \mathcal{F}$ be the smallest σ-algebra containing all the elementary cylinders

$$C = \{\omega : X_{t_1}(\omega) \in A_1, \ldots, X_{t_m}(\omega) \in A_m\},$$

where $t_1, \ldots, t_m \in T$, $k_1 \le t_1, \ldots, t_m \le k_2$, and $A_1, \ldots, A_m$ are Borel sets of the real line. A special role will be played by the σ-algebras $\mathcal{F}_{-\infty}^k$.

Definition 16.9. *A random process is called regular if the σ-algebra $\cap_k \mathcal{F}_{-\infty}^k$ contains only sets of measure one and zero.*

Remark 16.10. Let $\widetilde{\Omega}$ be the space of all functions defined on the parameter set of the process, with the σ-algebra generated by the cylindrical sets and the measure induced by the process. Let $\widetilde{\mathcal{F}}_{k_1}^{k_2}$ be the minimal σ-algebra which contains all the elementary cylindrical sets of the form

$$\{\widetilde{\omega} \in \widetilde{\Omega} : \widetilde{\omega}(t_1) \in A_1, \ldots, \widetilde{\omega}(t_m) \in A_m\},$$

where $t_1, \ldots, t_m \in T$, $k_1 \le t_1, \ldots, t_m \le k_2$, and $A_1, \ldots, A_m$ are Borel sets of the real line. Then the property of regularity can be equivalently formulated as follows. The process is regular if the σ-algebra $\cap_k \widetilde{\mathcal{F}}_{-\infty}^k$ contains only sets of measure one and zero. Therefore, the property of regularity depends only on the distribution of the process.

The σ-algebra $\cap_k \mathcal{F}_{-\infty}^k$ consists of events which depend on the behavior of the process in the infinite past. The property expressed in Definition 16.9 means that there is some loss of memory in the process. We shall need the following theorem by Doob.

Theorem 16.11 (Doob). *Let $\mathcal{H}^k$ be a decreasing sequence of σ-subalgebras of $\mathcal{F}$, $\mathcal{H}^{k+1} \subseteq \mathcal{H}^k$. If $\mathcal{H} = \cap_k \mathcal{H}^k$, then for any $C \in \mathcal{F}$,*

$$\lim_{k \to \infty} \mathrm{P}(C|\mathcal{H}^k) = \mathrm{P}(C|\mathcal{H}) \quad \text{almost surely}.$$

Proof. Let $H_k = L^2(\Omega, \mathcal{H}^k, \mathrm{P})$ be the Hilbert space of L^2 functions measurable with respect to the σ-algebra $\mathcal{H}^k$. Then, by Lemma 13.4, the function $\mathrm{P}(C|\mathcal{H}^k)$

is the projection of the indicator function χ_C onto H_k, while $P(C|\mathcal{H})$ is the projection of χ_C onto $H_\infty = \cap_k H_k$. By Lemma 15.21,

$$\lim_{k\to\infty} P(C|\mathcal{H}^k) = P(C|\mathcal{H}) \text{ in } L^2(\Omega, \mathcal{F}, \mathrm{P}).$$

We need to establish the convergence almost surely, however. Suppose that we do not have the convergence almost surely. Then there are a number $\varepsilon > 0$ and a set $A \in \mathcal{F}$ such that $\mathrm{P}(A) > 0$ and

$$\sup_{k\geq n} |P(C|\mathcal{H}^k)(\omega) - P(C|\mathcal{H})(\omega)| \geq \varepsilon \tag{16.8}$$

for all n and all $\omega \in A$. Take n so large that

$$\mathrm{E}|P(C|\mathcal{H}^n) - P(C|\mathcal{H})| < \frac{\mathrm{P}(A)\varepsilon}{2}.$$

Note that for any $m > n$ the sequence

$$(P(C|\mathcal{H}^m), \mathcal{H}^m)), (P(C|\mathcal{H}^{m-1}), \mathcal{H}^{m-1}), \ldots, (P(C|\mathcal{H}^n), \mathcal{H}^n)$$

is a martingale. Consequently, by the Doob Inequality (Theorem 13.22),

$$\mathrm{P}(\sup_{n\leq k\leq m} |P(C|\mathcal{H}^k) - P(C|\mathcal{H})| \geq \varepsilon)$$

$$\leq \frac{\mathrm{E}|P(C|\mathcal{H}^n) - P(C|\mathcal{H})|}{\varepsilon} < \frac{\mathrm{P}(A)}{2}.$$

Since m was arbitrary, we conclude that

$$\mathrm{P}(\sup_{k\geq n} |P(C|\mathcal{H}^k) - P(C|\mathcal{H})| \geq \varepsilon) \leq \frac{\mathrm{P}(A)}{2},$$

which contradicts (16.8). □

Using the Doob Theorem we shall prove the following statement.

Theorem 16.12. *If a stationary process X_t is regular, then it is mixing (and therefore ergodic).*

Proof. Let T be the shift transformation on $(\widetilde{\Omega}, \mathcal{B})$, and $\widetilde{\mathcal{F}}_{k_1}^{k_2}$ as in Remark 16.10. Let $\widetilde{\mathrm{P}}$ be the measure on $(\widetilde{\Omega}, \mathcal{B})$ induced by the process. We need to show that the relation (16.7) holds for any $B_1, B_2 \in \mathcal{B}$.

Let $\mathcal{G}$ be the collection of the elements of the σ-algebra $\mathcal{B}$ that can be well approximated by elements of $\widetilde{\mathcal{F}}_{-k}^k$ in the following sense: $B \in \mathcal{G}$ if for any $\varepsilon > 0$ there is a finite k and a set $C \in \widetilde{\mathcal{F}}_{-k}^k$ such that $\widetilde{\mathrm{P}}(B\Delta C) \leq \varepsilon$. Note that $\mathcal{G}$ is a Dynkin system. Since $\mathcal{G}$ contains all the cylindrical sets, by Lemma 4.13 it coincides with $\mathcal{B}$.

Therefore, it is sufficient to establish (16.7) for $B_1, B_2 \in \widetilde{\mathcal{F}}^k_{-k}$, where k is fixed. Since the shift transformation T is measure preserving and T^{-1} is measurable,

$$\widetilde{P}(B_1 \cap T^{-n}B_2) = \widetilde{P}(T^nB_1 \cap B_2).$$

It is easy to check that $T^nB_1 \in \widetilde{\mathcal{F}}^{k-n}_{-k-n} \subseteq \widetilde{\mathcal{F}}^{k-n}_{-\infty}$. Therefore,

$$\widetilde{P}(T^nB_1 \cap B_2) = \int_{T^nB_1} \widetilde{P}(B_2|\widetilde{\mathcal{F}}^{k-n}_{-\infty})d\widetilde{P}.$$

By the Doob Theorem and since the process is regular, $\lim_{n\to\infty} \widetilde{P}(B_2|\widetilde{\mathcal{F}}^{k-n}_{-\infty}) = \widetilde{P}(B_2)$ almost surely. Therefore,

$$\lim_{n\to\infty} \widetilde{P}(T^nB_1 \cap B_2) = \lim_{n\to\infty} \widetilde{P}(T^nB_1)\widetilde{P}(B_2) = \widetilde{P}(B_1)\widetilde{P}(B_2).$$

□

Thus, one of the ways to prove ergodicity or mixing of a stationary process is to prove its regularity, that is to prove that the intersection of the σ-algebras $\mathcal{F}^k_{-\infty}$ is the trivial σ-algebra. Statements of this type are sometimes called "zero-one laws", since, for a regular process, the probability of an event which belongs to all the σ-algebras $\mathcal{F}^k_{-\infty}$ is either zero or one. Let us prove the zero-one law for a sequence of independent random variables.

Theorem 16.13. *Let X_t, $t \in \mathbb{Z}$, be independent random variables. Then the process X_t is regular.*

Proof. As in the proof of Theorem 16.12, for an arbitrary $C \in \mathcal{F}^\infty_{-\infty}$, one can find $C_m \in \mathcal{F}^m_{-m}$ such that

$$\lim_{m\to\infty} P(C\Delta C_m) = 0. \qquad (16.9)$$

If, in addition, $C \in \cap_k \mathcal{F}^k_{-\infty}$, then

$$P(C \cap C_m) = P(C)P(C_m), \quad m \geq 1,$$

due to the independence of the σ-algebras $\cap_k \mathcal{F}^k_{-\infty}$ and $\mathcal{F}^m_{-m}$. This equality can be rewritten as

$$(P(C) + P(C_m) - P(C\Delta C_m))/2 = P(C)P(C_m), \quad m \geq 1. \qquad (16.10)$$

By (16.9), $\lim_{m\to\infty} P(C_m) = P(C)$ and therefore, by taking the limit as $m \to \infty$ in (16.10), we obtain

$$P(C) = P^2(C),$$

which implies that $P(C) = 0$ or 1. □

16.4 Stationary Processes with Continuous Time

In this section we shall modify the results on ergodicity, mixing, and regularity, to serve in the case of random processes with continuous time. Instead of a single transformation T, we now start with a measurable semi-group (or group) of transformations. By a measurable semi-group of transformations on a probability space $(\Omega, \mathcal{F}, P)$ preserving the measure P, we mean a family of mappings $T^t : \Omega \to \Omega$, $t \in \mathbb{R}^+$, with the following properties:

1. Each T^t is a measure preserving transformation.
2. For $\omega \in \Omega$ and $s, t \in \mathbb{R}^+$ we have $T^{s+t}\omega = T^s T^t \omega$.
3. The mapping $T^t(\omega) : \Omega \times \mathbb{R}^+ \to \mathbb{R}$ is measurable on the direct product $(\Omega, \mathcal{F}, P) \times (\mathbb{R}^+, \mathcal{B}, \lambda)$, where $\mathcal{B}$ is the σ-algebra of the Borel sets, and λ is the Lebesgue measure on $\mathbb{R}^+$.

For $f \in L^1(\Omega, \mathcal{F}, P)$, we define the time averages

$$A_t f = \frac{1}{t} \int_0^t f(T^s \omega) ds.$$

The Birkhoff Ergodic Theorem can be now formulated as follows. (We provide the statement in the continuous time case without a proof.)

Theorem 16.14 (Birkhoff Ergodic Theorem). *Let $(\Omega, \mathcal{F}, P)$ be a probability space, and T^t a measurable semi-group of transformations, preserving the measure P. Then, for any $f \in L^1(\Omega, \mathcal{F}, P)$, there exists $\overline{f} \in L^1(\Omega, \mathcal{F}, P)$ such that:*

1. *$A_t f \to \overline{f}$ both P-almost surely and in $L^1(\Omega, \mathcal{F}, P)$ as $t \to \infty$.*
2. *For every $t \in \mathbb{R}^+$, $\overline{f}(T^t \omega) = \overline{f}(\omega)$ almost surely.*
3. *$\int_\Omega \overline{f} dP = \int_\Omega f dP$.*

Definition 16.15. *A measurable semi-group of measure preserving transformations T^t is called ergodic if each function invariant (mod 0) for every T^t is a constant almost surely.*

In the ergodic case, the limit of time averages $\overline{f}$ given by the Birkhoff Ergodic Theorem is equal to a constant almost surely.

Definition 16.16. *A measurable semi-group of measure preserving transformations T^t is called mixing, if for any subsets $B_1, B_2 \in \mathcal{F}$ we have*

$$\lim_{t \to \infty} P(B_1 \cap T^{-t} B_2) = P(B_1)P(B_2). \tag{16.11}$$

As in the case of discrete time, mixing implies ergodicity.

Let us now relate measurable semi-groups of measure preserving transformations to stationary processes with continuous time. The definition of a stationary process (Definition 16.1) remains unchanged. Given a semi-group

T^t and an arbitrary measurable function $f : \Omega \to \mathbb{R}$, we can define a random process X_t, $t \in \mathbb{R}^+$, as

$$X_t = f(T^t\omega).$$

It is clear that X_t is a stationary measurable process. Conversely, if we start with a stationary process, we can define the semi-group of shift transformations $T^t : \widetilde{\Omega} \to \widetilde{\Omega}$ by

$$T^t\widetilde{\omega}(s) = \widetilde{\omega}(s + t).$$

This is a semi-group of measure-preserving transformations which, strictly speaking, is not measurable as a function from $\widetilde{\Omega} \times \mathbb{R}^+$, even if the process X_t is measurable. Nevertheless, the notions of ergodicity, mixing, and regularity still make sense for a stationary measurable process.

Definition 16.17. *A stationary process X_t is called ergodic if each measurable function $f : \widetilde{\Omega} \to \mathbb{R}$ which is invariant (mod 0) for the shift transformation T^t for every t is constant almost surely. The process X_t is called mixing if for any subsets $B_1, B_2 \in \mathcal{B}$, we have*

$$\lim_{t\to\infty} \mathrm{P}(B_1 \cap T^{-t}B_2) = \mathrm{P}(B_1)\mathrm{P}(B_2). \qquad (16.12)$$

It is clear that if a semi-group of measure-preserving transformations is ergodic (mixing), and the function f is fixed, then the corresponding stationary process is also ergodic (mixing). The definition of regularity is the same as in the case of discrete time.

Definition 16.18. *A random process X_t, $t \in \mathbb{R}$, is called regular if the σ-algebra $\cap_t \mathcal{F}^t_{-\infty}$ contains only sets of measure one and zero.*

It is possible to show that, for a stationary measurable process, regularity implies mixing which, in turn, implies ergodicity. The Birkhoff Ergodic Theorem can be applied to any stationary measurable L^1-valued process to conclude that the limit $\lim_{t\to\infty} \frac{1}{t} \int_0^t X_s(\omega)ds$ exists almost surely and in $L^1(\Omega, \mathcal{F}, \mathrm{P})$.

16.5 Problems

1. Show that a sequence of independent identically distributed random variables is a stationary random process.
2. Let $X = \{1, \ldots, r\}$, P be an $r \times r$ stochastic matrix, and π a distribution on X such that $\pi P = \pi$. Prove that a Markov chain with the stationary distribution π and the transition matrix P is a stationary random process.
3. Show that if a Gaussian random process is wide-sense stationary, then it is strictly stationary.
4. Let S be the unit circle in the complex plane, and θ a random variable with values in S. Assume that θ is uniformly distributed on S. Prove that $X_n = \theta e^{i\lambda n}$, $n \in \mathbb{Z}$, is a strictly stationary process.

5. Prove that a measure preserving transformation T is ergodic if and only if every T-invariant (mod 0) event has measure one or zero.

6. Prove that the mixing property is equivalent to the following: for any two bounded measurable functions f_1 and f_2

$$\lim_{n\to\infty} \int_\Omega f_1(\omega) f_2(T^n \omega) dP(\omega) = \int_\Omega f_1(\omega) dP(\omega) \int_\Omega f_2(\omega) dP(\omega).$$

7. Let T be the following transformation of the two-dimensional torus

$$T(x_1, x_2) = (\{x_1 + \alpha\}, \{x_2 + x_1\}),$$

where $\{x\}$ stands for the fractional part of x, and α is irrational. Prove that T preserves the Lebesgue measure on the torus, and that it is ergodic. Is T mixing?

8. Let X_n, $n \in \mathbb{Z}$, be a stationary random process such that $E|X_n| < \infty$. Prove that

$$\lim_{n\to\infty} \frac{X_n}{n} = 0$$

almost surely.

9. Let $\xi_1, \xi_2, \ldots$ be a sequence of independent identically distributed random variables with uniform distribution on $[0, 1]$. Prove that the limit

$$\lim_{n\to\infty} \frac{1}{n} \sum_{i=1}^n \sin(2\pi(\xi_{i+1} - \xi_i))$$

exists almost surely, and find its value.

10. Let X_n, $n \in \mathbb{Z}$, be a stationary Gaussian process. Prove that for almost every ω there is a constant $c(\omega)$ such that

$$\max_{0 \le i \le n} |X_i(\omega)| \le c(\omega) \ln n, \quad n = 1, 2, \ldots$$

11. Let $X = \{1, \ldots, r\}$, P be an $r \times r$ stochastic matrix, and π a distribution on X such that $\pi P = \pi$. Let X_t, $t \in \mathbb{Z}$, be a Markov chain with the stationary distribution π and the transition matrix P. Under what conditions on π and P is the Markov chain a regular process.

12. Let $\xi_1, \xi_2, \ldots$ be independent identically distributed integer valued random variables. Assume that the distribution of ξ_1 is symmetric in the sense that $P(\xi_1 = m) = P(\xi_1 = -m)$. Let $S_n = \sum_{i=1}^n \xi_i$. Show that

$$P(\lim_{n\to\infty} S_n = +\infty) = 0.$$

Generalized Random Processes

17.1 Generalized Functions and Generalized Random Processes

We start this section by recalling the definitions of test functions and generalized functions.[1] Thenwe shall introduce the notion of generalized random processes and see that they play the same role, when compared to ordinary random processes, as the generalized functions, when compared to ordinary functions.

As the space of test functions we shall consider the particular example of infinitely differentiable functions whose derivatives decay faster than any power. To simplify the notation we shall define test functions and generalized functions over $\mathbb{R}$, although the definitions can be easily replicated in the case of $\mathbb{R}^n$.

Definition 17.1. *The space S of test functions consists of infinitely differentiable complex-valued functions φ such that for any non-negative integers r and q,*

$$\max_{0 \le s \le r} \sup_{t \in \mathbb{R}} ((1 + t^2)^q |\varphi^{(s)}(t)|) = c_{q,r}(\varphi) < \infty.$$

Note that $c_{q,r}(\varphi)$ are norms on the space S, so that together with the collection of norms $c_{q,r}$, S is a countably-normed linear space. It is, therefore, a linear topological space with the basis of neighborhoods of zero given by the collection of sets $U_{q,r,\varepsilon} = \{\varphi : c_{q,r}(\varphi) < \varepsilon\}$.

Let us now consider the linear continuous functionals on the space S.

Definition 17.2. *The space S' of generalized functions consists of all the linear continuous functionals on the space S.*

---

[1] This chapter can be omitted during the first reading.

L. Koralov and Y.G. Sinai, *Theory of Probability and Random Processes*,
Universitext, DOI 10.1007/978-3-540-68829-7_17,
© Springer-Verlag Berlin Heidelberg 2012

The action of a generalized function $f \in S'$ on a test function φ will be denoted by $f(\varphi)$ or (f, φ). Our basic example of a generalized function is the following. Let $\mu(t)$ be a σ-finite measure on the real line such that the integral

$$\int_{-\infty}^{\infty} (1 + t^2)^{-q} d\mu(t)$$

converges for some q. Then the integral

$$(f, \varphi) = \int_{-\infty}^{\infty} \varphi(t) d\mu(t)$$

is defined for any $\varphi(t) \in S$ and is a continuous linear functional on the space of test functions. Similarly, if $g(t)$ is a continuous complex-valued function whose absolute value is bounded from above by a polynomial, then it defines a generalized function via

$$(f, \varphi) = \int_{-\infty}^{\infty} \varphi(t) \overline{g(t)} dt$$

(the complex conjugation is needed here if $g(t)$ is complex-valued). The space of generalized functions is closed under the operations of taking the derivative and Fourier transform. Namely, for $f \in S'$, we can define

$$(f', \varphi) = -(f, \varphi') \quad \text{and} \quad (\widehat{f}, \varphi) = (f, \widetilde{\varphi}),$$

where $\widetilde{\varphi}$ stands for the inverse Fourier transform of the test function φ. Note that the right-hand sides of these equalities are linear continuous functionals on the space S, and thus the functionals f' and $\widehat{f}$ belong to S'.

Since all the elements of S are bounded continuous functions, they can be considered as elements of S', that is $S \subset S'$. The operations of taking derivative and Fourier transform introduced above are easily seen to coincide with the usual derivative and Fourier transform for the elements of the space S.

Let us now introduce the notion of generalized random processes. From the physical point of view, the concept of a random process X_t is related to measurements of random quantities at certain moments of time, without taking the values at other moments of time into account. However, in many cases, it is impossible to localize the measurements to a single point of time. Instead, one considers the "average" measurements $\Phi(\varphi) = \int \varphi(t) X_t dt$, where φ is a test function. Such measurements should depend on φ linearly, and should not change much with a small change of φ.

This leads to the following definition of generalized random processes.

Definition 17.3. *Let $\Phi(\varphi)$ be a collection of complex-valued random variables on a common probability space $(\Omega, \mathcal{F}, P)$ indexed by the elements of the space of test functions $\varphi \in S$ with the following properties:*

1. *Linearity: $\Phi(a_1 \varphi_1 + a_2 \varphi_2) = a_1 \Phi(\varphi_1) + a_2 \Phi(\varphi_2)$ almost surely, for $a_1, a_2 \in \mathbb{C}$ and $\varphi_1, \varphi_2 \in S$.*

2. *Continuity: If $\psi_k^n \to \varphi_k$ in $\mathcal{S}$ as $n \to \infty$ for $k = 1, \ldots, m$, then the vector-valued random variables $(\Phi(\psi_1^n), \ldots, \Phi(\psi_m^n))$ converge in distribution to $(\Phi(\varphi_1), \ldots, \Phi(\varphi_m))$ as $n \to \infty$.*

Then $\Phi(\varphi)$ is called a generalized random process (over the space $\mathcal{S}$ of test functions).

Note that if $X_t(\omega)$ is an ordinary random process such that $X_t(\omega)$ is continuous in t for almost every ω, and $|X_t(\omega)| \leq p_\omega(t)$ for some polynomial $p_\omega(t)$, then $\Phi(\varphi) = \int \varphi(t)\overline{X_t}dt$ is a generalized random process. Alternatively, we could require that $X_t(\omega)$ be an ordinary random process continuous in t as a function from $\mathbb{R}$ to $L^2(\Omega, \mathcal{F}, \mathrm{P})$ and such that $\|X_t\|_{L^2} \leq p(t)$ for some polynomial $p(t)$.

As with generalized functions, we can define the derivative and Fourier transform of a generalized random process via

$$\Phi'(\varphi) = -\Phi(\varphi'), \quad \widehat{\Phi}(\varphi) = \Phi(\widetilde{\varphi}).$$

A generalized random process Φ is called strictly stationary if, for any $\varphi_1, \ldots, \varphi_n \in \mathcal{S}$ and any $h \in \mathbb{R}$, the random vector $(\Phi(\varphi_1(t+h)), \ldots, \Phi(\varphi_n(t+h)))$ has the same distribution as the vector $(\Phi(\varphi_1(t)), \ldots, \Phi(\varphi_n(t)))$.

We can consider the expectation and the covariance functional of the generalized random process. Namely, assuming that the right-hand side is a continuous functional, we define

$$m(\varphi) = \mathrm{E}\Phi(\varphi).$$

Assuming that the right-hand side is a continuous functional of each of the variables, we define

$$B(\varphi, \psi) = \mathrm{E}\Phi(\varphi)\overline{\Phi(\psi)}.$$

Clearly, the expectation and the covariance functional are linear and hermitian functionals respectively on the space $\mathcal{S}$ (hermitian meaning linear in the first argument and anti-linear in the second). The covariance functional is non-negative definite, that is $B(\varphi, \varphi) \geq 0$ for any φ. A generalized process is called wide-sense stationary if

$$m(\varphi(t)) = m(\varphi(t+h)), \quad B(\varphi(t), \psi(t)) = B(\varphi(t+h), \psi(t+h))$$

for any $h \in \mathbb{R}$. If an ordinary random process is strictly stationary or wide-sense stationary, then so too is the corresponding generalized random process. It is easily seen that the only linear continuous functionals on the space $\mathcal{S}$, which are invariant with respect to translations, are those of the form

$$m(\varphi) = a \int_{-\infty}^{\infty} \varphi(t)dt,$$

where a is a constant. The number a can also be referred to as the expectation of the wide-sense stationary generalized process.

The notions of spectral measure and random spectral measure can be extended to the case of generalized random processes which are wide-sense stationary. Consider a generalized random process with zero expectation. In order to define the notion of spectral measure, we need the following lemma, which we provide here without a proof. (See "Generalized Functions", Volume 4, by I.M. Gelfand and N.Y. Vilenkin.)

Lemma 17.4. *Let $B(\varphi, \psi)$ be a hermitian functional on S, which is continuous in each of the arguments, translation-invariant, and non-negative definite (that is $B(\varphi, \varphi) \geq 0$ for all $\varphi \in S$). Then there is a unique σ-finite measure ρ on the real line such that the integral*

$$\int_{-\infty}^{\infty} (1 + t^2)^{-q} d\rho(t)$$

converges for some $q \geq 0$, and

$$B(\varphi, \psi) = \int_{-\infty}^{\infty} \widehat{\varphi}(\lambda)\overline{\widehat{\psi}(\lambda)} d\rho(\lambda). \tag{17.1}$$

Note that the covariance functional satisfies all the requirements of the lemma. We can thus define the spectral measure as the measure ρ for which (17.1) holds, where B on the left-hand side is the covariance functional.

Furthermore, it can be shown that there exists a unique orthogonal random measure Z such that $E|Z(\Delta)|^2 = \rho(\Delta)$, and

$$\Phi(\varphi) = \int_{-\infty}^{\infty} \widehat{\varphi} dZ(\lambda). \tag{17.2}$$

Let μ_ρ be the generalized function corresponding to the measure ρ. Let $F = \widetilde{\mu}_\rho$ be its inverse Fourier transform in the sense of generalized functions. We can then rewrite (17.1) as

$$B(\varphi, \psi) = (F, \varphi * \psi^*),$$

where the convolution of two test functions is defined as

$$\varphi * \psi(t) = \int_{-\infty}^{\infty} \varphi(s)\psi(t - s) ds,$$

and $\psi^*(t) = \overline{\psi(-t)}$. For generalized processes which are wide-sense stationary, the generalized function F is referred to as the covariance function.

Let us assume that X_t is a stationary ordinary process with zero expectation, which is continuous in the L^2 sense. As previously mentioned, we can also consider it as a generalized process, $\Phi(\varphi) = \int \varphi(t)\overline{X_t} dt$. We have two sets of definitions of the covariance function, spectral measure, and the random orthogonal measure (one for the ordinary process X_t, and the other for the generalized process Φ). It would be natural if the two sets of definitions led to the same concepts of the covariance function, spectral measure, and the random orthogonal measure. This is indeed the case (we leave this statement as an exercise for the reader).

Finally, let us discuss the relationship between generalized random processes and measures on $\mathcal{S}'$. Given a Borel set $B \subseteq \mathbb{C}^n$ and n test functions $\varphi_1, \ldots, \varphi_n$, we define a cylindrical subset of $\mathcal{S}'$ as the set of elements $f \in \mathcal{S}'$ for which $(f(\varphi_1), \ldots, f(\varphi_n)) \in B$. The Borel σ-algebra $\mathcal{F}$ is defined as the minimal σ-algebra which contains all the cylindrical subsets of $\mathcal{S}'$. Any probability measure P on $\mathcal{F}$ defines a generalized process, since $f(\varphi)$ is a random variable on $(\mathcal{S}', \mathcal{F}, P)$ for any $\varphi \in \mathcal{S}$ and all the conditions of Definition 17.3 are satisfied. The converse statement is also true. We formulate it here as a theorem. The proof is non-trivial and we do not provide it here. (See "Generalized Functions", Volume 4, by I.M. Gelfand and N.Y. Vilenkin.)

Theorem 17.5. *Let $\Phi(\varphi)$ be a generalized random process on $\mathcal{S}$. Then there exists a unique probability measure P on $\mathcal{S}'$ such that for any n and any $\varphi_1, \ldots, \varphi_n \in \mathcal{S}$ the random vectors $(f(\varphi_1), \ldots, f(\varphi_n))$ and $(\Phi(\varphi_1), \ldots, \Phi(\varphi_n))$ have the same distributions.*

17.2 Gaussian Processes and White Noise

A generalized random process Φ is called Gaussian if for any test functions $\varphi_1, \ldots, \varphi_k$, the random vector $(\Phi(\varphi_1), \ldots, \Phi(\varphi_k))$ is Gaussian. To simplify the notation, let us consider Gaussian processes with zero expectation. We shall also assume that the process is real-valued, meaning that $\Phi(\varphi)$ is real, whenever φ is a real-valued element of $\mathcal{S}$.

The covariance matrix of the vector $(\Phi(\varphi_1), \ldots, \Phi(\varphi_k))$ is simply $B_{ij} = E(\Phi(\varphi_i)\overline{\Phi}(\varphi_j)) = B(\varphi_i, \varphi_j)$. Therefore, all the finite-dimensional distributions with $\varphi_1, \ldots, \varphi_k$ real are determined by the covariance functional. We shall say that a hermitian form is real if $B(\varphi, \psi)$ is real whenever φ and ψ are real.

Recall that the covariance functional of any generalized random process is a non-negative definite hermitian form which is continuous in each of the variables. We also have the converse statement.

Theorem 17.6. *Let $B(\varphi, \psi)$ be a real non-negative definite hermitian form which is continuous in each of the variables. Then there is a real-valued Gaussian generalized process with zero expectation with $B(\varphi, \psi)$ as its covariance functional.*

To prove this theorem we shall need the following important fact from the theory of countably normed spaces. We provide it here without a proof.

Lemma 17.7. *If a hermitian functional $B(\varphi, \psi)$ on the space $\mathcal{S}$ is continuous in each of the variables separately, then it is continuous in the pair of the variables, that is $\lim_{(\varphi,\psi)\to(\varphi_0,\psi_0)} B(\varphi, \psi) = B(\varphi_0, \psi_0)$ for any (φ_0, ψ_0).*

Proof of Theorem 17.6. Let $\mathcal{S}_r$ be the set of real-valued elements of $\mathcal{S}$. Let Ω be the space of all functions (not necessarily linear) defined on $\mathcal{S}_r$. Let $\mathcal{B}$ be the smallest σ-algebra containing all the cylindrical subsets of Ω, that is the sets of the form

$$\{\omega : (\omega(\varphi_1),\ldots,\omega(\varphi_k)) \in A\},$$

where $\varphi_1,\ldots,\varphi_k \in \mathcal{S}_r$ and A is a Borel subset of $\mathbb{R}^k$. Let $\mathcal{B}_{\varphi_1,\ldots,\varphi_k}$ be the smallest σ-algebra which contains all such sets, where A is allowed to vary but $\varphi_1,\ldots,\varphi_k$ are fixed. We define the measure $\mathrm{P}_{\varphi_1,\ldots,\varphi_k}$ on $\mathcal{B}_{\varphi_1,\ldots,\varphi_k}$ by

$$\mathrm{P}_{\varphi_1,\ldots,\varphi_k}(\{\omega : (\omega(\varphi_1),\ldots,\omega(\varphi_k)) \in A\}) = \eta(A),$$

where η is a Gaussian distribution with the covariance matrix $B_{ij} = B(\varphi_i,\varphi_j)$. The measures $\mathrm{P}_{\varphi_1,\ldots,\varphi_k}$ clearly satisfy the assumptions of Kolmogorov's Consistency Theorem and, therefore, there exists a unique measure P on $\mathcal{B}$ whose restriction to each $\mathcal{B}_{\varphi_1,\ldots,\varphi_k}$ coincides with $\mathrm{P}_{\varphi_1,\ldots,\varphi_k}$.

We define $\Phi(\varphi)$, where $\varphi \in \mathcal{S}_r$ for now, simply by putting $\Phi(\varphi)(\omega) = \omega(\varphi)$. Let us show that $\Phi(\varphi)$ is the desired generalized process. By construction, $\mathrm{E}(\Phi(\varphi)\overline{\Phi}(\psi)) = B(\varphi,\psi)$. Next, let us show that $\Phi(a\varphi + b\psi) = a\Phi(\varphi) + b\Phi(\psi)$ almost surely with respect to the measure P, when $\varphi, \psi \in \mathcal{S}_r$ and $a, b \in \mathbb{R}$. Note that we defined Ω as the set of all functions on $\mathcal{S}_r$, not just the linear ones. To prove the linearity of Φ, note that the variance of $\Phi(a\varphi+b\psi)-a\Phi(\varphi)-b\Phi(\psi)$ is equal to zero. Therefore $\Phi(a\varphi + b\psi) = a\Phi(\varphi) + b\Phi(\psi)$ almost surely.

We also need to demonstrate the continuity of $\Phi(\varphi)$. If $\psi_k^n \to \varphi_k$ in $\mathcal{S}_r$ as $n \to \infty$ for $k = 1,\ldots,m$, then the covariance matrix of the vector $(\Phi(\psi_1^n),\ldots,\Phi(\psi_m^n))$ is $B_{ij}^n = B(\psi_i^n,\psi_j^n)$, while the covariance matrix of the vector $(\Phi(\varphi_1),\ldots,\Phi(\varphi_m))$ is equal to $B_{ij} = B(\varphi_i,\varphi_j)$. If $\psi_k^n \to \varphi_k$ in $\mathcal{S}_r$ as $n \to \infty$ for $k = 1,\ldots,m$, then $\lim_{n\to\infty} B_{ij}^n = B_{ij}$ due to Lemma 17.7. Since the vectors are Gaussian, the convergence of covariance matrices implies the convergence in distribution.

Finally, for $\varphi = \varphi_1 + i\varphi_2$, where φ_1 and φ_2 are real, we define $\Phi(\varphi) = \Phi(\varphi_1) + i\Phi(\varphi_2)$. Clearly, $\Phi(\varphi)$ is the desired generalized random process. $\square$

We shall say that a generalized function F is non-negative definite if $(F, \varphi * \varphi^*) \geq 0$ for any $\varphi \in \mathcal{S}$. There is a one-to-one correspondence between non-negative definite generalized functions and continuous translation-invariant non-negative definite hermitian forms. Namely, given a generalized function F, we can define the form $B(\varphi,\psi) = (F, \varphi * \psi^*)$. Conversely, the existence of the non-negative definite generalized function corresponding to a form is guaranteed by Lemma 17.4. Theorem 17.6 can now be applied in the translation-invariant case to obtain the following statement.

Lemma 17.8. *For any non-negative definite generalized function F, there is a real-valued stationary Gaussian generalized process with zero expectation for which F is the covariance function.*

Let us introduce an important example of a generalized process. Note that the delta-function (the generalized function defined as $(\delta, \varphi) = \varphi(0)$) is non-negative definite.

Definition 17.9. *A real-valued stationary Gaussian generalized process with zero expectation and covariance function equal to delta-function is called white noise.*

Let us examine what happens to the covariance functional of a generalized process when we take the derivative of the process. If B_Φ is the covariance functional of the process Φ and $B_{\Phi'}$ is the covariance functional of Φ', then

$$B_{\Phi'}(\varphi, \psi) = \mathrm{E}(\Phi'(\varphi)\overline{\Phi}'(\psi)) = \mathrm{E}(\Phi(\varphi')\overline{\Phi}(\psi')) = B_\Phi(\varphi', \psi').$$

If the process Φ is stationary, and F_Φ and $F_{\Phi'}$ are the covariance functions of Φ and Φ' respectively, we obtain

$$(F_{\Phi'}, \varphi * \psi^*) = (F_\Phi, \varphi' * (\psi')^*).$$

Since $\varphi' * (\psi')^* = -(\varphi * \psi^*)''$,

$$(F_{\Phi'}, \varphi * \psi^*) = (-F_\Phi'', \varphi * \psi^*).$$

Therefore, the generalized functions $F_{\Phi'}$ and $-F_\Phi''$ agree on all test functions of the form $\varphi * \psi^*$. It is not difficult to show that such test functions are dense in S. Therefore, $F_{\Phi'} = -F_\Phi''$.

In Chap. 18 we shall study Brownian motion (also called Wiener process). It is a real Gaussian process, denoted by W_t, whose covariance functional is given by the formula

$$B_W(\varphi, \psi) = \int_{-\infty}^{\infty} \int_{-\infty}^{\infty} k(s, t)\varphi(s)\overline{\psi}(t)dsdt,$$

where

$$k(s, t) = \begin{cases} \min(|s|, |t|) & \text{if } s \text{ and } t \text{ have the same sign,} \\ 0 & \text{otherwise.} \end{cases}$$

Although the Wiener process itself is not stationary, its derivative is, as will be seen below. Indeed, by using integration by parts,

$$\int_{-\infty}^{\infty} \int_{-\infty}^{\infty} k(s, t)\varphi'(s)\overline{\psi}'(t)dsdt = \int_{-\infty}^{\infty} \varphi(t)\overline{\psi}(t)dt.$$

Therefore, the covariance functional of the derivative of the Wiener process is equal to

$$B_{W'}(\varphi, \psi) = B_W(\varphi', \psi') = \int_{-\infty}^{\infty} \varphi(t)\overline{\psi}(t)dt = (\delta, \varphi * \psi^*).$$

Since the derivative of a Gaussian process is a (generalized) Gaussian process, and the distributions of a Gaussian process are uniquely determined by its covariance function, we see that the derivative of the Wiener process is a white noise.

18

Brownian Motion

18.1 Definition of Brownian Motion

The term Brownian motion comes from the name of the botanist R. Brown, who described the irregular motion of minute particles suspended in water, while the water itself remained seemingly still. It is now known that this motion is due to the cumulative effect of water molecules hitting the particle at various angles.

The rigorous definition and the first mathematical proof of the existence of Brownian motion are due to N. Wiener, who studied Brownian motion in the 1920s, almost a century after it was observed by R. Brown. Wiener process is another term for Brownian motion, both terms being used equally often. Brownian motion and more general diffusion processes are extremely important in physics, economics, finance, and many branches of mathematics beyond probability theory.

We start by defining one-dimensional Brownian motion as a process (with a certain list of properties) on an abstract probability space $(\Omega, \mathcal{F}, P)$. We shall then discuss the space $C([0, \infty))$ of continuous functions, show that it carries a probability measure (Wiener measure) corresponding to Brownian motion, and that $C([0, \infty))$ can be taken as the underlying probability space Ω in the definition of Brownian motion.

Definition 18.1. *A process W_t on a probability space $(\Omega, \mathcal{F}, P)$ is called a one-dimensional Brownian motion if:*

1. *Sample paths $W_t(\omega)$ are continuous functions of t for almost all ω.*
2. *For any $k \geq 1$ and $0 \leq t_1 \leq \ldots \leq t_k$, the random vector $(W_{t_1}, \ldots, W_{t_k})$ is Gaussian with zero mean and covariance matrix $B(t_i, t_j) = \mathrm{E}(W_{t_i} W_{t_j}) = t_i \wedge t_j$, where $1 \leq i, j \leq k$.*

Since the matrix $B(t_i, t_j) = t_i \wedge t_j$ is non-negative definite for any k, and $0 \leq t_1 \leq \ldots \leq t_k$, by the Kolmogorov Consistency Theorem there exists a probability measure on the space $\widetilde{\Omega}$ of all functions such that the process

L. Koralov and Y.G. Sinai, *Theory of Probability and Random Processes*,
Universitext, DOI 10.1007/978-3-540-68829-7_18,
© Springer-Verlag Berlin Heidelberg 2012

$\widetilde{W}_t(\widetilde{\omega}) = \widetilde{\omega}(t)$ is Gaussian with the desired covariance matrix. Since $C([0,\infty))$ is not a measurable set in $\widetilde{\Omega}$, however, we can not simply restrict this measure to the space of continuous functions. This does not preclude us from trying to define another process with the desired properties. We shall prove the existence of a Brownian motion in two different ways in the following sections.

Here is another list of conditions that characterize Brownian motion.

Lemma 18.2. *A process W_t on a probability space $(\Omega, \mathcal{F}, \mathrm{P})$ is a Brownian motion if and only if:*

1. *Sample paths $W_t(\omega)$ are continuous functions of t for almost all ω.*
2. *$W_0(\omega) = 0$ for almost all ω.*
3. *For $0 \le s \le t$, the increment $W_t - W_s$ is a Gaussian random variable with zero mean and variance $t - s$.*
4. *Random variables $W_{t_0}, W_{t_1} - W_{t_0}, \ldots, W_{t_k} - W_{t_{k-1}}$ are independent for every $k \ge 1$ and $0 = t_0 \le t_1 \le \ldots \le t_k$.*

Proof. Assume that W_t is a Brownian motion. Then $\mathrm{E}W_0^2 = 0 \wedge 0 = 0$, which implies that $W_0 = 0$ almost surely.

Let $0 \le s \le t$. Since the vector (W_s, W_t) is Gaussian, so is the random variable $W_t - W_s$. Its variance is equal to

$$\mathrm{E}(W_t - W_s)^2 = t \wedge t + s \wedge s - 2s \wedge t = t - s.$$

Let $k \ge 1$ and $0 = t_0 \le t_1 \le \ldots \le t_k$. Since $(W_{t_0}, \ldots, W_{t_k})$ is a Gaussian vector, so is $(W_{t_0}, W_{t_1} - W_{t_0}, \ldots, W_{t_k} - W_{t_{k-1}})$. In order to verify that its components are independent, it is enough to show that they are uncorrelated. If $1 \le i < j \le k$, then

$$\mathrm{E}[(W_{t_i} - W_{t_{i-1}})(W_{t_j} - W_{t_{j-1}})] = t_i \wedge t_j + t_{i-1} \wedge t_{j-1}$$

$$-t_i \wedge t_{j-1} - t_{i-1} \wedge t_j = t_i + t_{i-1} - t_i - t_{i-1} = 0.$$

Thus a Brownian motion satisfies all the conditions of Lemma 18.2. The converse statement can be proved similarly, so we leave it as an exercise for the reader. $\square$

Sometimes it is important to consider Brownian motion in conjunction with a filtration.

Definition 18.3. *A process W_t on a probability space $(\Omega, \mathcal{F}, \mathrm{P})$ adapted to a filtration $(\mathcal{F}_t)_{t \in \mathbb{R}^+}$ is called a Brownian motion relative to the filtration $\mathcal{F}_t$ if*

1. *Sample paths $W_t(\omega)$ are continuous functions of t for almost all ω.*
2. *$W_0(\omega) = 0$ for almost all ω.*
3. *For $0 \le s \le t$, the increment $W_t - W_s$ is a Gaussian random variable with zero mean and variance $t - s$.*
4. *For $0 \le s \le t$, the increment $W_t - W_s$ is independent of the σ-algebra $\mathcal{F}_s$.*

If we are given a Brownian motion W_t, but no filtration is specified, then we can consider the filtration generated by the process, $\mathcal{F}_t^W = \sigma(W_s, s \leq t)$. Let us show that W_t is a Brownian motion relative to the filtration $\mathcal{F}_t^W$.

Lemma 18.4. *Let X_t, $t \in \mathbb{R}^+$, be a random process such that $X_{t_0}, X_{t_1} - X_{t_0}, \ldots, X_{t_k} - X_{t_{k-1}}$ are independent random variables for every $k \geq 1$ and $0 = t_0 \leq t_1 \leq \ldots \leq t_k$. Then for $0 \leq s \leq t$, the increment $X_t - X_s$ is independent of the σ-algebra $\mathcal{F}_s^X$.*

Proof. For fixed $n \geq 1$ and $0 = t_0 \leq t_1 \leq \ldots \leq t_k \leq s$, the σ-algebra $\sigma(X_{t_0}, X_{t_1}, \ldots, X_{t_k}) = \sigma(X_{t_0}, X_{t_1} - X_{t_0}, \ldots, X_{t_k} - X_{t_{k-1}})$ is independent of $X_t - X_s$. Let $\mathcal{K}$ be the union of all such σ-algebras. It forms a collection of sets closed under pair-wise intersections, and is thus a π-system.

Let $\mathcal{G}$ be the collection of sets which are independent of $X_t - X_s$. Then $A \in \mathcal{G}$ implies that $\Omega \backslash A \in \mathcal{G}$. Furthermore, $A_1, A_2, \ldots \in \mathcal{G}$, $A_n \cap A_m = \emptyset$ for $n \neq m$ imply that $\bigcup_{n=1}^{\infty} A_n \in \mathcal{G}$. Therefore $\mathcal{F}_s^X = \sigma(\mathcal{K}) \subseteq \mathcal{G}$ by Lemma 4.13. $\square$

Let us also define d-dimensional Brownian motion. For a process X_t defined on a probability space $(\Omega, \mathcal{F}, \mathrm{P})$, let $\mathcal{F}^X$ be the σ-algebra generated by X_t, that is $\mathcal{F}^X = \sigma(X_t, t \in T)$. Recall that the processes $X_t^1, \ldots, X_t^d$ defined on a common probability space are said to be independent if the σ-algebras $\mathcal{F}^{X^1}, \ldots, \mathcal{F}^{X^d}$ are independent.

Definition 18.5. *An $\mathbb{R}^d$-valued process $W_t = (W_t^1, \ldots, W_t^d)$ is said to be a (standard) d-dimensional Brownian motion if its components $W_t^1, \ldots, W_t^d$ are independent one-dimensional Brownian motions.*

An $\mathbb{R}^d$-valued process $W_t = (W_t^1, \ldots, W_t^d)$ is said to be a (standard) d-dimensional Brownian motion relative to a filtration $\mathcal{F}_t$ if its components are independent one-dimensional Brownian motions relative to the filtration $\mathcal{F}_t$.

As in the one-dimensional case, if W_t is a d-dimensional Brownian motion, we can consider the filtration $\mathcal{F}_t^W = \sigma(W_s^i, s \leq t, 1 \leq i \leq d)$. Then W_t is a d-dimensional Brownian motion relative to the filtration $\mathcal{F}_t^W$.

18.2 The Space $C([0, \infty))$

Definition 18.6. *The space $C([0, \infty))$ is the metric space which consists of all continuous real-valued functions $\omega = \omega(t)$ on $[0, \infty)$ with the metric*

$$d(\omega_1, \omega_2) = \sum_{n=1}^{\infty} \frac{1}{2^n} \min(\sup_{0 \leq t \leq n} |\omega_1(t) - \omega_2(t)|, 1).$$

Remark 18.7. One can also consider the space $C([0, T])$ of continuous real-valued functions $\omega = \omega(t)$ on $[0, T]$ with the metric of uniform convergence

$$d(\omega_1, \omega_2) = \sup_{0 \le t \le T} |\omega_1(t) - \omega_2(t)|.$$

Convergence in the metric of $C([0, \infty))$ is equivalent to uniform convergence on each finite interval $[0, T]$. This, however, does not imply uniform convergence on the entire axis $[0, \infty)$. It can be easily checked that $C([0, T])$ and $C([0, \infty))$ are complete, separable metric spaces. Note that $C([0, T])$ is in fact a Banach space with the norm $\|\omega\| = \sup_{0 \le t \le T} |\omega(t)|$.

We can consider cylindrical subsets of $C([0, \infty))$. Namely, given a finite collection of points $t_1, \ldots, t_k \in \mathbb{R}^+$ and a Borel set $A \in \mathcal{B}(\mathbb{R}^k)$, we define a cylindrical subset of $C([0, \infty))$ as

$$\{\omega : (\omega(t_1), \ldots, \omega(t_k)) \in A\}.$$

Denote by $\mathcal{B}$ the minimal σ-algebra that contains all the cylindrical sets (for all choices of $k, t_1, \ldots, t_k$, and A).

Lemma 18.8. *The minimal σ-algebra $\mathcal{B}$ that contains all the cylindrical sets is the σ-algebra of Borel sets of $C([0, \infty))$.*

Proof. Let us first show that all cylindrical sets are Borel sets. All cylindrical sets belong to the minimal σ-algebra which contains all sets of the form $B = \{\omega : \omega(t) \in A\}$, where $t \in \mathbb{R}^+$ and A is open in $\mathbb{R}$. But B is open in $C([0, \infty))$ since, together with any $\overline{\omega} \in B$, it contains a sufficiently small ball $B(\overline{\omega}, \varepsilon) = \{\omega : d(\omega, \overline{\omega}) < \varepsilon\}$. Therefore, all cylindrical sets are Borel sets. Consequently, $\mathcal{B}$ is contained in the Borel σ-algebra.

To prove the converse inclusion, note that any open set is a countable union of open balls, since the space $C([0, \infty))$ is separable. We have

$$B(\overline{\omega}, \varepsilon) = \{\omega : \sum_{n=1}^{\infty} \frac{1}{2^n} \min(\sup_{0 \le t \le n} |\omega(t) - \overline{\omega}(t)|, 1) < \varepsilon\}$$

$$= \{\omega : \sum_{n=1}^{\infty} \frac{1}{2^n} \min(\sup_{0 \le t \le n, t \in \mathbb{Q}} |\omega(t) - \overline{\omega}(t)|, 1) < \varepsilon\},$$

where $\mathbb{Q}$ is the set of rational numbers.

The function $f(\omega) = \sup_{0 \le t \le n, t \in \mathbb{Q}} |\omega(t) - \overline{\omega}(t)|$ defined on $C([0, \infty))$ is measurable with respect to the σ-algebra $\mathcal{B}$ generated by the cylindrical sets and, therefore, $B(\overline{\omega}, \varepsilon)$ belongs to $\mathcal{B}$. We conclude that all open sets and, therefore, all Borel sets belong to the minimal σ-algebra which contains all cylindrical sets. □

This lemma shows, in particular, that any random process $X_t(\omega)$ with continuous realizations defined on a probability space $(\Omega, \mathcal{F}, \mathrm{P})$ can be viewed as a measurable function from $(\Omega, \mathcal{F})$ to $(C([0, \infty)), \mathcal{B})$, and thus induces a probability measure on $\mathcal{B}$.

Conversely, given a probability measure P on $(C([0, \infty)), \mathcal{B})$, we can define the random process on the probability space $(C([0, \infty)), \mathcal{B}, P)$ which is simply

$$X_t(\omega) = \omega(t).$$

Given a finite collection of points $t_1, \ldots, t_k \in \mathbb{R}^+$, we can define the projection mapping $\pi_{t_1, \ldots, t_k} : C([0, \infty)) \to \mathbb{R}^k$ as

$$\pi_{t_1, \ldots, t_k}(\omega) = (\omega(t_1), \ldots, \omega(t_k)).$$

This mapping is continuous and thus measurable. The cylindrical sets of $C([0, \infty))$ are exactly the pre-images of Borel sets under the projection mappings. Given a measure P on $C([0, \infty))$, we can consider $\widetilde{P}(A) = P(\pi_{t_1, \ldots, t_k}^{-1}(A))$, a measure on $\mathbb{R}^k$ that is the push-forward of P under the projection mapping. These measures will be referred to as finite-dimensional measures or finite-dimensional distributions of P.

Let P_n be a sequence of probability measures on $C([0, \infty))$, and $\widetilde{P}_n$ their finite-dimensional distributions for given $t_1, \ldots, t_k$. If f is a bounded continuous function from $\mathbb{R}^k$ to $\mathbb{R}$, then $f(\pi_{t_1, \ldots, t_k}) : C([0, \infty)) \to \mathbb{R}$ is also bounded and continuous. Therefore, if the P_n converge to P weakly, then

$$\int_{\mathbb{R}^k} f d\widetilde{P}_n = \int_{C([0, \infty))} f(\pi_{t_1, \ldots, t_k}) dP_n \to \int_{C([0, \infty))} f(\pi_{t_1, \ldots, t_k}) dP = \int_{\mathbb{R}^k} f d\widetilde{P},$$

that is, the finite-dimensional distributions also converge weakly. Conversely, the convergence of the finite-dimensional distributions implies the convergence of the measures on $C([0, \infty))$, provided the sequence of measures P_n is tight.

Lemma 18.9. *A sequence of probability measures on $(C([0, \infty)), \mathcal{B})$ converges weakly if and only if it is tight and all of its finite-dimensional distributions converge weakly.*

Remark 18.10. When we state that convergence of finite-dimensional distributions and tightness imply weak convergence, we do not require that all the finite-dimensional distributions converge to the finite-dimensional distributions of the same measure on $C([0, \infty))$. The fact that they do converge to the finite-dimensional distributions of the same measure follows from the proof of the lemma.

Proof. If P_n is a sequence of probability measures converging weakly to a measure P, then it is weakly compact, and therefore tight by the Prokhorov Theorem. The convergence of the finite-dimensional distributions of P_n to those of P was justified above.

To prove the converse statement, assume that a sequence of measures is tight, and the finite-dimensional distributions converge weakly. For each $k \geq 1$ and $t_1, \ldots, t_k$, let $\widetilde{P}_n^{t_1, \ldots, t_k}$ be the finite-dimensional distribution of the measure P_n, and $\mu_{t_1, \ldots, t_k}$ be the measure on $\mathbb{R}^k$ such that $\widetilde{P}_n^{t_1, \ldots, t_k} \to \mu_{t_1, \ldots, t_k}$ weakly.

Again by the Prokhorov Theorem, there is a subsequence P'_n of the original sequence converging weakly to a measure P. If a different subsequence P''_n converges weakly to a measure Q, then P and Q have the same finite-dimensional distributions (namely $\mu_{t_1,\ldots,t_k}$) and, therefore, must coincide. Let us demonstrate that the original sequence P_n converges to the same limit. If this is not the case, there exists a bounded continuous function f on $C([0,\infty))$ and a subsequence $\overline{P}_n$ such that $\int f d\overline{P}_n$ do not converge to $\int f dP$. Then one can find a subsequence $\overline{\overline{P}}_n$ of $\overline{P}_n$ such that $|\int f d\overline{\overline{P}}_n - \int f dP| > \varepsilon$ for some $\varepsilon > 0$ and all n. On the other hand, the sequence $\overline{\overline{P}}_n$ is tight and contains a subsequence which converges to P. This leads to a contradiction, and therefore P_n converges to P. □

We shall now work towards formulating a useful criterion for tightness of a sequence of probability measures on $C([0,\infty))$. We define the modulus of continuity of a function $\omega \in C([0,\infty))$ on the interval $[0,T]$ by

$$m^T(\omega, \delta) = \sup_{|t-s| \le \delta, \ 0 \le s,t \le T} |\omega(t) - \omega(s)|.$$

Note that the function $m^T(\omega, \delta)$ is continuous in ω in the metric of $C([0,\infty))$. This implies that the set $\{\omega : m^T(\omega, \delta) < \varepsilon\}$ is open for any $\varepsilon > 0$. Also note that $\lim_{\delta \to 0} m^T(\omega, \delta) = 0$ for any ω.

Definition 18.11. *A set of functions $A \subseteq C([0,\infty))$ (or $A \subseteq C([0,T])$) is called equicontinuous on the interval $[0,T]$ if*

$$\lim_{\delta \to 0} \sup_{\omega \in A} m^T(\omega, \delta) = 0.$$

It is called uniformly bounded on the interval $[0,T]$ if it is bounded in the $C([0,T])$ norm, that is

$$\sup_{\omega \in A} \sup_{0 \le t \le T} |\omega(t)| < \infty.$$

Theorem 18.12 (Arzela-Ascoli Theorem). *A set $A \subseteq C([0,\infty))$ has compact closure if and only if it is uniformly bounded and equicontinuous on every interval $[0,T]$.*

Proof. Let us assume that A is uniformly bounded and equicontinuous on every interval $[0,T]$. In order to prove that the closure of A is compact, it is sufficient to demonstrate that every sequence $(\omega_n)_{n \ge 1} \subseteq A$ has a convergent subsequence.

Let $(q_1, q_2, \ldots)$ be an enumeration of $\mathbb{Q}^+$ (the set of non-negative rational numbers). Since the sequence $(\omega_n(q_1))_{n \ge 1}$ is bounded, we can select a subsequence of functions $(\omega_{1,n})_{n \ge 1}$ from the sequence $(\omega_n)_{n \ge 1}$ such that the numeric sequence $(\omega_{1,n}(q_1))_{n \ge 1}$ converges to a limit. From the sequence

$(\omega_{1,n})_{n \geq 1}$ we can select a subsequence $(\omega_{2,n})_{n \geq 1}$ such that $(\omega_{2,n}(q_2))_{n \geq 1}$ converges to a limit. We can continue this process, and then consider the diagonal sequence $(\overline{\omega}_n)_{n \geq 1} = (\omega_{n,n})_{n \geq 1}$, which is a subsequence of the original sequence and has the property that $(\overline{\omega}_n(q))_{n \geq 1}$ converges for all $q \in \mathbb{Q}^+$.

Let us demonstrate that, for each T, the sequence $(\overline{\omega}_n)_{n \geq 1}$ is a Cauchy sequence in the metric of uniform convergence on $[0, T]$. This will imply that it converges uniformly to a continuous function on each finite interval, and therefore converges in the metric of $C([0, \infty))$. Given $\varepsilon > 0$, we take $\delta > 0$ such that

$$\sup_{\omega \in A} m^T(\omega, \delta) < \frac{\varepsilon}{3}.$$

Let S be a finite subset of $\mathbb{Q}^+$ such that $\text{dist}(t, S) < \delta$ for every $t \in [0, T]$. Let us take n_0 such that $|\overline{\omega}_n(q) - \overline{\omega}_m(q)| < \frac{\varepsilon}{3}$ for $m, n \geq n_0$ for all $q \in S$. Then $\sup_{t \in [0, T]} |\overline{\omega}_n(t) - \overline{\omega}_m(t)| < \varepsilon$ if $m, n \geq n_0$. Indeed, for any $t \in [0, T]$ we can find $q \in S$ with $\text{dist}(t, S) < \delta$ and

$$|\overline{\omega}_n(t) - \overline{\omega}_m(t)| \leq |\overline{\omega}_n(t) - \overline{\omega}_n(q)| + |\overline{\omega}_n(q) - \overline{\omega}_m(q)| + |\overline{\omega}_m(t) - \overline{\omega}_m(q)| < \varepsilon.$$

Thus $(\overline{\omega}_n)_{n \geq 1}$ is a Cauchy sequence, and the set A has compact closure.

Conversely, let us assume that A has compact closure. Let $T > 0$ be fixed. To show that A is uniformly bounded on $[0, T]$, we introduce the sets $U_k = \{\omega : \sup_{0 \leq t \leq T} |\omega(t)| < k\}$. Clearly these sets are open in the metric of $C([0, \infty))$, and $C([0, \infty)) = \bigcup_{k=1}^{\infty} U_k$. Therefore $A \subseteq U_k$ for some k, which shows that A is uniformly bounded on $[0, T]$.

Let $\varepsilon > 0$ be fixed. Consider the sets $V_\delta = \{\omega : m^T(\omega, \delta) < \varepsilon\}$. These sets are open, and $C([0, \infty)) = \bigcup_{\delta > 0} V_\delta$. Therefore $A \subseteq V_\delta$ for some $\delta > 0$, which shows that $\sup_{\omega \in A} m^T(\omega, \delta) \leq \varepsilon$. Since $\varepsilon > 0$ was arbitrary, this shows that A is equicontinuous on $[0, T]$. $\qquad \square$

With the help of the Arzela-Ascoli Theorem we can now prove the following criterion for tightness of a sequence of probability measures.

Theorem 18.13. *A sequence P_n of probability measures on $(C([0, \infty)), \mathcal{B})$ is tight if and only if the following two conditions hold:*

(a) For any $T > 0$ and $\eta > 0$, there is $a > 0$ such that

$$P_n(\{\omega : \sup_{0 \leq t \leq T} |\omega(t)| > a\}) \leq \eta, \quad n \geq 1.$$

(b) For any $T > 0$, $\eta > 0$, and $\varepsilon > 0$, there is $\delta > 0$ such that

$$P_n(\{\omega : m^T(\omega, \delta) > \varepsilon\}) \leq \eta, \quad n \geq 1.$$

Proof. Assume first that the sequence P_n is tight. Given $\eta > 0$, we can find a compact set K with $P_n(K) \geq 1 - \eta$ for all n. Let $T > 0$ and $\varepsilon > 0$ be also given. By the Arzela-Ascoli Theorem, there exist $a > 0$ and $\delta > 0$ such that

$$\sup_{\omega \in K} \sup_{0 \le t \le T} |\omega(t)| < a \quad \text{and} \quad \sup_{\omega \in K} m^T(\omega, \delta) < \varepsilon.$$

This proves that conditions (a) and (b) are satisfied.

Let us now assume that (a) and (b) are satisfied. For a given $\eta > 0$ and every positive integers T and m, we find $a_T > 0$ and $\delta_{m,T} > 0$ such that

$$P_n(\{\omega : \sup_{0 \le t \le T} |\omega(t)| > a_T\}) \le \frac{\eta}{2^{T+1}}, \quad n \ge 1,$$

and

$$P_n(\{\omega : m^T(\omega, \delta_{m,T}) > \frac{1}{m}\}) \le \frac{\eta}{2^{T+1+m}}, \quad n \ge 1.$$

The sets $A_T = \{\omega : \sup_{0 \le t \le T} |\omega(t)| \le a_T\}$ and $B_{m,T} = \{\omega : m^T(\omega, \delta_{m,T}) \le \frac{1}{m}\}$ are closed and satisfy

$$P_n(A_T) \ge 1 - \frac{\eta}{2^{T+1}}, \quad P_n(B_{m,T}) \ge 1 - \frac{\eta}{2^{T+1+m}}, \quad n \ge 1.$$

Therefore,

$$P_n((\bigcap_{T=1}^{\infty} A_T) \bigcap (\bigcap_{m,T=1}^{\infty} B_{m,T})) \ge 1 - \sum_{T=1}^{\infty} \frac{\eta}{2^{T+1}} - \sum_{m,T=1}^{\infty} \frac{\eta}{2^{T+1+m}} = 1 - \eta.$$

The set $K = (\bigcap_{T=1}^{\infty} A_T) \bigcap (\bigcap_{m,T=1}^{\infty} B_{m,T})$ is compact by the Arzela-Ascoli Theorem. We have thus exhibited a compact set K such that $P_n(K) \ge 1 - \eta$ for all n. This implies tightness since η was an arbitrary positive number. $\square$

18.3 Existence of the Wiener Measure, Donsker Theorem

Definition 18.14. *A probability measure $\mathcal{W}$ on $(C([0,\infty)), \mathcal{B})$ is called the Wiener measure if the coordinate process $W_t(\omega) = \omega(t)$ on $(C([0,\infty)), \mathcal{B}, \mathcal{W})$ is a Brownian motion relative to the filtration $\mathcal{F}_t^W$.*

In this section we shall give a constructive proof of the existence of the Wiener measure. By Lemma 18.4, in order to show that $\mathcal{W}$ is the Wiener measure, it is sufficient to show that the increments of the coordinate process $W_t - W_s$ are independent Gaussian variables with respect to $\mathcal{W}$, with zero mean, variance $t - s$, and $W_0 = 0$ almost surely. Also note that a measure which has these properties is unique.

Let $\xi_1, \xi_2, \dots$ be a sequence of independent identically distributed random variables on a probability space $(\Omega, \mathcal{F}, P)$. We assume that the expectation of each of the variables is equal to zero and the variance is equal to one. Let S_n be the partial sums, that is $S_0 = 0$ and $S_n = \sum_{i=1}^{n} \xi_i$ for $n \geq 1$. We define a sequence of measurable functions $X_t^n : \Omega \to C([0, \infty))$ via

$$X_t^n(\omega) = \frac{1}{\sqrt{n}} S_{[nt]}(\omega) + (nt - [nt]) \frac{1}{\sqrt{n}} \xi_{[nt]+1}(\omega),$$

where $[t]$ stands for the integer part of t. One can think of X_t^n as a random walk with steps of order $\frac{1}{\sqrt{n}}$ and time steps of size $\frac{1}{n}$. In between the consecutive steps of the random walk the value of X_t^n is obtained by linear interpolation. The following theorem is due to Donsker.

Theorem 18.15 (Donsker). *The measures on $C([0, \infty))$ induced by X_t^n converge weakly to the Wiener measure.*

The proof of the Donsker Theorem will rely on a sequence of lemmas.

Lemma 18.16. *For $0 \leq t_1 \leq \ldots \leq t_k$,*

$$\lim_{n \to \infty} (X_{t_1}^n, \ldots, X_{t_k}^n) = (\eta_{t_1}, \ldots, \eta_{t_k}) \quad in \quad distribution,$$

where $(\eta_{t_1}, \ldots, \eta_{t_k})$ is a Gaussian vector with zero mean and the covariance matrix $\mathrm{E}\eta_{t_i}\eta_{t_j} = t_j \wedge t_i$.

Proof. It is sufficient to demonstrate that the vector $(X_{t_1}^n, X_{t_2}^n - X_{t_1}^n, \ldots, X_{t_k}^n - X_{t_{k-1}}^n)$ converges to a vector of independent Gaussian variables with variances $t_1, t_2 - t_1, \ldots, t_k - t_{k-1}$. Since the term $(nt - [nt]) \frac{1}{\sqrt{n}} \xi_{[nt]+1}$ converges to zero in probability for every t, it is sufficient to establish the convergence to a Gaussian vector for

$$(V_1^n, \ldots, V_k^n) = (\frac{1}{\sqrt{n}} S_{[nt_1]}, \frac{1}{\sqrt{n}} S_{[nt_2]} - \frac{1}{\sqrt{n}} S_{[nt_1]}, \ldots, \frac{1}{\sqrt{n}} S_{[nt_k]} - \frac{1}{\sqrt{n}} S_{[nt_{k-1}]}).$$

Each of the components converges to a Gaussian random variable by the Central Limit Theorem for independent identically distributed random variables. Let us write $\xi_j = \lim_{n \to \infty} V_j^n$, and let $\varphi_j(\lambda_j)$ be the characteristic function of ξ_j. Thus, $\varphi_1(\lambda_1) = e^{-\frac{t_1 \lambda_1^2}{2}}$, $\varphi_2(\lambda_2) = e^{-\frac{(t_2 - t_1)\lambda_2^2}{2}}$, etc.

In order to show that the vector $(V_1^n, \ldots, V_k^n)$ converges to a Gaussian vector, it is sufficient to consider the characteristic function $\varphi^n(\lambda_1, \ldots, \lambda_k) = \mathrm{E}e^{i(\lambda_1 V_1^n + \ldots + \lambda_k V_k^n)}$. Due to independence of the components of the vector $(V_1^n, \ldots, V_k^n)$, the characteristic function $\varphi^n(\lambda_1, \ldots, \lambda_k)$ is equal to the product of the characteristic functions of the components, and thus converges to $\varphi_1(\lambda_1) \cdot \ldots \cdot \varphi_k(\lambda_k)$, which is the characteristic function of a Gaussian vector with independent components. $\square$

Let us now prove that the family of measures induced by X_t^n is tight. First, we use Theorem 18.13 to prove the following lemma.

Lemma 18.17. *A sequence* P_n *of probability measures on* $(C([0,\infty)), \mathcal{B})$ *is tight if the following two conditions hold:*

(a) For any $\eta > 0$, *there is* $a > 0$ *such that*

$$P_n(\{\omega : |\omega(0)| > a\}) \leq \eta, \quad n \geq 1.$$

(b) For any $T > 0$, $\eta > 0$, *and* $\varepsilon > 0$, *there are* $0 < \delta < 1$ *and an integer* n_0 *such that, for all* $t \in [0,T]$, *we have*

$$P_n(\{\omega : \sup_{t \leq s \leq \min(t+\delta, T)} |\omega(s) - \omega(t)| > \varepsilon\}) \leq \delta\eta, \quad n \geq n_0.$$

Proof. Let us show that assumption (b) in this lemma implies assumption (b) of Theorem 18.13. For fixed δ we denote

$$A_t = \{\omega : \sup_{t \leq s \leq \min(t+2\delta, T)} |\omega(s) - \omega(t)| > \frac{\varepsilon}{2}\}.$$

By the second assumption of the lemma, we can take δ and n_0 such that $P_n(A_t) \leq \frac{\delta\eta}{T}$ for all t and $n \geq n_0$.

Consider $[\frac{T}{\delta}]$ overlapping intervals, $I_0 = [0, 2\delta], I_1 = [\delta, 3\delta], \ldots, I_{[\frac{T}{\delta}]-1} = [([\frac{T}{\delta}] - 1)\delta, T]$. If $|s - t| \leq \delta$, there is at least one interval such that both s and t belong to it. Therefore,

$$P_n(\{\omega : m^T(\omega, \delta) > \varepsilon\}) \leq P_n(\bigcup_{i=0}^{[\frac{T}{\delta}]-1} A_{i\delta}) \leq \sum_{i=0}^{[\frac{T}{\delta}]-1} P_n(A_{i\delta}) \leq \frac{T}{\delta} \frac{\delta\eta}{T} = \eta.$$

Thus, we have justified that assumption (b) of Theorem 18.13 holds for $n \geq n_0$. Since a finite family of measures on $(C([0,\infty)), \mathcal{B})$ is always tight, we can take a smaller δ, if needed, to make sure that (b) of Theorem 18.13 holds for $n \geq 1$. This, together with assumption (a) of this lemma, immediately imply that assumption (a) of Theorem 18.13 holds. $\square$

We now wish to apply Lemma 18.17 to the sequence of measures induced by X_t^n. Since $X_0^n = 0$ almost surely, we only need to verify the second assumption of the lemma. We need to show that for any $T > 0$, $\eta > 0$, and $\varepsilon > 0$, there are $0 < \delta < 1$ and an integer n_0 such that for all $t \in [0, T]$ we have

$$P(\{\omega : \sup_{t \leq s \leq \min(t+\delta, T)} |X_s^n - X_t^n| > \varepsilon\}) \leq \delta\eta, \quad n \geq n_0.$$

Since the value of X_t^n changes linearly when t is between integer multiples of $\frac{1}{n}$, and the interval $[t, t+\delta]$ is contained inside the interval $[\frac{k}{n}, \frac{k+[n\delta+2]}{n}]$ for some integer k, it is sufficient to check that for $T > 0$, $\eta > 0$, and $\varepsilon > 0$, there are $0 < \delta < 1$ and an integer n_0 such that

$$P(\{\omega : \max_{k \leq i \leq k+[n\delta+2]} \frac{1}{\sqrt{n}}|S_i - S_k| > \frac{\varepsilon}{2}\}) \leq \delta\eta, \quad n \geq n_0$$

for all k. Obviously, we can replace $\frac{\varepsilon}{2}$ by ε and $[n\delta + 2]$ by $[n\delta]$. Thus, it is sufficient to show that

$$P(\{\omega : \max_{k \leq i \leq k+[n\delta]} \frac{1}{\sqrt{n}}|S_i - S_k| > \varepsilon\}) \leq \delta\eta, \quad n \geq n_0. \tag{18.1}$$

Lemma 18.18. *For any $\varepsilon > 0$, there is $\lambda > 1$ such that*

$$\limsup_{n\to\infty} P(\max_{i \leq n}|S_i| > \lambda\sqrt{n}) \leq \frac{\varepsilon}{\lambda^2}.$$

Before proving this lemma, let us employ it to justify (18.1). Suppose that $\eta > 0$ and $0 < \varepsilon < 1$ are given. By the lemma, there exist $\lambda > 1$ and n_1 such that

$$P(\max_{i \leq n}|S_i| > \lambda\sqrt{n}) \leq \frac{\eta\varepsilon^2}{\lambda^2}, \quad n \geq n_1.$$

Let $\delta = \frac{\varepsilon^2}{\lambda^2}$. Then $0 < \delta < 1$ since $0 < \varepsilon < 1$ and $\lambda > 1$. Take $n_0 = [\frac{n_1}{\delta}] + 1$. Then $n \geq n_0$ implies that $[n\delta] \geq n_1$, and therefore

$$P(\max_{i \leq [n\delta]}|S_i| > \lambda\sqrt{[n\delta]}) \leq \frac{\eta\varepsilon^2}{\lambda^2}.$$

This implies (18.1) with $k = 0$, since $\lambda\sqrt{[n\delta]} \leq \varepsilon\sqrt{n}$ and $\frac{\eta\varepsilon^2}{\lambda^2} = \delta\eta$. Finally, note that the probability on the left-hand side of (18.1) does not depend on k, since the variables $\xi_1, \xi_2, \ldots$ are independent and identically distributed. We have thus established that Lemma 18.18 implies the tightness of the sequence of measures induced by X_t^n.

Proof of Lemma 18.18. Let us first demonstrate that

$$P(\max_{i \leq n}|S_i| > \lambda\sqrt{n}) \leq 2P(|S_n| \geq (\lambda - \sqrt{2})\sqrt{n}) \quad \text{for } \lambda \geq \sqrt{2}. \tag{18.2}$$

Consider the events

$$A_i = \{\max_{j < i}|S_j| < \lambda\sqrt{n} \leq |S_i|\}, \quad 1 \leq i \leq n.$$

Then

$$P(\max_{i \leq n}|S_i| > \lambda\sqrt{n}) \leq P(|S_n| \geq (\lambda - \sqrt{2})\sqrt{n})$$
$$+ \sum_{i=1}^{n-1} P(A_i \cap \{|S_n| < (\lambda - \sqrt{2})\sqrt{n}\}). \tag{18.3}$$

Note that

$$A_i \cap \{|S_n| < (\lambda - \sqrt{2})\sqrt{n}\} \subseteq A_i \cap \{|S_n - S_i| \geq \sqrt{2n}\}.$$

The events A_i and $\{|S_n - S_i| \geq \sqrt{2n}\}$ are independent, while the probability of the latter can be estimated using the Chebyshev Inequality and the fact that $\xi_1, \xi_2, \dots$ is a sequence of independent random variables with variances equal to one,

$$P(|S_n - S_i| \geq \sqrt{2n}) \leq \frac{n-i}{2n} \leq \frac{1}{2}.$$

Therefore,

$$\sum_{i=1}^{n-1} P(A_i \cap \{|S_n| < (\lambda - \sqrt{2})\sqrt{n}\}) \leq \frac{1}{2} \sum_{i=1}^{n-1} P(A_i) \leq \frac{1}{2} P(\max_{i \leq n} |S_i| > \lambda\sqrt{n}).$$

This and (18.3) imply that (18.2) holds. For $\lambda > 2\sqrt{2}$, from (18.2) we obtain

$$P(\max_{i \leq n} |S_i| > \lambda\sqrt{n}) \leq 2P(|S_n| \geq \frac{1}{2}\lambda\sqrt{n}).$$

By the Central Limit Theorem,

$$\lim_{n \to \infty} P(|S_n| \geq \frac{1}{2}\lambda\sqrt{n}) = \frac{1}{\sqrt{2\pi}} \int_{\frac{1}{2}\lambda}^{\infty} e^{-\frac{1}{2}t^2} dt \leq \frac{\varepsilon}{2\lambda^2},$$

where the last inequality holds for all sufficiently large λ. Therefore,

$$\limsup_{n \to \infty} P(\max_{i \leq n} |S_i| > \lambda\sqrt{n}) \leq \frac{\varepsilon}{\lambda^2}.$$

$\square$

We now have all the ingredients needed for the proof of the Donsker Theorem.

Proof of Theorem 18.15. We have demonstrated that the sequence of measures induced by X_t^n is tight. The finite-dimensional distributions converge by Lemma 18.16. Therefore, by Lemma 18.9, the sequence of measures induced by X_t^n converges weakly to a probability measure, which we shall denote by $\mathcal{W}$. By Lemma 18.16 and the discussion following Definition 18.14, the limiting measure $\mathcal{W}$ satisfies the requirements of Definition 18.14. $\square$

18.4 Kolmogorov Theorem

In this section we provide an alternative proof of the existence of Brownian motion. It relies on an important theorem which, in particular, shows that almost all the sample paths of Brownian motion are locally Holder continuous with any exponent $\gamma < 1/2$.

Theorem 18.19 (Kolmogorov). *Let X_t, $t \in \mathbb{R}^+$, be a random process on a probability space $(\Omega, \mathcal{F}, P)$. Suppose that there are positive constants α and β, and for each $T \geq 0$ there is a constant $c(T)$ such that*

$$E|X_t - X_s|^\alpha \leq c(T)|t - s|^{1+\beta} \quad \text{for } 0 \leq s, t \leq T. \tag{18.4}$$

Then there is a continuous modification Y_t of the process X_t such that for every $\gamma \in (0, \beta/\alpha)$ and $T > 0$ there is $\delta > 0$, and for each ω there is $h(\omega) > 0$ such that

$$P(\{\omega : |Y_t(\omega) - Y_s(\omega)| \leq \delta|t - s|^\gamma \text{ for all } s, t \in [0, T], \; 0 \leq t - s < h(\omega)\}) = 1. \tag{18.5}$$

Proof. Let us first construct a process Y_t^1 with the desired properties, with the parameter t taking values in the interval $[0, 1]$. Let $c = c(1)$.

We introduce the finite sets $D_n = \{k/2^n, k = 0, 1, \ldots, 2^n\}$ and the countable set $D = \bigcup_{n=1}^\infty D_n$.

From the Chebyshev Inequality and (18.4) it follows that for any $\varepsilon > 0$ and $0 \leq s, t \leq 1$ we have

$$P(|X_t - X_s| \geq \varepsilon) \leq \frac{E|X_t - X_s|^\alpha}{\varepsilon^\alpha} \leq c|t - s|^{1+\beta}\varepsilon^{-\alpha}. \tag{18.6}$$

In particular, using this inequality with $\varepsilon = 2^{-\gamma n}$ and $k = 1, \ldots, 2^n$, we obtain

$$P(|X_{k/2^n} - X_{(k-1)/2^n}| \geq 2^{-\gamma n}) \leq c2^{-n(1+a)},$$

where $a = \beta - \alpha\gamma > 0$. By taking the union of events on the left-hand side over all values of k, we obtain

$$P(\{\omega : \max_{1 \leq k \leq n} |X_{k/2^n}(\omega) - X_{(k-1)/2^n}(\omega)| \geq 2^{-\gamma n}\}) \leq c2^{-na}.$$

Since the series $\sum_{n=1}^\infty 2^{-na}$ converges, by the first Borel-Cantelli Lemma there exist an event $\Omega' \in \mathcal{F}$ with $P(\Omega') = 1$ and, for each $\omega \in \Omega'$, an integer $n'(\omega)$ such that

$$\max_{1 \leq k \leq n} |X_{k/2^n}(\omega) - X_{(k-1)/2^n}(\omega)| < 2^{-\gamma n} \quad \text{for } \omega \in \Omega' \text{ and } n \geq n'(\omega). \tag{18.7}$$

Let us show that if $\omega \in \Omega'$ is fixed, the function $X_t(\omega)$ is uniformly Holder continuous in $t \in D$ with exponent γ. Take $h(\omega) = 2^{-n'(\omega)}$. Let $s, t \in D$ be such that $0 < t - s < h(\omega)$. Let us take n such that $2^{-(n+1)} \leq t - s < 2^{-n}$. (Note that $n \geq n'(\omega)$ here.) Take m large enough, so that $s, t \in D_m$. Clearly, the interval $[s, t]$ can be represented as a finite union of intervals of the form $[(k-1)/2^j, k/2^j]$ with $n + 1 \leq j \leq m$, with no more than two such intervals for each j. From (18.7) we conclude that

$$|X_t(\omega) - X_s(\omega)| \leq 2 \sum_{j=n+1}^m 2^{-\gamma j} \leq \frac{2}{1 - 2^\gamma} 2^{-\gamma(n+1)} \leq \delta|t - s|^\gamma,$$

where $\delta = 2/(1 - 2^{-\gamma})$.

Let us now define the process Y_t^1. First, we define it for $\omega \in \Omega'$. Since $X_t(\omega)$ is uniformly continuous as a function of t on the set D, and D is dense in $[0,1]$, the following limit is defined for all $\omega \in \Omega'$ and $t \in [0,1]$:

$$Y_t^1(\omega) = \lim_{s \to t, s \in D} X_s(\omega).$$

In particular, $Y_t^1(\omega) = X_t(\omega)$ for $\omega \in \Omega'$, $t \in D$. The function $Y_t^1(\omega)$ is also Holder continuous, that is $|Y_t^1(\omega) - Y_s^1(\omega)| \leq \delta|t-s|^\gamma$ for all $\omega \in \Omega'$ and $s, t \in [0,1]$, $|t-s| < h(\omega)$. For $\omega \in \Omega \setminus \Omega'$ we define $Y_t^1(\omega) = 0$ for all $t \in [0,1]$.

Let us show that Y_t^1 is a modification of X_t. For any $t \in [0,1]$,

$$Y_t^1 = \lim_{s \to t, s \in D} X_s \quad \text{almost surely}$$

(namely, for all $\omega \in \Omega'$), while $X_t = \lim_{s \to t, s \in D} X_s$ in probability due to (18.6). Therefore, $Y_t^1(\omega) = X_t(\omega)$ almost surely for any $t \in [0,1]$.

We defined the process Y_t^1 with the parameter t taking values in the interval $[0,1]$. We can apply the same arguments to construct a process Y_t^m on the interval $[0, 2^m]$. The main difference in the construction is that now the sets D_n are defined as $D_n = \{k/2^n, k = 0, 1, \ldots, 2^{mn}\}$. If t is of the form $t = k/2^n$ for some integers k and n, and belongs to the parameter set of both $Y_t^{m_1}$ and $Y_t^{m_2}$, then $Y_t^{m_1} = Y_t^{m_2} = X_t$ almost surely by the construction. Therefore, the set

$$\widetilde{\Omega} = \{\omega : Y_t^{m_1}(\omega) = Y_t^{m_2}(\omega) \text{ for all } m_1, m_2 \text{ and } t = k/2^n, \, t \leq \min(2^{m_1}, 2^{m_2})\}$$

has measure one. Any two processes $Y_t^{m_1}(\omega)$ and $Y_t^{m_2}(\omega)$ must coincide for all $\omega \in \widetilde{\Omega}$ on the intersection of their parameter sets, since they are both continuous. Therefore, for fixed t and $\omega \in \widetilde{\Omega}$, we can define $Y_t(\omega)$ as any of the processes $Y_t^m(\omega)$ with sufficiently large m. We can define $Y_t(\omega)$ to be equal to zero for all t if $\omega \notin \widetilde{\Omega}$.

By construction, Y_t satisfies (18.5) for each T.   □

Remark 18.20. The function $h(\omega)$ can be taken to be a measurable function of ω, since the same is true for the function $n'(\omega)$ defined in the proof. Clearly, for each T the constant δ can be taken to be arbitrarily small.

The assumptions of the Kolmogorov Theorem are particularly easy to verify if X_t is a Gaussian process.

Theorem 18.21. *Let X_t, $t \in \mathbb{R}^+$, be a real-valued Gaussian random process with zero mean on a probability space $(\Omega, \mathcal{F}, \mathrm{P})$. Let $B(s,t) = \mathrm{E}(X_t X_s)$ be the covariance function of the process. Suppose there is a positive constant r and for each $T \geq 0$ there is a constant $c(T)$ such that*

$$B(t,t) + B(s,s) - 2B(s,t) \leq c(T)|t-s|^r \quad \text{for } 0 \leq s, t \leq T. \tag{18.8}$$

Then there is a continuous modification Y_t of the process X_t such that for every $\gamma \in (0, r/2)$ and every $T > 0$ there is $\delta > 0$, and for each ω there is $h(\omega) > 0$ such that (18.5) holds.

Proof. Let us examine the quantity $\mathrm{E}|X_t - X_s|^{2n}$, where n is a positive integer. The random variable $X_t - X_s$ is Gaussian, with zero mean and variance equal to the expression on the left-hand side of (18.8). For a Gaussian random variable ξ with zero mean and variance σ^2, we have

$$\mathrm{E}\xi^{2n} = \frac{1}{\sqrt{2\pi}\sigma} \int_{-\infty}^{\infty} x^{2n} \exp(\frac{-x^2}{2\sigma^2}) dx = k(n)\sigma^{2n},$$

where $k(n) = (1/\sqrt{2\pi}) \int_{-\infty}^{\infty} x^{2n} e^{-x^2/2} dx$. Thus we obtain

$$\mathrm{E}|X_t - X_s|^{2n} \leq k(n)(c(T)|t - s|^r)^n \leq c'(n, T)|t - s|^{rn}$$

for $0 \leq s, t \leq T$ and some constant $c'(n, T)$. By Theorem 18.19, (18.5) holds for any $\gamma \in (0, (rn - 1)/2n)$. Since we can take n to be arbitrarily large, this means that (18.5) holds for any $\gamma \in (0, r/2)$.   □

Remark 18.22. If the process X_t is stationary with the covariance function $b(t) = B(t, 0) = \mathrm{E}(X_t X_0)$, then condition (18.8) is reduced to

$$b(0) - b(t) \leq c(T)|t|^r \quad \text{for } |t| \leq T.$$

Let us now use Theorem 18.21 to justify the existence of Brownian motion. First, we note that the Kolmogorov Consistency Theorem guarantees the existence of a process X_t on some probability space $(\Omega, \mathcal{F}, \mathrm{P})$ with the following properties:

1. $X_0(\omega) = 0$ for almost all ω.
2. For $0 \leq s \leq t$ the increment $X_t - X_s$ is a Gaussian random variable with mean zero and variance $t - s$.
3. The random variables $X_{t_0}, X_{t_1} - X_{t_0}, \ldots, X_{t_k} - X_{t_{k-1}}$ are independent for every $k \geq 1$ and $0 = t_0 \leq t_1 \leq \ldots \leq t_k$.

To see that the assumptions of the Kolmogorov Consistency Theorem are satisfied, it is sufficient to note that conditions 1–3 are equivalent to the following: for any $t_1, t_2, \ldots, t_k \in \mathbb{R}^+$, the vector $(X_{t_1}, X_{t_2}, \ldots, X_{t_k})$ is Gaussian with the covariance matrix $B_{ij} = \mathrm{E}(X_{t_i} X_{t_j}) = t_i \wedge t_j$.

The assumptions of Theorem 18.21 are satisfied for the process X_t with $r = 1$. Indeed, $B(s, t) = s \wedge t$ and the expression on the left-hand side of (18.8) is equal to $|t - s|$. Therefore, there exists a continuous modification of X_t, which we shall denote by W_t, such that (18.5) holds for any $\gamma \in (0, 1/2)$ (with W_t instead of Y_t).

If we consider the filtration $\mathcal{F}_t^W = \sigma(W_s, s \leq t)$ generated by the process W_t, then Lemma 18.4 implies that W_t is a Brownian motion relative to the filtration $\mathcal{F}_t^W$.

It is easy to show that for fixed γ, δ, and T, the set of functions $x(t) \in C([0, \infty))$ satisfying (18.5) with $x(t)$ instead of $Y_t(\omega)$, is a measurable subset of $(C([0, \infty)), \mathcal{B})$ (see Problem 7). Therefore we have the following.

Lemma 18.23. *If W_t is a Brownian motion and $\gamma < 1/2$, then almost every trajectory $W_t(\omega)$ is a locally Holder continuous function of t with exponent γ.*

18.5 Some Properties of Brownian Motion

Scaling and Symmetry. If W_t is a Brownian motion and c is a positive constant, then the process

$$X_t = \frac{1}{\sqrt{c}} W_{ct}, \quad t \in \mathbb{R}^+,$$

is also a Brownian motion, which follows from the definition. Similarly, if W_t is a Brownian motion, then so is the process $X_t = -W_t$.

Strong Law of Large Numbers. Let W_t be a Brownian motion. We shall demonstrate that

$$\lim_{t \to \infty} \frac{W_t}{t} = 0 \text{ almost surely.} \tag{18.9}$$

For $c > 0$, let $A_c = \{\omega : \lim \sup_{t \to \infty} (W_t(\omega)/t) > c\}$. It is not difficult to show that A_c is measurable. Let us prove that $P(A_c) = 0$. Consider the events

$$B_c^n = \{\omega : \sup_{2^{n-1} \leq t \leq 2^n} W_t(\omega) > c 2^{n-1}\}.$$

It is clear that, in order for ω to belong to A_c, it must belong to B_c^n for infinitely many n. By the Doob Inequality (Theorem 13.30),

$$P(B_c^n) \leq P(\sup_{0 \leq t \leq 2^n} W_t(\omega) > c 2^{n-1}) \leq \frac{E W_{2^n}^2}{(c 2^{n-1})^2} = \frac{1}{c^2 2^{n-2}}.$$

Therefore $\sum_{n=1}^{\infty} P(B_c^n) < \infty$. By the first Borel-Cantelli Lemma, this implies that $P(A_c) = 0$. Since c was an arbitrary positive number, this means that $\lim \sup_{t \to \infty} (W_t(\omega)/t) \leq 0$ almost surely. After replacing W_t by $-W_t$, we see that $\lim \inf_{t \to \infty} (W_t(\omega)/t) \geq 0$ almost surely, and thus (18.9) holds.

Time Inversion. Let us show that, if W_t is a Brownian motion, then so is the process

$$X_t = \begin{cases} t W_{1/t} & \text{if } 0 < t < \infty \\ 0 & \text{if } t = 0. \end{cases}$$

Clearly, X_t has the desired finite-dimensional distributions and almost all realizations of X_t are continuous for $t > 0$. It remains to show that X_t is continuous at zero. By the Law of Large Numbers,

$$\lim_{t \to 0} X_t = \lim_{t \to 0} t W_{1/t} = \lim_{s \to \infty} (W_s/s) = 0 \text{ almost surely,}$$

that is X_t is almost surely continuous at $t = 0$.

Invariance Under Rotations and Reflections. Let $W_t = (W_t^1, \ldots, W_t^d)$ be a d-dimensional Brownian motion, and T a $d \times d$ orthogonal matrix. Let us show that $X_t = TW_t$ is also a d-dimensional Brownian motion.

Clearly, X_t is a Gaussian $\mathbb{R}^d$-values process, that is $(X_{t_1}^{i_1}, \ldots, X_{t_k}^{i_k})$ is a Gaussian vector for any $1 \leq i_1, \ldots, i_k \leq d$ and $t_1, \ldots, t_k \in \mathbb{R}^+$. The trajectories of X_t are continuous almost surely. Let us examine its covariance function. If $s, t \in \mathbb{R}^+$, then

$$\mathrm{E}(X_s^i X_t^j) = \sum_{k=1}^{d} \sum_{l=1}^{d} T_{ik} T_{jl} \mathrm{E}(W_s^k W_t^l) = (s \wedge t) \sum_{k=1}^{d} T_{ik} T_{jk}$$

$$= (s \wedge t)(TT^*)_{ij} = \delta_{ij},$$

since T is an orthogonal matrix. Since X_t is a Gaussian process and $\mathrm{E}(X_s^i X_t^j) = 0$ for $i \neq j$, the processes $X_t^1, \ldots, X_t^d$ are independent (see Problem 1), while the covariance function of X_t^i is $s \wedge t$, which proves that X_t^i is a Brownian motion for each i. Thus we have shown that X_t is a d-dimensional Brownian motion.

Convergence of Quadratic Variations. Let f be a function defined on an interval $[a, b]$ of the real line. Let $\sigma = \{t_0, t_1, \ldots, t_n\}$, $a = t_0 \leq t_1 \leq \ldots \leq t_n = b$, be a partition of the interval $[a, b]$ into n subintervals. We denote the length of the largest interval by $\delta(\sigma) = \max_{1 \leq i \leq n}(t_i - t_{i-1})$. Recall that the p-th variation (with $p > 0$) of the function f over the partition σ is defined as

$$V_{[a,b]}^p(f, \sigma) = \sum_{i=1}^{n} |f(t_i) - f(t_{i-1})|^p.$$

Let us consider a Brownian motion on an interval $[0, t]$. We shall prove that the quadratic variation of the Brownian motion over a partition σ converges to t in L^2 as the mesh of the partition gets finer.

Lemma 18.24. *Let W_t be a Brownian motion on a probability space $(\Omega, \mathcal{F}, \mathrm{P})$. Then*

$$\lim_{\delta(\sigma) \to 0} V_{[0,t]}^2(W_s(\omega), \sigma) = t \quad in \quad L^2(\Omega, \mathcal{F}, \mathrm{P}).$$

Proof. By the definition of $V_{[0,t]}^2$,

$$\mathrm{E}(V_{[0,t]}^2(W_s(\omega), \sigma) - t)^2 = \mathrm{E}(\sum_{i=1}^{n} [(W_{t_i} - W_{t_{i-1}})^2 - (t_i - t_{i-1})])^2$$

$$= \sum_{i=1}^{n} \mathrm{E}[(W_{t_i} - W_{t_{i-1}})^2 - (t_i - t_{i-1})]^2 \leq \sum_{i=1}^{n} \mathrm{E}(W_{t_i} - W_{t_{i-1}})^4 + \sum_{i=1}^{n}(t_i - t_{i-1})^2$$

$$= 4 \sum_{i=1}^{n}(t_i - t_{i-1})^2 \leq 4 \max_{1 \leq i \leq n}(t_i - t_{i-1}) \sum_{i=1}^{n}(t_i - t_{i-1}) = 4t\delta(\sigma),$$

where the second equality is justified by

$$E((W_{t_i} - W_{t_{i-1}})^2 - (t_i - t_{i-1}))((W_{t_j} - W_{t_{j-1}})^2 - (t_j - t_{j-1})) = 0 \text{ if } i \neq j.$$

Therefore, $\lim_{\delta(\sigma) \to 0} E(V^2_{[0,t]}(W_s(\omega), \sigma) - t)^2 = 0.$ □

Remark 18.25. This lemma does not imply that the quadratic variation of Brownian motion exists almost surely. In fact, the opposite is the case:

$$\limsup_{\delta(\sigma) \to 0} V^2_{[0,t]}(W_s(\omega), \sigma) = \infty \quad \text{almost surely.}$$

(See "Diffusion Processes and their Sample Paths" by K. Ito and H. McKean).

Law of Iterated Logarithm. Let W_t be a Brownian motion on a probability space $(\Omega, \mathcal{F}, P)$. For a fixed t, the random variable W_t is Gaussian with variance t, and therefore we could expect a typical value for W_t to be of order $\sqrt{t}$ if t is large. In fact, the running maximum of a Brownian motion grows slightly faster than $\sqrt{t}$. Namely, we have the following theorem, which we state without a proof.

Theorem 18.26 (Law of Iterated Logarithm). *If W_t is a Brownian motion, then*

$$\limsup_{t \to \infty} \frac{W_t}{\sqrt{2t \ln \ln t}} = 1 \quad \text{almost surely.}$$

Bessel Processes. Let $W_t = (W_t^1, \ldots, W_t^d)$, $d \geq 2$, be a d-dimensional Brownian motion on a probability space $(\Omega, \mathcal{F}, P)$, and let $x \in \mathbb{R}^d$. Consider the process with values in $\mathbb{R}^+$ on the same probability space defined by

$$R_t = ||W_t + x||.$$

Due to the rotation invariance of the Brownian motion, the law of R_t depends on x only through $r = ||x||$. We shall refer to the process R_t as the Bessel process with dimension d starting at r. Let us note a couple of properties of the Bessel process. (Their proof can be found in "Brownian Motion and Stochastic Calculus" by I. Karatzas and S. Shreve, for example.)

First, the Bessel process in dimension $d \geq 2$ starting at $r \geq 0$ almost surely never reaches the origin for $t > 0$, that is

$$P(R_t = 0 \text{ for some } t > 0) = 0.$$

Second, the Bessel process in dimension $d \geq 2$ starting at $r \geq 0$ almost surely satisfies the following integral equation:

$$R_t = r + \int_0^t \frac{d-1}{2R_s} ds + B_t, \quad t \geq 0,$$

where B_t is a one-dimensional Brownian motion. The integral on the right-hand side is finite almost surely.

18.6 Problems

1. Let $X_t = (X_t^1, \ldots, X_t^d)$ be a Gaussian $\mathbb{R}^d$-valued process (i.e., the vector $(X_{t_1}^1, \ldots, X_{t_1}^d, \ldots, X_{t_k}^1, \ldots, X_{t_k}^d)$ is Gaussian for any $t_1, \ldots, t_k$). Show that if $E(X_s^i X_t^j) = 0$ for any $i \neq j$ and $s, t \in \mathbb{R}^+$, then the processes $X_t^1, \ldots, X_t^d$ are independent.

2. Let W_t be a one-dimensional Brownian motion. Find the distribution of the random variable $\int_0^1 W_t dt$.

3. Let $W_t = (W_t^1, W_t^2)$ be a two-dimensional Brownian motion. Find the distribution of $\|W_t\| = \sqrt{(W_t^1)^2 + (W_t^2)^2}$.

4. The characteristic functional of a random process X_t, $T = \mathbb{R}$ (or $T = \mathbb{R}^+$), is defined by
$$L(\varphi) = E \exp(i \int_T \varphi(t) X_t dt),$$
where φ is an infinitely differentiable function with compact support. Find the characteristic functional of the Brownian motion.

5. Let W_t be a one-dimensional Brownian motion. Find all a and b for which the process $X_t = \exp(aW_t + bt)$, $t \geq 0$, is a martingale relative to the filtration $\mathcal{F}_t^W$.

6. Let W_t be a one-dimensional Brownian motion, and $a, b \in \mathbb{R}$. Show that the measure on $C([0, \infty])$ induced by the process $at + bW_t$ can be viewed as a weak limit of measures corresponding to certain random walks.

7. Prove that for fixed γ, δ, and T, the set of functions $x(t) \in C([0, \infty))$ satisfying (18.5) with $x(t)$ instead of $Y_t(\omega)$ is a measurable subset of $(C([0, \infty)), \mathcal{B})$.

8. Let $b : \mathbb{R} \to \mathbb{R}$ be a nonnegative-definite function that is $2k$ times differentiable. Assume that there are constants $r, c > 0$ such that
$$|b^{(2k)}(t) - b^{(2k)}(0)| \leq c|t|^r$$
for all t. Prove that there is a Gaussian process with zero mean and covariance function b such that all of its realizations are k times continuously differentiable.

9. Let P_1 and P_2 be two measures on $C([0, 1])$ induced by the processes $c_1 W_t$ and $c_2 W_t$, where $0 < c_1 < c_2$, and W_t is a Brownian motion. Prove that P_1 and P_2 are mutually singular (i.e., there are two measurable subsets A_1 and A_2 of $C([0, 1])$ such that $P_1(A_1) = 1$, $P_2(A_2) = 1$, while $A_1 \cap A_2 = \emptyset$).

10. Let W_t be a one-dimensional Brownian motion on a probability space $(\Omega, \mathcal{F}, P)$. Prove that, for any $a, b > 0$, one can find an event $A \in \mathcal{F}$ with $P(A) = 1$ and a function $t(\omega)$ such that $\inf_{s \geq t(\omega)}(as + bW_s(\omega)) \geq 0$ for $\omega \in A$.

11. Let $X_t = aW_t + bt$, where W_t is a one-dimensional Brownian motion on a probability space $(\Omega, \mathcal{F}, P)$, and a and b are some constants. Find the following limit

$$\lim_{\delta(\sigma) \to 0} V^2_{[0,t]}(X_s(\omega), \sigma)$$

in $L^2(\Omega, \mathcal{F}, P)$.

12. Let W_t be a one-dimensional Brownian motion and σ_n the partition of the interval $[0, t]$ into 2^n subintervals of equal length. Prove that

$$\lim_{n \to \infty} V^2_{[0,t]}(W_s, \sigma_n) = t \quad \text{almost surely.}$$

13. Let W_t be a one-dimensional Brownian motion on a probability space $(\Omega, \mathcal{F}, P)$. For $\delta > 0$, let Ω_δ be the event $\Omega_\delta = \{\omega \in \Omega : |W_1(\omega)| \le \delta\}$. Let $\mathcal{F}_\delta$ be defined by: $A \in \mathcal{F}_\delta$ if $A \in \mathcal{F}$ and $A \subseteq \Omega_\delta$. Define the measure P_δ on $(\Omega_\delta, \mathcal{F}_\delta)$ as

$$P_\delta(A) = P(A)/P(\Omega_\delta).$$

Let W^δ_t, $t \in [0, 1]$, be the process on the probability space $(\Omega_\delta, \mathcal{F}_\delta, P_\delta)$ defined simply by

$$W^\delta_t(\omega) = W_t(\omega).$$

Prove that there is a process B_t, $t \in [0, 1]$, with continuous realizations, such that W^δ_t converge to B_t as $\delta \downarrow 0$, i.e., the measures on $C([0, 1])$ induced by the processes W^δ_t weakly converge to the measure induced by B_t. Such a process B_t is called a Brownian Bridge.

Prove that a Brownian Bridge is a Gaussian process and find its covariance function.

19

Markov Processes and Markov Families

19.1 Distribution of the Maximum of Brownian Motion

Let W_t be a one-dimensional Brownian motion relative to a filtration $\mathcal{F}_t$ on a probability space $(\Omega, \mathcal{F}, \mathrm{P})$. We denote the maximum of W_t on the interval $[0, T]$ by M_T,

$$M_T(\omega) = \sup_{0 \leq t \leq T} W_t(\omega).$$

In this section we shall use intuitive arguments in order to find the distribution of M_T. Rigorous arguments will be provided later in this chapter, after we introduce the notion of a strong Markov family. Thus, the problem at hand may serve as a simple example motivating the study of the strong Markov property.

For a non-negative constant c, define the stopping time τ_c as the first time the Brownian motion reaches the level c if this occurs before time T, and otherwise as T, that is

$$\tau_c(\omega) = \min(\inf\{t \geq 0 : W_t(\omega) = c\}, T).$$

Since the probability of the event $W_T = c$ is equal to zero,

$$\mathrm{P}(M_T \geq c) = \mathrm{P}(\tau_c < T) = \mathrm{P}(\tau_c < T, W_T < c) + \mathrm{P}(\tau_c < T, W_T > c).$$

The key observation is that the probabilities of the events $\{\tau_c < T, W_T < c\}$ and $\{\tau_c < T, W_T > c\}$ are the same. Indeed, the Brownian motion is equally likely to be below c and above c at time T under the condition that it reaches level c before time T. This intuitive argument hinges on our ability to stop the process at time τ_c and then "start it anew" in such a way that the increment $W_T - W_{\tau_c}$ has symmetric distribution and is independent of $\mathcal{F}_{\tau_c}$.

Since $\tau_c < T$ almost surely on the event $\{W_T > c\}$,

$$\mathrm{P}(M_T \geq c) = 2\mathrm{P}(\tau_c < T, W_T > c) = 2\mathrm{P}(W_T > c) = \frac{\sqrt{2}}{\sqrt{\pi T}} \int_c^\infty e^{-\frac{x^2}{2T}} \, dx.$$

L. Koralov and Y.G. Sinai, *Theory of Probability and Random Processes*, 273
Universitext, DOI 10.1007/978-3-540-68829-7_19,
© Springer-Verlag Berlin Heidelberg 2012

Therefore,

$$P(M_T \leq c) = 1 - P(M_T \geq c) = 1 - \frac{\sqrt{2}}{\sqrt{\pi T}} \int_c^\infty e^{-\frac{x^2}{2T}} dx,$$

which is the desired expression for the distribution of the maximum of Brownian motion.

19.2 Definition of the Markov Property

Let $(X, \mathcal{G})$ be a measurable space. In Chap. 5 we defined a Markov chain as a measure on the space of sequences with elements in X which is generated by a Markov transition function. In this chapter we use a different approach, defining a Markov process as a random process with certain properties, and a Markov family as a family of such random processes. We then reconcile the two points of view by showing that a Markov family defines a transition function. In turn, by using a transition function and an initial distribution we can define a measure on the space of realizations of the process.

For the sake of simplicity of notation, we shall primarily deal with the time-homogeneous case. Let us assume that the state space is $\mathbb{R}^d$ with the σ-algebra of Borel sets, that is $(X, \mathcal{G}) = (\mathbb{R}^d, \mathcal{B}(\mathbb{R}^d))$. Let $(\Omega, \mathcal{F}, P)$ be a probability space with a filtration $\mathcal{F}_t$.

Definition 19.1. *Let μ be a probability measure on $\mathcal{B}(\mathbb{R}^d)$. An adapted process X_t with values in $\mathbb{R}^d$ is called a Markov process with initial distribution μ if:*

(1) $P(X_0 \in \Gamma) = \mu(\Gamma)$ for any $\Gamma \in \mathcal{B}(\mathbb{R}^d)$.
(2) If $s, t \geq 0$ and $\Gamma \subseteq \mathbb{R}^d$ is a Borel set, then

$$P(X_{s+t} \in \Gamma | \mathcal{F}_s) = P(X_{s+t} \in \Gamma | X_s) \quad almost \ surely. \tag{19.1}$$

Definition 19.2. *Let X_t^x, $x \in \mathbb{R}^d$, be a family of processes with values in $\mathbb{R}^d$ which are adapted to a filtration $\mathcal{F}_t$. This family of processes is called a time-homogeneous Markov family if:*

(1) The function $p(t, x, \Gamma) = P(X_t^x \in \Gamma)$ is Borel-measurable as a function of $x \in \mathbb{R}^d$ for any $t \geq 0$ and any Borel set $\Gamma \subseteq \mathbb{R}^d$.
(2) $P(X_0^x = x) = 1$ for any $x \in \mathbb{R}^d$.
(3) If $s, t \geq 0$, $x \in \mathbb{R}^d$, and $\Gamma \subseteq \mathbb{R}^d$ is a Borel set, then

$$P(X_{s+t}^x \in \Gamma | \mathcal{F}_s) = p(t, X_s^x, \Gamma) \quad almost \ surely.$$

The function $p(t, x, \Gamma)$ is called the transition function for the Markov family X_t^x. It has the following properties:

(1') For fixed $t \geq 0$ and $x \in \mathbb{R}^d$, the function $p(t, x, \Gamma)$, as a function of Γ, is a probability measure, while for fixed t and Γ it is a measurable function of x.

(2') $p(0, x, \{x\}) = 1$.
(3') If $s, t \geq 0$, $x \in \mathbb{R}^d$, and $\Gamma \subseteq \mathbb{R}^d$ is a Borel set, then

$$p(s + t, x, \Gamma) = \int_{\mathbb{R}^d} p(s, x, dy) p(t, y, \Gamma).$$

The first two properties are obvious. For the third one it is sufficient to write

$$p(s + t, x, \Gamma) = P(X^x_{s+t} \in \Gamma) = EP(X^x_{s+t} \in \Gamma | \mathcal{F}_s)$$

$$= Ep(t, X^x_s, \Gamma) = \int_{\mathbb{R}^d} p(s, x, dy) p(t, y, \Gamma),$$

where the last equality follows by Theorem 3.14.

Now assume that we are given a function $p(t, x, \Gamma)$ with properties (1')–(3') and a measure μ on $\mathcal{B}(\mathbb{R}^d)$. As we shall see below, this pair can be used to define a measure on the space of all functions $\widetilde{\Omega} = \{\widetilde{\omega} : \mathbb{R}^+ \to \mathbb{R}^d\}$ in such a way that $\widetilde{\omega}(t)$ is a Markov process. Recall that in Chap. 5 we defined a Markov chain as the measure corresponding to a Markov transition function and an initial distribution (see the discussion following Definition 5.17).

Let $\widetilde{\Omega}$ be the set of all functions $\widetilde{\omega} : \mathbb{R}^+ \to \mathbb{R}^d$. Take a finite collection of points $0 \leq t_1 \leq \ldots \leq t_k < \infty$, and Borel sets $A_1, \ldots, A_k \in \mathcal{B}(\mathbb{R}^d)$. For an elementary cylinder $B = \{\widetilde{\omega} : \widetilde{\omega}(t_1) \in A_1, \ldots, \widetilde{\omega}(t_k) \in A_k\}$, we define the finite-dimensional measure $P^\mu_{t_1, \ldots, t_k}(B)$ via

$$P^\mu_{t_1, \ldots, t_k}(B) = \int_{\mathbb{R}^d} \mu(dx) \int_{A_1} p(t_1, x, dy_1) \int_{A_2} p(t_2 - t_1, y_1, dy_2) \ldots$$

$$\int_{A_{k-1}} p(t_{k-1} - t_{k-2}, y_{k-2}, dy_{k-1}) \int_{A_k} p(t_k - t_{k-1}, y_{k-1}, dy_k).$$

The family of finite-dimensional probability measures $P^\mu_{t_1, \ldots, t_k}$ is consistent and, by the Kolmogorov Theorem, defines a measure P^μ on $\mathcal{B}$, the σ-algebra generated by all the elementary cylindrical sets. Let $\mathcal{F}_t$ be the σ-algebra generated by the elementary cylindrical sets $B = \{\widetilde{\omega} : \widetilde{\omega}(t_1) \in A_1, \ldots, \widetilde{\omega}(t_k) \in A_k\}$ with $0 \leq t_1 \leq \ldots \leq t_k \leq t$, and $X_t(\widetilde{\omega}) = \widetilde{\omega}(t)$. We claim that X_t is a Markov process on $(\widetilde{\Omega}, \mathcal{B}, P^\mu)$ relative to the filtration $\mathcal{F}_t$. Clearly, the first property in Definition 19.1 holds. To verify the second property, it is sufficient to show that

$$P^\mu(B \bigcap \{X_{s+t} \in \Gamma\}) = \int_B p(t, X_s, \Gamma) dP^\mu \qquad (19.2)$$

for any $B \in \mathcal{F}_s$, since the integrand on the right-hand side is clearly $\sigma(X_s)$-measurable. When $B = \{\widetilde{\omega} : \widetilde{\omega}(t_1) \in A_1, \ldots, \widetilde{\omega}(t_k) \in A_k\}$ with $0 \leq t_1 \leq \ldots \leq t_k \leq s$, both sides of (19.2) are equal to

$$\int_{\mathbb{R}^d} \mu(dx) \int_{A_1} p(t_1, x, dy_1) \int_{A_2} p(t_2 - t_1, y_1, dy_2) \ldots$$

$$\int_{A_k} p(t_k - t_{k-1}, y_{k-1}, dy_k) \int_{\mathbb{R}^d} p(s - t_k, y_k, dy)p(t, y, \Gamma).$$

Since such elementary cylindrical sets form a π-system, it follows from Lemma 4.13 that (19.2) holds for all $B \in \mathcal{F}_s$.

Let $\widetilde{\Omega}$ be the space of all functions from $\mathbb{R}^+$ to $\mathbb{R}^d$ with the σ-algebra $\mathcal{B}$ generated by cylindrical sets. We can define a family of shift transformations $\theta_s : \widetilde{\Omega} \to \widetilde{\Omega}$, $s \geq 0$, which act on functions $\widetilde{\omega} \in \widetilde{\Omega}$ via

$$(\theta_s \widetilde{\omega})(t) = \widetilde{\omega}(s + t).$$

If X_t is a random process with realizations denoted by $X.(\omega)$, we can apply θ_s to each realization to get a new process, whose realizations will be denoted by $X_{s+.}(\omega)$.

If $f : \widetilde{\Omega} \to \mathbb{R}$ is a bounded measurable function and X_t^x, $x \in \mathbb{R}^d$, is a Markov family, we can define the function $\varphi_f(x) : \mathbb{R}^d \to \mathbb{R}$ as

$$\varphi_f(x) = \mathrm{E}f(X_.^x).$$

Now we can formulate an important consequence of the Markov property.

Lemma 19.3. *Let X_t^x, $x \in \mathbb{R}^d$, be a Markov family of processes relative to a filtration $\mathcal{F}_t$. If $f : \widetilde{\Omega} \to \mathbb{R}$ is a bounded measurable function, then*

$$\mathrm{E}(f(X_{s+.}^x)|\mathcal{F}_s) = \varphi_f(X_s^x) \quad almost \ surely. \tag{19.3}$$

Proof. Let us show that for any bounded measurable function $g : \mathbb{R}^d \to \mathbb{R}$ and $s, t \geq 0$,

$$\mathrm{E}(g(X_{s+t}^x)|\mathcal{F}_s) = \int_{\mathbb{R}^d} g(y)p(t, X_s^x, dy) \quad almost \ surely. \tag{19.4}$$

Indeed, if g is the indicator function of a Borel set $\Gamma \subseteq \mathbb{R}^d$, this statement is part of the definition of a Markov family. By linearity, it also holds for finite linear combinations of indicator functions. Therefore, (19.4) holds for all bounded measurable functions, since they can be uniformly approximated by finite linear combinations of indicator functions.

To prove (19.3), we first assume that f is the indicator function of an elementary cylindrical set, that is $f = \chi_A$, where

$$A = \{\widetilde{\omega} : \widetilde{\omega}(t_1) \in A_1, \ldots, \widetilde{\omega}(t_k) \in A_k\}$$

with $0 \leq t_1 \leq \ldots \leq t_k$ and some Borel sets $A_1, \ldots, A_k \subseteq \mathbb{R}^d$. In this case the left-hand side of (19.3) is equal to $\mathrm{P}(X_{s+t_1}^x \in A_1, \ldots, X_{s+t_k}^x \in A_k|\mathcal{F}_s)$. We can transform this expression by inserting conditional expectations with respect to $\mathcal{F}_{s+t_{k-1}}, \ldots, \mathcal{F}_{s+t_1}$ and applying (19.4) repeatedly. We thus obtain

$$P(X^x_{s+t_1} \in A_1, \ldots, X^x_{s+t_k} \in A_k | \mathcal{F}_s)$$

$$= E(\chi_{\{X^x_{s+t_1} \in A_1\}} \cdots \chi_{\{X^x_{s+t_k} \in A_k\}} | \mathcal{F}_s)$$

$$= E(\chi_{\{X^x_{s+t_1} \in A_1\}} \cdots \chi_{\{X^x_{s+t_{k-1}} \in A_{k-1}\}} E(\chi_{\{X^x_{s+t_k} \in A_k\}} | \mathcal{F}_{s+t_{k-1}}) | \mathcal{F}_s)$$

$$= E(\chi_{\{X^x_{s+t_1} \in A_1\}} \cdots \chi_{\{X^x_{s+t_{k-1}} \in A_{k-1}\}} p(t_k - t_{k-1}, X^x_{s+t_{k-1}}, A_k) | \mathcal{F}_s)$$

$$= E(\chi_{\{X^x_{s+t_1} \in A_1\}} \cdots \chi_{\{X^x_{s+t_{k-2}} \in A_{k-2}\}} E(\chi_{\{X^x_{s+t_{k-1}} \in A_{k-1}\}}$$
$$p(t_k - t_{k-1}, X^x_{s+t_{k-1}}, A_k) | \mathcal{F}_{s+t_{k-2}}) | \mathcal{F}_s)$$

$$= E(\chi_{\{X^x_{s+t_1} \in A_1\}} \cdots \chi_{\{X^x_{s+t_{k-2}} \in A_{k-2}\}}$$

$$\int_{A_{k-1}} p(t_{k-1} - t_{k-2}, X^x_{s+t_{k-2}}, dy_{k-1}) p(t_k - t_{k-1}, y_{k-1}, A_k) | \mathcal{F}_s) = \ldots$$

$$= \int_{A_1} p(t_1 - s, X^x_s, dy_1) \int_{A_2} p(t_2 - t_1, y_1, dy_2) \ldots$$

$$\int_{A_{k-1}} p(t_{k-1} - t_{k-2}, y_{k-2}, dy_{k-1}) p(t_k - t_{k-1}, y_{k-1}, A_k).$$

Note that $\varphi_f(x)$ is equal to $P(X^x_{t_1} \in A_1, \ldots, X^x_{t_k} \in A_k)$. If we insert conditional expectations with respect to $\mathcal{F}_{t_{k-1}}, \ldots, \mathcal{F}_{t_1}, \mathcal{F}_0$ and apply (19.4) repeatedly,

$$P(X^x_{t_1} \in A_1, \ldots, X^x_{t_k} \in A_k) = \int_{A_1} p(t_1 - s, x, dy_1) \int_{A_2} p(t_2 - t_1, y_1, dy_2) \ldots$$

$$\int_{A_{k-1}} p(t_{k-1} - t_{k-2}, y_{k-2}, dy_{k-1}) p(t_k - t_{k-1}, y_{k-1}, A_k).$$

If we replace x with X^x_s, we see that the right-hand side of (19.3) coincides with the left-hand side if f is an indicator function of an elementary cylinder.

Next, let us show that (19.3) holds if $f = \chi_A$ is an indicator function of any set $A \in \mathcal{B}$. Indeed, elementary cylinders form a π-system, while the collection of sets A for which (19.3) is true with $f = \chi_A$ is a Dynkin system. By Lemma 4.13, formula (19.3) holds for $f = \chi_A$, where A is any element from the σ-algebra generated by the elementary cylinders, that is $\mathcal{B}$.

Finally, any bounded measurable function f can be uniformly approximated by finite linear combinations of indicator functions. □

Remark 19.4. If we assume that X^x_t are continuous processes, Lemma 19.3 applies in the case when f is a bounded measurable function on $C([0, \infty))$.

Remark 19.5. The arguments in the proof of the lemma imply that φ_f is a measurable function for any bounded measurable f. It is enough to take $s = 0$.

It is sometimes useful to formulate the third condition of Definition 19.2 in a slightly different way. Let g be a bounded measurable function $g : \mathbb{R}^d \to \mathbb{R}$. Then we can define a new function $\psi_g : \mathbb{R}^+ \times \mathbb{R}^d \to \mathbb{R}$ by

$$\psi_g(t, x) = \mathrm{E}g(X_t^x).$$

Note that $\psi_g(t, x) = \varphi_f(x)$, if we define $f : \widetilde{\Omega} \to \mathbb{R}$ by $f(\widetilde{\omega}) = g(\widetilde{\omega}(t))$.

Lemma 19.6. *If conditions (1) and (2) of Definition 19.2 are satisfied, then condition (3) is equivalent to the following:*

(3′) If $s, t \geq 0$, $x \in \mathbb{R}^d$, and $g : \mathbb{R}^d \to \mathbb{R}$ is a bounded continuous function, then

$$\mathrm{E}(g(X_{s+t}^x)|\mathcal{F}_s) = \psi_g(t, X_s^x) \quad \text{almost surely.}$$

Proof. Clearly, (3) implies (3′) as a particular case of Lemma 19.3. Conversely, let $s, t \geq 0$ and $x \in \mathbb{R}^d$ be fixed, and assume that $\Gamma \subseteq \mathbb{R}^d$ is a closed set. In this case we can find a sequence of non-negative bounded continuous functions g_n such that $g_n(x) \downarrow \chi_\Gamma(x)$ for all $x \in \mathbb{R}^d$. By taking the limit as $n \to \infty$ in the equality

$$\mathrm{E}(g_n(X_{s+t}^x)|\mathcal{F}_s) = \psi_{g_n}(t, X_s^x) \quad \text{almost surely,}$$

we obtain

$$\mathrm{P}(X_{s+t}^x \in \Gamma|\mathcal{F}_s) = p(t, X_s^x, \Gamma) \quad \text{almost surely} \qquad (19.5)$$

for closed sets Γ. The collection of all closed sets is a π-system, while the collection of all sets Γ for which (19.5) holds is a Dynkin system. Therefore (19.5) holds for all Borel sets Γ by Lemma 4.13. $\qquad \square$

19.3 Markov Property of Brownian Motion

Let W_t be a d-dimensional Brownian motion relative to a filtration $\mathcal{F}_t$. Consider the family of processes $W_t^x = x + W_t$. Let us show that W_t^x is a time-homogeneous Markov family relative to the filtration $\mathcal{F}_t$.

Since W_t^x is a Gaussian vector for fixed t, there is an explicit formula for $\mathrm{P}(W_t^x \in \Gamma)$. Namely,

$$p(t, x, \Gamma) = \mathrm{P}(W_t^x \in \Gamma) = (2\pi t)^{-\frac{d}{2}} \int_\Gamma \exp(-||y - x||^2/2t)dy \qquad (19.6)$$

if $t > 0$. As a function of x, $p(0, x, \Gamma)$ is simply the indicator function of the set Γ. Therefore, $p(t, x, \Gamma)$ is a Borel-measurable function of x for any $t \geq 0$ and any Borel set Γ.

Clearly, the second condition of Definition 19.2 is satisfied by the family of processes W_t^x.

In order to verify the third condition, let us assume that $t > 0$, since otherwise the condition is satisfied. For a Borel set $S \subseteq \mathbb{R}^{2d}$ and $x \in \mathbb{R}^d$, let

$$S_x = \{y \in \mathbb{R}^d : (x, y) \in S\}.$$

Let us show that

$$P((W_s^x, W_{s+t}^x - W_s^x) \in S|\mathcal{F}_s) = (2\pi t)^{-\frac{d}{2}} \int_{S_{W_s^x}} \exp(-||y||^2/2t) dy. \quad (19.7)$$

First, assume that $S = A \times B$, where A and B are Borel subsets of $\mathbb{R}^d$. In this case,

$$P(W_s^x \in A, W_{s+t}^x - W_s^x \in B|\mathcal{F}_s) = \chi_{\{W_s^x \in A\}} P(W_{s+t}^x - W_s^x \in B|\mathcal{F}_s)$$

$$= \chi_{\{W_s^x \in A\}} P(W_{s+t}^x - W_s^x \in B) = \chi_{\{W_s^x \in A\}} (2\pi t)^{-\frac{d}{2}} \int_B \exp(-||y||^2/2t) dy,$$

since $W_{s+t}^x - W_s^x$ is independent of $\mathcal{F}_s$. Thus, (19.7) holds for sets of the form $S = A \times B$. The collection of sets that can be represented as such a direct product is a π-system. Since the collection of sets for which (19.7) holds is a Dynkin system, we can apply Lemma 4.13 to conclude that (19.7) holds for all Borel sets. Finally, let us apply (19.7) to the set $S = \{(x, y) : x + y \in \Gamma\}$. Then,

$$P(W_{s+t}^x \in \Gamma|\mathcal{F}_s) = (2\pi t)^{-\frac{d}{2}} \int_\Gamma \exp(-||y - W_s^x||^2/2t) dy = p(t, W_s^x, \Gamma).$$

This proves that the third condition of Definition 19.2 is satisfied, and that W_t^x is a Markov family.

19.4 The Augmented Filtration

Let W_t be a d-dimensional Brownian motion on a probability space $(\Omega, \mathcal{F}, P)$. We shall exhibit a probability space and a filtration satisfying the usual conditions such that W_t is a Brownian motion relative to this filtration.

Recall that $\mathcal{F}_t^W = \sigma(W_s, s \leq t)$ is the filtration generated by the Brownian motion, and $\mathcal{F}^W = \sigma(W_s, s \in \mathbb{R}^+)$ is the σ-algebra generated by the Brownian motion. Let $\mathcal{N}$ be the collection of all P-negligible sets relative to $\mathcal{F}^W$, that is $A \in \mathcal{N}$ if there is an event $B \in \mathcal{F}^W$ such that $A \subseteq B$ and $P(B) = 0$. Define the new filtration $\widetilde{\mathcal{F}}_t^W = \sigma(\mathcal{F}_t^W \bigcup \mathcal{N})$, called the augmentation of $\mathcal{F}_t^W$, and the new σ-algebra $\widetilde{\mathcal{F}}^W = \sigma(\mathcal{F}^W \bigcup \mathcal{N})$.

Now consider the process W_t on the probability space $(\Omega, \widetilde{\mathcal{F}}^W, P)$, and note that it is a Brownian motion relative to the filtration $\widetilde{\mathcal{F}}_t^W$.

Lemma 19.7. *The augmented filtration $\widetilde{\mathcal{F}}_t^W$ satisfies the usual conditions.*

Proof. It is clear that $\widetilde{\mathcal{F}}_0^W$ contains all the P-negligible events from $\widetilde{\mathcal{F}}^W$. It remains to prove that $\widetilde{\mathcal{F}}_t^W$ is right-continuous.

Our first observation is that $W_t - W_s$ is independent of the σ-algebra $\mathcal{F}_{s+}^W$ if $0 \leq s \leq t$. Indeed, assuming that $s < t$, the variable $W_t - W_{s+\delta}$ is independent of $\mathcal{F}_{s+}^W$ for all positive δ. Then, as $\delta \downarrow 0$, the variable $W_t - W_{s+\delta}$ tends to $W_t - W_s$ almost surely, which implies that $W_t - W_s$ is also independent of $\mathcal{F}_{s+}^W$.

Next, we claim that $\mathcal{F}_{s+}^W \subseteq \widetilde{\mathcal{F}}_s^W$. Indeed, let $t_1, \ldots, t_k \geq s$ for some positive integer k, and let $B_1, \ldots, B_k$ be Borel subsets of $\mathbb{R}^d$. By Lemma 19.3, the random variable $P(W_{t_1} \in B_1, \ldots, W_{t_k} \in B_k | \mathcal{F}_s^W)$ has a $\sigma(W_s)$-measurable version. The same remains true if we replace $\mathcal{F}_s^W$ by $\mathcal{F}_{s+}^W$. Indeed, in the statement of the Markov property for the Brownian motion, we can replace $\mathcal{F}_s^W$ by $\mathcal{F}_{s+}^W$, since in the arguments of Sect. 19.3 we can use that $W_t - W_s$ is independent of $\mathcal{F}_{s+}^W$.

Let $s_1, \ldots, s_{k_1} \leq s \leq t_1, \ldots, t_{k_2}$ for some positive integers k_1 and k_2, and let $A_1, \ldots, A_{k_1}, B_1, \ldots, B_{k_2}$ be Borel subsets of $\mathbb{R}^d$. Then,

$$P(W_{s_1} \in A_1, \ldots, W_{s_{k_1}} \in A_{k_1}, W_{t_1} \in B_1, \ldots, W_{t_{k_2}} \in B_{k_2} | \mathcal{F}_{s+}^W)$$

$$= \chi_{\{W_{s_1} \in A_1, \ldots, W_{s_{k_1}} \in A_{k_1}\}} P(W_{t_1} \in B_1, \ldots, W_{t_{k_2}} \in B_{k_2} | \mathcal{F}_{s+}^W),$$

which has a $\mathcal{F}_s^W$-measurable version. The collection of sets $A \in \mathcal{F}^W$, for which $P(A|\mathcal{F}_{s+}^W)$ has a $\mathcal{F}_s^W$-measurable version, forms a Dynkin system. Therefore, by Lemma 4.13, $P(A|\mathcal{F}_{s+}^W)$ has a $\mathcal{F}_s^W$-measurable version for each $A \in \mathcal{F}^W$. This easily implies our claim that $\mathcal{F}_{s+}^W \subseteq \widetilde{\mathcal{F}}_s^W$.

Finally, let us show that $\widetilde{\mathcal{F}}_{s+}^W \subseteq \widetilde{\mathcal{F}}_s^W$. Let $A \in \widetilde{\mathcal{F}}_{s+}^W$. Then $A \in \widetilde{\mathcal{F}}_{s+\frac{1}{n}}^W$ for every positive integer n. We can find sets $A_n \in \mathcal{F}_{s+\frac{1}{n}}^W$ such that $A \Delta A_n \in \mathcal{N}$. Define

$$B = \bigcap_{m=1}^{\infty} \bigcup_{n=m}^{\infty} A_n.$$

Then $B \in \mathcal{F}_{s+}^W$, since $B \in \mathcal{F}_{s+\frac{1}{m}}^W$ for any m. It remains to show that $A \Delta B \in \mathcal{N}$. Indeed,

$$B \setminus A \subseteq \bigcup_{n=1}^{\infty} (A_n \setminus A) \in \mathcal{N},$$

while

$$A \setminus B = A \bigcap \left(\bigcup_{m=1}^{\infty} \bigcap_{n=m}^{\infty} (\Omega \setminus A_n) \right) = \bigcup_{m=1}^{\infty} \left(A \bigcap \left(\bigcap_{n=m}^{\infty} (\Omega \setminus A_n) \right) \right)$$

$$\subseteq \bigcup_{m=1}^{\infty} (A \bigcap (\Omega \setminus A_m)) = \bigcup_{m=1}^{\infty} (A \setminus A_m) \in \mathcal{N}.$$

$\square$

Lemma 19.8 (Blumenthal Zero-One Law). *If* $A \in \widetilde{\mathcal{F}}_0^W$, *then either* $P(A) = 0$ *or* $P(A) = 1$.

Proof. For $A \in \widetilde{\mathcal{F}}_0^W$, there is a set $A_0 \in \mathcal{F}_0^W$ such that $A \Delta A_0 \in \mathcal{N}$. The set A_0 can be represented as $\{\omega \in \Omega : W_0(\omega) \in B\}$, where B is a Borel subset of $\mathbb{R}^d$. Now it is clear that $P(A_0)$ is equal to either 0 or 1, depending on whether the set B contains the origin. Since $P(A) = P(A_0)$, we obtain the desired result. $\square$

19.5 Definition of the Strong Markov Property

It is sometimes necessary in the formulation of the Markov property to replace $\mathcal{F}_s$ by a σ-algebra $\mathcal{F}_\sigma$, where σ is a stopping time. This leads to the notions of a strong Markov process and a strong Markov family. First, we need the following definition.

Definition 19.9. *A random process* X_t *is called progressively measurable with respect to a filtration* $\mathcal{F}_t$ *if* $X_s(\omega)$ *is* $\mathcal{F}_t \times \mathcal{B}([0, t])$-*measurable as a function of* $(\omega, s) \in \Omega \times [0, t]$ *for each fixed* $t \geq 0$.

For example, any progressively measurable process is adapted, and any continuous adapted process is progressively measurable (see Problem 1). If X_t is progressively measurable and τ is a stopping time, then $X_{t \wedge \tau}$ is also progressively measurable, and X_τ is $\mathcal{F}_\tau$-measurable (see Problem 2).

Definition 19.10. *Let* μ *be a probability measure on* $\mathcal{B}(\mathbb{R}^d)$. *A progressively measurable process* X_t *(with respect to filtration* $\mathcal{F}_t$*) with values in* $\mathbb{R}^d$ *is called a strong Markov process with initial distribution* μ *if:*

(1) $P(X_0 \in \Gamma) = \mu(\Gamma)$ *for any* $\Gamma \in \mathcal{B}(\mathbb{R}^d)$.
(2) If $t \geq 0$, σ *is a stopping time of* $\mathcal{F}_t$, *and* $\Gamma \subseteq \mathbb{R}^d$ *is a Borel set, then*

$$P(X_{\sigma+t} \in \Gamma | \mathcal{F}_\sigma) = P(X_{\sigma+t} \in \Gamma | X_\sigma) \quad \text{almost surely.} \quad (19.8)$$

Definition 19.11. *Let* X_t^x, $x \in \mathbb{R}^d$, *be a family of progressively measurable processes with values in* $\mathbb{R}^d$. *This family of processes is called a time-homogeneous strong Markov family if:*

(1) The function $p(t, x, \Gamma) = P(X_t^x \in \Gamma)$ *is Borel-measurable as a function of* $x \in \mathbb{R}^d$ *for any* $t \geq 0$ *and any Borel set* $\Gamma \subseteq \mathbb{R}^d$.
(2) $P(X_0^x = x) = 1$ *for any* $x \in \mathbb{R}^d$.
(3) If $t \geq 0$, σ *is a stopping time of* $\mathcal{F}_t$, $x \in \mathbb{R}^d$, *and* $\Gamma \subseteq \mathbb{R}^d$ *is a Borel set, then*

$$P(X_{\sigma+t}^x \in \Gamma | \mathcal{F}_\sigma) = p(t, X_\sigma^x, \Gamma) \quad \text{almost surely.}$$

We have the following analog of Lemmas 19.3 and 19.6.

Lemma 19.12. *Let X_t^x, $x \in \mathbb{R}^d$, be a strong Markov family of processes relative to a filtration $\mathcal{F}_t$. If $f : \widetilde{\Omega} \to \mathbb{R}$ is a bounded measurable function and σ is a stopping time of $\mathcal{F}_t$, then*

$$\mathrm{E}(f(X_{\sigma+\cdot}^x)|\mathcal{F}_\sigma) = \varphi_f(X_\sigma^x) \quad almost\ surely, \qquad (19.9)$$

where $\varphi_f(x) = \mathrm{E}f(X^x)$.

Remark 19.13. If we assume that X_t^x are continuous processes, Lemma 19.12 applies in the case when f is a bounded measurable function on $C([0,\infty))$.

Lemma 19.14. *If conditions (1) and (2) of Definition 19.11 are satisfied, then condition (3) is equivalent to the following:*

(3′) If $t \geq 0$, σ is a stopping time of $\mathcal{F}_t$, $x \in \mathbb{R}^d$, and $g : \mathbb{R}^d \to \mathbb{R}$ is a bounded continuous function, then

$$\mathrm{E}(g(X_{\sigma+t}^x)|\mathcal{F}_\sigma) = \psi_g(t, X_\sigma^x) \quad almost\ surely,$$

where $\psi_g(t, x) = \mathrm{E}g(X_t^x)$.

We omit the proofs of these lemmas since they are analogous to those in Sect. 19.3. Let us derive another useful consequence of the strong Markov property.

Lemma 19.15. *Let X_t^x, $x \in \mathbb{R}^d$, be a strong Markov family of processes relative to a filtration $\mathcal{F}_t$. Assume that X_t^x is right-continuous for every $x \in \mathbb{R}^d$. Let σ and τ be stopping times of $\mathcal{F}_t$ such that $\sigma \leq \tau$ and τ is $\mathcal{F}_\sigma$-measurable. Then for any bounded measurable function $g : \mathbb{R}^d \to \mathbb{R}$,*

$$\mathrm{E}(g(X_\tau^x)|\mathcal{F}_\sigma) = \psi_g(\tau - \sigma, X_\sigma^x) \quad almost\ surely,$$

where $\psi_g(t, x) = \mathrm{E}g(X_t^x)$.

Remark 19.16. The function $\psi_g(t, x)$ is jointly measurable in (t, x) if X_t^x is right-continuous. Indeed, if g is continuous, then $\psi_g(t, x)$ is right-continuous in t. This is sufficient to justify the joint measurability, since it is measurable in x for each fixed t. Using arguments similar to those in the proof of Lemma 19.6, one can show that $\psi_g(t, x)$ is jointly measurable when g is an indicator function of a measurable set. Approximating an arbitrary bounded measurable function by finite linear combinations of indicator functions justifies the statement in the case of an arbitrary bounded measurable g.

Proof of Lemma 19.15. First assume that g is a continuous function, and that $\tau - \sigma$ takes a finite or countable number of values. Then we can write $\Omega = A_1 \cup A_2 \cup \ldots$, where $\tau(\omega) - \sigma(\omega) = t_k$ for $\omega \in A_k$, and all t_k are distinct. Thus,

$$\mathrm{E}(g(X_\tau^x)|\mathcal{F}_\sigma) = \mathrm{E}(g(X_{\sigma+t_k}^x)|\mathcal{F}_\sigma) \quad almost\ surely\ on\ A_k,$$

since $g(X^x_\tau) = g(X^x_{\sigma+t_k})$ on A_k, and $A_k \in \mathcal{F}_\sigma$. Therefore,

$$E(g(X^x_\tau)|\mathcal{F}_\sigma) = E(g(X^x_{\sigma+t_k})|\mathcal{F}_\sigma) = \psi_g(t_k, X^x_\sigma) = \psi_g(\tau - \sigma, X^x_\sigma) \quad \text{a.s. on } A_k,$$

which implies

$$E(g(X^x_\tau)|\mathcal{F}_\sigma) = \psi_g(\tau - \sigma, X^x_\sigma) \quad \text{almost surely.} \tag{19.10}$$

If the distribution of $\tau - \sigma$ is not necessarily discrete, it is possible to find a sequence of stopping times τ_n such that $\tau_n - \sigma$ takes at most a countable number of values for each n, $\tau_n \downarrow \tau$, and each τ_n is $\mathcal{F}_\sigma$-measurable. For example, we can take $\tau_n(\omega) = \sigma(\omega) + k/2^n$ for all ω such that $(k-1)/2^n \leq \tau(\omega) - \sigma(\omega) < k/2^n$, where $k \geq 1$. Thus,

$$E(g(X^x_{\tau_n})|\mathcal{F}_\sigma) = \psi_g(\tau_n - \sigma, X^x_\sigma) \quad \text{almost surely.}$$

Clearly, $\psi_g(\tau_n - \sigma, x)$ is a Borel-measurable function of x. Since g is bounded and continuous, and X_t is right-continuous,

$$\lim_{n \to \infty} \psi_g(\tau_n - \sigma, x) = \psi_g(\tau - \sigma, x).$$

Therefore, $\lim_{n \to \infty} \psi_g(\tau_n - \sigma, X^x_\sigma) = \psi_g(\tau - \sigma, X^x_\sigma)$ almost surely. By the Dominated Convergence Theorem for conditional expectations,

$$\lim_{n \to \infty} E(g(X^x_{\tau_n})|\mathcal{F}_\sigma) = E(g(X^x_\tau)|\mathcal{F}_\sigma),$$

which implies that (19.10) holds for all σ and τ satisfying the assumptions of the theorem.

As in the proof of Lemma 19.6, we can show that (19.10) holds if g is an indicator function of a measurable set. Since a bounded measurable function can be uniformly approximated by finite linear combinations of indicator functions, (19.10) holds for all bounded measurable g.   □

19.6 Strong Markov Property of Brownian Motion

As before, let W_t be a d-dimensional Brownian motion relative to a filtration $\mathcal{F}_t$, and $W^x_t = x + W_t$. In this section we show that W^x_t is a time-homogeneous strong Markov family relative to the filtration $\mathcal{F}_t$.

Since the first two conditions of Definition 19.11 were verified in Sect. 19.3, it remains to verify condition (3') from Lemma 19.14. Let σ be a stopping time of $\mathcal{F}_t$, $x \in \mathbb{R}^d$, and $g : \mathbb{R}^d \to \mathbb{R}$ be a bounded continuous function. The case when $t = 0$ is trivial, therefore we can assume that $t > 0$. In this case, $\psi_g(t, x) = Eg(W^x_t)$ is a bounded continuous function of x.

First, assume that σ takes a finite or countable number of values. Then we can write $\Omega = A_1 \cup A_2 \cup \ldots$, where $\sigma(\omega) = s_k$ for $\omega \in A_k$, and all s_k are distinct. Since a set $B \subseteq A_k$ belongs to $\mathcal{F}_\sigma$ if and only if it belongs to $\mathcal{F}_{s_k}$, and $g(W^x_{\sigma+t}) = g(W^x_{s_k+t})$ on A_k,

$$\mathrm{E}(g(W^x_{\sigma+t})|\mathcal{F}_\sigma) = \mathrm{E}(g(W^x_{s_k+t})|\mathcal{F}_{s_k}) \quad \text{almost surely on } A_k.$$

Therefore,

$$\mathrm{E}(g(W^x_{\sigma+t})|\mathcal{F}_\sigma) = \mathrm{E}(g(W^x_{s_k+t})|\mathcal{F}_{s_k}) = \psi_g(t, W^x_{s_k}) = \psi_g(t, W^x_\sigma) \quad \text{a.s. on } A_k,$$

which implies that

$$\mathrm{E}(g(W^x_{\sigma+t})|\mathcal{F}_\sigma) = \psi_g(t, W^x_\sigma) \quad \text{almost surely.} \tag{19.11}$$

If the distribution of σ is not necessarily discrete, we can find a sequence of stopping times σ_n, each taking at most a countable number of values, such that $\sigma_n(\omega) \downarrow \sigma(\omega)$ for all ω. We wish to derive (19.11) starting from

$$\mathrm{E}(g(W^x_{\sigma_n+t})|\mathcal{F}_{\sigma_n}) = \psi_g(t, W^x_{\sigma_n}) \quad \text{almost surely.} \tag{19.12}$$

Since the realizations of Brownian motion are continuous almost surely, and $\psi_g(t, x)$ is a continuous function of x,

$$\lim_{n\to\infty} \psi_g(t, W^x_{\sigma_n}) = \psi_g(t, W^x_\sigma) \quad \text{almost surely.}$$

Let $\mathcal{F}^+ = \bigcap_{n=1}^\infty \mathcal{F}_{\sigma_n}$. By the Doob Theorem (Theorem 16.11),

$$\lim_{n\to\infty} \mathrm{E}(g(W^x_{\sigma+t})|\mathcal{F}_{\sigma_n}) = \mathrm{E}(g(W^x_{\sigma+t})|\mathcal{F}^+).$$

We also need to estimate the difference $\mathrm{E}(g(W^x_{\sigma_n+t})|\mathcal{F}_{\sigma_n}) - \mathrm{E}(g(W^x_{\sigma+t})|\mathcal{F}_{\sigma_n})$. Since the sequence $g(W^x_{\sigma_n+t}) - g(W^x_{\sigma+t})$ tends to zero almost surely, and $g(W^x_{\sigma_n+t})$ is uniformly bounded, it is easy to show that $\mathrm{E}(g(W^x_{\sigma_n+t})|\mathcal{F}_{\sigma_n}) - \mathrm{E}(g(W^x_{\sigma+t})|\mathcal{F}_{\sigma_n})$ tends to zero in probability. (We leave this statement as an exercise for the reader.) Therefore, upon taking the limit as $n \to \infty$ in (19.12),

$$\mathrm{E}(g(W^x_{\sigma+t})|\mathcal{F}^+) = \psi_g(t, W^x_\sigma) \quad \text{almost surely.}$$

Since $\mathcal{F}_\sigma \subseteq \mathcal{F}^+$, and W^x_σ is $\mathcal{F}_\sigma$-measurable, we can take conditional expectations with respect to $\mathcal{F}_\sigma$ on both sides of this equality to obtain (19.11). This proves that W^x_t is a strong Markov family.

Let us conclude this section with several examples illustrating the use of the strong Markov property.

Example. Let us revisit the problem on the distribution of the maximum of Brownian motion. We use the same notation as in Sect. 19.1. Since W^x_t is a strong Markov family, we can apply Lemma 19.15 with $\sigma = \tau_c$, $\tau = T$, and $g = \chi_{(c,\infty)}$. Since $\mathrm{P}(W_T > c|\mathcal{F}_{\tau_c}) = 0$ on the event $\{\tau_c \geq T\}$,

$$\mathrm{P}(W_T > c|\mathcal{F}_{\tau_c}) = \chi_{\{\tau_c < T\}}\mathrm{P}(W^c_t > c)|_{t=T-\tau_c}.$$

Since $P(W_t^c > c) = 1/2$ for all t,

$$P(W_T > c | \mathcal{F}_{\tau_c}) = \frac{1}{2}\chi_{\{\tau_c < T\}},$$

and, after taking expectation on both sides,

$$P(W_T > c) = \frac{1}{2}P(\tau_c < T). \tag{19.13}$$

Since the event $\{W_T > c\}$ is contained in the event $\{\tau_c < T\}$, (19.13) implies $P(\tau_c < T, W_T < c) = P(\tau_c < T, W_T > c)$, thus justifying the arguments of Sect. 19.1.

Example. Let W_t be a Brownian motion relative to a filtration $\mathcal{F}_t$, and σ be a stopping time of $\mathcal{F}_t$. Define the process $\widetilde{W}_t = W_{\sigma+t} - W_\sigma$. Let us show that $\widetilde{W}_t$ is a Brownian motion independent of $\mathcal{F}_\sigma$.

Let Γ be a Borel subset of $\mathbb{R}^d$, $t \geq 0$, and let $f : \widetilde{\Omega} \to \mathbb{R}$ be the indicator function of the set $\{\widetilde{\omega} : \widetilde{\omega}(t) - \widetilde{\omega}(0) \in \Gamma\}$. By Lemma 19.12,

$$P(\widetilde{W}_t \in \Gamma | \mathcal{F}_\sigma) = E(f(W_{\sigma+\cdot}) | \mathcal{F}_\sigma) = \varphi_f(W_\sigma) \text{ almost surely,}$$

where $\varphi_f(x) = Ef(W^x) = P(W_t^x - W_0^x \in \Gamma) = P(W_t \in \Gamma)$, thus showing that $\varphi_f(x)$ does not depend on x. Therefore, $P(\widetilde{W}_t \in \Gamma | \mathcal{F}_\sigma) = P(\widetilde{W}_t \in \Gamma)$. Since Γ was an arbitrary Borel set, $\widetilde{W}_t$ is independent of $\mathcal{F}_\sigma$.

Now let $k \geq 1$, $t_1, \ldots, t_k \in \mathbb{R}^+$, B be a Borel subset of $\mathbb{R}^{dk}$, and $f : \widetilde{\Omega} \to \mathbb{R}$ the indicator function of the set $\{\widetilde{\omega} : (\widetilde{\omega}(t_1) - \widetilde{\omega}(0), \ldots, \widetilde{\omega}(t_k) - \widetilde{\omega}(0)) \in B\}$. By Lemma 19.12,

$$P((\widetilde{W}_{t_1}, \ldots, \widetilde{W}_{t_k}) \in B | \mathcal{F}_\sigma) = E(f(W_{\sigma+\cdot}) | \mathcal{F}_\sigma) = \varphi_f(W_\sigma) \text{ almost surely,} \tag{19.14}$$

where

$$\varphi_f(x) = Ef(W^x) = P((W_{t_1}^x - W_0^x, \ldots, W_{t_k}^x - W_0^x) \in B)$$
$$= P((W_{t_1}, \ldots, W_{t_k}) \in B),$$

which does not depend on x. Taking expectation on both sides of (19.14) gives

$$P((\widetilde{W}_{t_1}, \ldots, \widetilde{W}_{t_k}) \in B) = P((W_{t_1}, \ldots, W_{t_k}) \in B),$$

which shows that $\widetilde{W}_t$ has the finite-dimensional distributions of a Brownian motion. Clearly, the realizations of $\widetilde{W}_t$ are continuous almost surely, that is $\widetilde{W}_t$ is a Brownian motion.

Example. Let W_t be a d-dimensional Brownian motion and $W_t^x = x + W_t$. Let D be a bounded open domain in $\mathbb{R}^d$, and f a bounded measurable function defined on ∂D. For a point $x \in D$, we define τ^x to be the first time the process W_t^x reaches the boundary of D, that is

$$\tau^x(\omega) = \inf\{t \geq 0 : W_t^x(\omega) \in \partial D\}.$$

Since D is a bounded domain, the stopping time τ^x is finite almost surely. Let us follow the process W_t^x till it reaches ∂D and evaluate f at the point $W_{\tau^x(\omega)}^x(\omega)$. Let us define

$$u(x) = Ef(W_{\tau^x}^x) = \int_{\partial D} f(y)d\mu_x(y),$$

where $\mu_x(A) = P(W_{\tau^x}^x \in A)$ is the measure on ∂D induced by the random variable $W_{\tau^x}^x$ and $A \in \mathcal{B}(\partial D)$. Let us show that $u(x)$ is a harmonic function, that is $\Delta u(x) = 0$ for $x \in D$.

Let B^x be a ball in $\mathbb{R}^d$ centered at x and contained in D. Let σ^x be the first time the process W_t^x reaches the boundary of B^x, that is

$$\sigma^x(\omega) = \inf\{t \geq 0 : W_t^x(\omega) \in \partial B^x\}.$$

For a continuous function $\widetilde{\omega} \in \widetilde{\Omega}$, denote by $\tau(\widetilde{\omega})$ the first time $\widetilde{\omega}$ reaches ∂D, and put $\tau(\widetilde{\omega})$ equal to infinity if $\widetilde{\omega}$ never reaches ∂D, that is

$$\tau(\widetilde{\omega}) = \begin{cases} \inf\{t \geq 0 : \widetilde{\omega}(t) \in \partial D\} & \text{if } \widetilde{\omega}(t) \in \partial D \text{ for some } t \in \mathbb{R}^+, \\ \infty & \text{otherwise.} \end{cases}$$

Define the function $\widetilde{f}$ on the space $\widetilde{\Omega}$ via

$$\widetilde{f}(\widetilde{\omega}) = \begin{cases} f(\widetilde{\omega}(\tau(\widetilde{\omega}))) & \text{if } \widetilde{\omega}(t) \in \partial D \text{ for some } t \in \mathbb{R}^+, \\ 0 & \text{otherwise.} \end{cases}$$

Let us apply Lemma 19.12 to the family of processes W_t^x, the function $\widetilde{f}$, and the stopping time σ^x:

$$E(\widetilde{f}(W_{\sigma^x+\cdot}^x)|\mathcal{F}_{\sigma^x}) = \varphi_{\widetilde{f}}(W_{\sigma^x}^x) \text{ almost surely,}$$

where $\varphi_{\widetilde{f}}(x) = E\widetilde{f}(W^x) = Ef(W_{\tau^x}^x) = u(x)$. The function $u(x)$ is measurable by Remark 19.5. Note that $\widetilde{f}(\widetilde{\omega}) = \widetilde{f}(\widetilde{\omega}(s + \cdot))$ if $s < \tau(\widetilde{\omega})$, and therefore the above equality can be rewritten as

$$E(f(W_{\tau^x}^x)|\mathcal{F}_{\sigma^x}) = u(W_{\sigma^x}^x) \text{ almost surely.}$$

After taking expectation on both sides,

$$u(x) = Ef(W_{\tau^x}^x) = Eu(W_{\sigma^x}^x) = \int_{\partial B^x} u(y)d\nu^x(y),$$

where ν^x is the measure on ∂B^x induced by the random variable $W_{\sigma^x}^x$. Due to the spherical symmetry of Brownian motion, the measure ν^x is the uniform measure on the sphere ∂B^x. Thus $u(x)$ is equal to the average value of $u(y)$

over the sphere ∂B^x. For a bounded measurable function u, this property, when valid for all x and all the spheres centered at x and contained in the domain D (which is the case here), is equivalent to u being harmonic (see "Elliptic Partial Differential Equations of Second Order" by D. Gilbarg and N. Trudinger, for example). We shall further discuss the properties of the function $u(x)$ in Sect. 21.2.

19.7 Problems

1. Prove that any right-continuous adapted process is progressively measurable. (Hint: see the proof of Lemma 12.3.)
2. Prove that if a process X_t is progressively measurable with respect to a filtration $\mathcal{F}_t$, and τ is a stopping time of the same filtration, then $X_{t \wedge \tau}$ is also progressively measurable and X_τ is $\mathcal{F}_\tau$-measurable.
3. Let W_t be a one-dimensional Brownian motion. For a positive constant c, define the stopping time τ_c as the first time the Brownian motion reaches the level c, that is

$$\tau_c(\omega) = \inf\{t \geq 0 : W_t(\omega) = c\}.$$

   Prove that $\tau_c < \infty$ almost surely, and find the distribution function of τ_c. Prove that $\mathrm{E}\tau_c = \infty$.
4. Let W_t be a one-dimensional Brownian motion. Prove that one can find positive constants c and λ such that

$$\mathrm{P}(\sup_{1 \leq s \leq 2^t} \frac{|W_s|}{\sqrt{s}} \leq 1) \leq c e^{-\lambda t}, \quad t \geq 1.$$

5. Let W_t be a one-dimensional Brownian motion and $V_t = \int_0^t W_s ds$. Prove that the pair (W_t, V_t) is a two-dimensional Markov process.
6. Let W_t be a one-dimensional Brownian motion. Find $\mathrm{P}(\sup_{0 \leq t \leq 1} |W_t| \leq 1)$.
7. Let $W_t = (W_t^1, W_t^2)$ be a standard two-dimensional Brownian motion. Let τ_1 be the first time when $W_t^1 = 1$, that is

$$\tau_1(\omega) = \inf\{t \geq 0 : W_t^1(\omega) = 1\}.$$

   Find the distribution of $W_{\tau_1}^2$.
8. Let W_t be a one-dimensional Brownian motion. Prove that with probability one the set $S = \{t : W_t = 0\}$ is unbounded.

Stochastic Integral and the Ito Formula

20.1 Quadratic Variation of Square-Integrable Martingales

In this section we shall apply the Doob-Meyer Decomposition to submartingales of the form X_t^2, where X_t is a square-integrable martingale with continuous sample paths. This decomposition will be essential in the construction of the stochastic integral in the next section.

We shall call two random processes equivalent if they are indistinguishable. We shall often use the same notation for a process and the equivalence class it represents.

Definition 20.1. *Let $\mathcal{F}_t$ be a filtration on a probability space $(\Omega, \mathcal{F}, P)$. Let $\mathcal{M}_2^c$ denote the space of all equivalence classes of square-integrable martingales which start at zero, and whose sample paths are continuous almost surely. That is, $X_t \in \mathcal{M}_2^c$ if $(X_t, \mathcal{F}_t)$ is a square-integrable martingale, $X_0 = 0$ almost surely, and X_t is continuous almost surely.*

We shall always assume that the filtration $\mathcal{F}_t$ satisfies the usual conditions (as is the case, for example, if $\mathcal{F}_t$ is the augmented filtration for a Brownian motion).

Let us consider the process X_t^2. Since it is equal to a convex function (namely x^2) applied to the martingale X_t, the process X_t^2 is a submartingale. Let S_a be the set of all stopping times bounded by a. If $\tau \in S_a$, by the Optional Sampling Theorem

$$\int_{\{X_\tau^2 > \lambda\}} X_\tau^2 dP \leq \int_{\{X_\tau^2 > \lambda\}} X_a^2 dP.$$

By the Chebyshev Inequality,

$$P(X_\tau^2 > \lambda) \leq \frac{EX_\tau^2}{\lambda} \leq \frac{EX_a^2}{\lambda} \to 0 \quad \text{as} \quad \lambda \to \infty.$$

L. Koralov and Y.G. Sinai, *Theory of Probability and Random Processes*, Universitext, DOI 10.1007/978-3-540-68829-7_20, © Springer-Verlag Berlin Heidelberg 2012

Since the integral is an absolutely continuous function of sets,

$$\lim_{\lambda \to \infty} \sup_{\tau \in S_a} \int_{\{X_\tau^2 > \lambda\}} X_\tau^2 d\mathrm{P} = 0,$$

that is, the set of random variables $\{X_\tau\}_{\tau \in S_a}$ is uniformly integrable.

Therefore, we can apply the Doob-Meyer Decomposition (Theorem 13.26) to conclude that there are unique (up to indistinguishability) processes M_t and A_t, whose paths are continuous almost surely, such that $X_t^2 = M_t + A_t$, where $(M_t, \mathcal{F}_t)$ is a martingale, and A_t is an adapted non-decreasing process, and $M_0 = A_0 = 0$ almost surely.

Definition 20.2. *The process A_t in the above decomposition $X_t^2 = M_t + A_t$ of the square of the martingale $X_t \in \mathcal{M}_2^c$ is called the quadratic variation of X_t and is denoted by $\langle X \rangle_t$.*

Example. Let us prove that $\langle W \rangle_t = t$. Indeed for $s \le t$,

$$\mathrm{E}(W_t^2|\mathcal{F}_s) = \mathrm{E}((W_t - W_s)^2|\mathcal{F}_s) + 2\mathrm{E}(W_t W_s|\mathcal{F}_s) - \mathrm{E}(W_s^2|\mathcal{F}_s) = W_s^2 + t - s.$$

Therefore, $W_t^2 - t$ is a martingale, and $\langle W \rangle_t = t$ due to the uniqueness of the Doob-Meyer Decomposition.

Example. Let $X_t \in \mathcal{M}_2^c$ and τ be a stopping time of the filtration $\mathcal{F}_t$ (here τ is allowed to take the value ∞ with positive probability). Thus, the process $Y_t = X_{t \wedge \tau}$ also belongs to $\mathcal{M}_2^c$. Indeed, it is a continuous martingale by Lemma 13.29. It is square-integrable since $Y_t \chi_{\{t < \tau\}} = X_t \chi_{\{t < \tau\}}$, while

$$Y_t \chi_{\{\tau \le t\}} = X_\tau \chi_{\{\tau \le t\}} = \mathrm{E}(X_t \chi_{\{\tau \le t\}}|\mathcal{F}_\tau) \in L^2(\Omega, \mathcal{F}, \mathrm{P}).$$

Since $X_t^2 - \langle X \rangle_t$ is a continuous martingale, the process $X_{t \wedge \tau}^2 - \langle X \rangle_{t \wedge \tau}$ is also a martingale by Lemma 13.29. Since $\langle X \rangle_{t \wedge \tau}$ is an adapted non-decreasing process, we conclude from the uniqueness of the Doob-Meyer Decomposition that $\langle Y \rangle_t = \langle X \rangle_{t \wedge \tau}$.

Lemma 20.3. *Let $X_t \in \mathcal{M}_2^c$. Let τ be a stopping time such that $\langle X \rangle_\tau = 0$ almost surely. Then $X_t = 0$ for all $0 \le t \le \tau$ almost surely.*

Proof. Since $\langle X \rangle_t$ is non-decreasing, $\langle X \rangle_{t \wedge \tau} = 0$ almost surely for each t. By Lemma 13.29, the process $X_{t \wedge \tau}^2 - \langle X \rangle_{t \wedge \tau}$ is a martingale. Therefore, since the expectation of a martingale is a constant,

$$\mathrm{E}X_{t \wedge \tau}^2 = \mathrm{E}(X_{t \wedge \tau}^2 - \langle X \rangle_{t \wedge \tau}) = 0$$

for each $t \ge 0$, that is $X_{t \wedge \tau} = 0$ almost surely. Since X_t is continuous, $X_t = 0$ for all $0 \le t \le \tau$ almost surely. $\square$

Clearly, the linear combinations of elements of $\mathcal{M}_2^c$ are also elements of $\mathcal{M}_2^c$.

Definition 20.4. *Let two processes X_t and Y_t belong to $\mathcal{M}_2^c$. We define their cross-variation as*

$$\langle X, Y \rangle_t = \frac{1}{4}(\langle X + Y \rangle_t - \langle X - Y \rangle_t). \tag{20.1}$$

Clearly, $X_t Y_t - \langle X, Y \rangle_t$ is a continuous martingale, the cross-variation is bi-linear and symmetric in X and Y, and $|\langle X, Y \rangle_t|^2 \le \langle X \rangle_t \langle Y \rangle_t$.

Let us introduce a metric which will turn $\mathcal{M}_2^c$ into a complete metric space.

Definition 20.5. *For $X, Y \in \mathcal{M}_2^c$ and $0 \le t < \infty$, we define*

$$\|X\|_t = \sqrt{EX_t^2}, \quad \text{and} \quad d_{\mathcal{M}}(X, Y) = \sum_{n=1}^{\infty} \frac{1}{2^n} \min(\|X - Y\|_n, 1).$$

In order to prove that $d_{\mathcal{M}}$ is a metric, we need to show, in particular, that $d_{\mathcal{M}}(X, Y) = 0$ implies that $X_t - Y_t$ is indistinguishable from zero. If $d_{\mathcal{M}}(X, Y) = 0$, then $X_n - Y_n = 0$ almost surely for every positive integer n. Since $X_t - Y_t$ is a martingale, $X_t - Y_t = E(X_n - Y_n | \mathcal{F}_t) = 0$ almost surely for every $0 \le t \le n$. Therefore,

$$P(\{\omega : X_t(\omega) - Y_t(\omega) = 0 \text{ for all rational } t\}) = 1.$$

This implies that $X_t - Y_t$ is indistinguishable from zero, since it is continuous almost surely. It is clear that $d_{\mathcal{M}}$ has all the other properties required of a metric. Let us show that the space $\mathcal{M}_2^c$ is complete, which will be essential in the construction of the stochastic integral.

Lemma 20.6. *The space $\mathcal{M}_2^c$ with the metric $d_{\mathcal{M}}$ is complete.*

Proof. Let X_t^m be a Cauchy sequence in $\mathcal{M}_2^c$. Then X_n^m is a Cauchy sequence in $L^2(\Omega, \mathcal{F}_n, P)$ for each n. If $t \le n$, then $E|X_t^{m_1} - X_t^{m_2}|^2 \le E|X_n^{m_1} - X_n^{m_2}|^2$ for all m_1 and m_2, since $|X_t^{m_1} - X_t^{m_2}|^2$ is a submartingale. This proves that X_t^m is a Cauchy sequence in $L^2(\Omega, \mathcal{F}_t, P)$ for each t. Let X_t be defined for each t as the limit of X_t^m in $L^2(\Omega, \mathcal{F}_t, P)$. Let $0 \le s \le t$, and $A \in \mathcal{F}_s$. Then,

$$\int_A X_t dP = \lim_{m \to \infty} \int_A X_t^m dP = \lim_{m \to \infty} \int_A X_s^m dP = \int_A X_s dP,$$

where the middle equality follows from X_t^m being a martingale, and the other two are due to the L^2 convergence. This shows that $(X_t, \mathcal{F}_t)$ is a martingale. By Lemma 13.25, we can choose a right-continuous modification of X_t. We can therefore apply the Doob Inequality (Theorem 13.30) to the submartingale $|X_t^m - X_t|^2$ to obtain

$$P(\sup_{0 \le s \le t} |X_s^m - X_s| \ge \lambda) \le \frac{1}{\lambda^2} E|X_t^m - X_t|^2 \to 0 \quad \text{as } m \to \infty$$

for any t. We can, therefore, extract a subsequence m_k such that

$$P(\sup_{0 \le s \le t} |X_s^{m_k} - X_s| \ge \frac{1}{k}) \le \frac{1}{2^k} \text{ for } k \ge 1.$$

The First Borel-Cantelli Lemma implies that $X_t^{m_k}$ converges to X_t uniformly on $[0, t]$ for almost all ω. Since t was arbitrary, this implies that X_t is continuous almost surely, and thus $\mathcal{M}_2^c$ is complete. $\qquad\square$

Next we state a lemma which explains the relation between the quadratic variation of a martingale (as in Definition 20.1) and the second variation of the martingale over a partition (as in Sect. 3.2).

More precisely, let f be a function defined on an interval $[a, b]$ of the real line. Let $\sigma = \{t_0, t_1, \ldots, t_n\}$, $a = t_0 \le t_1 \le \ldots \le t_n = b$, be a partition of the interval $[a, b]$ into n subintervals. We denote the length of the largest interval by $\delta(\sigma) = \max_{1 \le i \le n}(t_i - t_{i-1})$. Let $V_{[a,b]}^2(f, \sigma) = \sum_{i=1}^{n} |f(t_i) - f(t_{i-1})|^2$ be the second variation of the function f over the partition σ.

Lemma 20.7. *Let $X_t \in \mathcal{M}_2^c$ and $t \ge 0$ be fixed. Then, for any $\varepsilon > 0$*

$$\lim_{\delta(\sigma) \to 0} P(|V_{[0,t]}^2(X_s, \sigma) - \langle X \rangle_t| > \varepsilon) = 0.$$

We omit the proof of this lemma, instead referring the reader to "Brownian Motion and Stochastic Calculus" by I. Karatzas and S. Shreve. Note, however, that Lemma 18.24 contains a stronger statement (convergence in L^2 instead of convergence in probability) when the martingale X_t is a Brownian motion.

Corollary 20.8. *Assume that $V_{[0,t]}^1(X_s(\omega)) < \infty$ for almost all $\omega \in \Omega$, where $X_t \in \mathcal{M}_2^c$ and $t \ge 0$ is fixed. Then $X_s(\omega) = 0$, $s \in [0, t]$, for almost all $\omega \in \Omega$.*

Proof. Let us assume the contrary. Then, by Lemma 20.3, there is a positive constant c_1 and an event $A' \subseteq \Omega$ with $P(A') > 0$ such that $\langle X \rangle_t(\omega) \ge c_1$ for almost all $\omega \in A'$. Since $V_{[0,t]}^1(X_s(\omega)) < \infty$ for almost all $\omega \in A'$, we can find a constant c_2 and a subset $A'' \subseteq A'$ with $P(A'') > 0$ such that $V_{[0,t]}^1(X_s(\omega)) \le c_2$ for almost all $\omega \in A''$.

Let σ_n be a sequence of partitions of $[0, t]$ into 2^n intervals of equal length. By Lemma 20.7, we can assume, without loss of generality, that $V_{[0,t]}^2(X_s(\omega), \sigma_n) \neq 0$ for large enough n almost surely on A''. Since a continuous function is also uniformly continuous,

$$\lim_{n \to \infty} \frac{V_{[0,t]}^1(X_s(\omega), \sigma_n)}{V_{[0,t]}^2(X_s(\omega), \sigma_n)} = \infty \text{ almost surely on } A''.$$

This, however, contradicts $V_{[0,t]}^2(X_s(\omega), \sigma_n) \to \langle X \rangle_t(\omega) \ge c_1$ (in probability), while $\lim_{n \to \infty} V_{[0,t]}^1(X_s(\omega), \sigma_n) = V_{[0,t]}^1(X_s(\omega)) \le c_2$ for almost all $\omega \in A''$. $\qquad\square$

Lemma 20.9. *Let $X_t, Y_t \in \mathcal{M}_2^c$. There is a unique (up to indistinguishability) adapted continuous process of bounded variation A_t such that $A_0 = 0$ almost surely and $X_t Y_t - A_t$ is a martingale. In fact, $A_t = \langle X, Y \rangle_t$.*

Proof. The existence part was demonstrated above. Suppose there are two processes A_t^1 and A_t^2 with the desired properties. Then $M_t = A_t^1 - A_t^2$ is a continuous martingale with bounded variation. Define the sequence of stopping times $\tau_n = \inf\{t \geq 0 : |M_t| = n\}$, where the infimum of an empty set is equal to $+\infty$. This is a non-decreasing sequence, which tends to infinity almost surely. Note that $M_t^{(n)} = M_{t \wedge \tau_n}$ is a square-integrable martingale for each n (by Lemma 13.29), and that $M_t^{(n)}$ is also a process of bounded variation. By Corollary 20.8, $M_t^{(n)} = 0$ for all t almost surely. Since $\tau_n \to \infty$, A_t^1 and A_t^2 are indistinguishable. □

An immediate consequence of this result is the following lemma.

Lemma 20.10. *Let $X_t, Y_t \in \mathcal{M}_2^c$ with the filtration $\mathcal{F}_t$, and let τ be a stopping time for $\mathcal{F}_t$. Then $\langle X, Y \rangle_{t \wedge \tau}$ is the cross-variation of the processes $X_{t \wedge \tau}$ and $Y_{t \wedge \tau}$.*

20.2 The Space of Integrands for the Stochastic Integral

Let $(M_t, \mathcal{F}_t)_{t \in \mathbb{R}^+}$ be a continuous square-integrable martingale on a probability space $(\Omega, \mathcal{F}, P)$, and let X_t be an adapted process. In this chapter we shall define the stochastic integral $\int_0^t X_s dM_s$, also denoted by $I_t(X)$.

We shall carefully state additional assumptions on X_t in order to make sense of the integral. Note that the above expression cannot be understood as the Lebesgue-Stieltjes integral defined for each ω, unless $\langle M \rangle_t(\omega) = 0$. Indeed, the function $M_s(\omega)$ has unbounded first variation on the interval $[0, t]$ if $\langle M \rangle_t(\omega) \neq 0$, as discussed in the previous section.

While the stochastic integral could be defined for a general square integrable martingale M_t (by imposing certain restrictions on the process X_t), we shall stick to the assumption that $M_t \in \mathcal{M}_2^c$. Our prime example is $M_t = W_t$.

Let us now discuss the assumptions on the integrand X_t. We introduce a family of measures μ_t, $0 \leq t < \infty$, associated to the process M_t, on the product space $\Omega \times [0, t]$ with the σ-algebra $\mathcal{F} \times \mathcal{B}([0, t])$.

Namely, let $\mathcal{K}$ be the collection of sets of the form $A = B \times [a, b]$, where $B \in \mathcal{F}$ and $[a, b] \subseteq [0, t]$. Let $\mathcal{G}$ be the collection of measurable sets $A \in \mathcal{F} \times \mathcal{B}([0, t])$ for which $\int_0^t \chi_A(\omega, s) d\langle M \rangle_s(\omega)$ exists for almost all ω and is a measurable function of ω. Note that $\mathcal{K} \subseteq \mathcal{G}$, that $\mathcal{K}$ is a π-system, and that $\mathcal{G}$ is closed under unions of non-intersecting sets and complements in $\Omega \times [0, t]$. Therefore, $\mathcal{F} \times \mathcal{B}([0, t]) = \sigma(\mathcal{K}) = \mathcal{G}$, where the second equality is due to Lemma 4.12.

We can now define

$$\mu_t(A) = \mathrm{E} \int_0^t \chi_A(\omega, s) d\langle M\rangle_s(\omega),$$

where $A \in \mathcal{F} \times \mathcal{B}([0,t])$. The expectation exists since the integral is a measurable function of ω bounded from above by $\langle M\rangle_t$. The fact that μ_t is σ-additive (that is a measure) follows from the Levi Convergence Theorem. If f is defined on $\Omega \times [0,t]$ and is measurable with respect to the σ-algebra $\mathcal{F} \times \mathcal{B}([0,t])$, then

$$\int_{\Omega \times [0,t]} f d\mu_t = \mathrm{E} \int_0^t f(\omega, s) d\langle M\rangle_s(\omega).$$

(If the function f is non-negative, and the expression on one side of the equality is defined, then the expression on the other side is also defined. If the function f is not necessarily non-negative, and the expression on the left-hand side is defined, then the expression on the right-hand side is also defined). Indeed, this formula is true for indicator functions of measurable sets, and therefore, for simple functions with a finite number of values. It also holds for non-negative functions since they can be approximated by monotonic sequences of simple functions with a finite number of values. Furthermore, any function can be represented as a difference of two non-negative functions, and thus, if the expression on the left-hand side is defined, so is the one on the right-hand side.

We can also consider the σ-finite measure μ on the product space $\Omega \times \mathbb{R}^+$ with the σ-algebra $\mathcal{F} \times \mathcal{B}(\mathbb{R}^+)$ whose restriction to $\Omega \times [0,t]$ coincides with μ_t for each t. For example, if $M_t = W_t$, then $d\langle M\rangle_t(\omega)$ is the Lebesgue measure for each ω, and μ is equal to the product of the measure P and the Lebesgue measure on the half-line.

Let $\mathcal{H}_t = L^2(\Omega \times [0,t], \mathcal{F} \times \mathcal{B}([0,t]), \mu_t)$, and $|| \cdot ||_{\mathcal{H}_t}$ be the L^2 norm on this space. We define $\mathcal{H}$ as the space of classes of functions on $\Omega \times \mathbb{R}^+$ whose restrictions to $\Omega \times [0,t]$ belong to $\mathcal{H}_t$ for every $t \geq 0$. Two functions f and g belong to the same class, and thus correspond to the same element of $\mathcal{H}$, if $f = g$ almost surely with respect to the measure μ. We can define the metric on $\mathcal{H}$ by

$$d_{\mathcal{H}}(f, g) = \sum_{n=1}^{\infty} \frac{1}{2^n} \min(||f - g||_{\mathcal{H}_n}, 1).$$

It is easy to check that this turns $\mathcal{H}$ into a complete metric space.

We shall define the stochastic integral $I_t(X)$ for all progressively measurable processes X_t such that $X_t \in \mathcal{H}$. We shall see that $I_t(X)$ is indistinguishable from $I_t(Y)$ if X_t and Y_t coincide as elements of $\mathcal{H}$. The set of elements of $\mathcal{H}$ which have a progressively measurable representative will be denoted by $\mathcal{L}^*$ or $\mathcal{L}^*(M)$, whenever it is necessary to stress the dependence on the martingale M_t. It can be also viewed as a metric space with the metric $d_{\mathcal{H}}$, and it can be shown that this space is also complete (although we will not use this fact).

Lemma 20.11. *Let X_t be a progressively measurable process and A_t a continuous adapted processes such that $A_t(\omega)$ almost surely has bounded variation on any finite interval and*

$$Y_t(\omega) = \int_0^t X_s(\omega) dA_s(\omega) < \infty \quad almost \quad surely.$$

Then Y_t is progressively measurable.

Proof. As before, X_t can be approximated by simple functions from below, proving $\mathcal{F}_t$-measurability of Y_t for fixed t. The process Y_t is progressively measurable since it is continuous. □

20.3 Simple Processes

In this section we again assume that we have a probability space $(\Omega, \mathcal{F}, P)$ and a continuous square-integrable martingale $M_t \in \mathcal{M}_2^c$.

Definition 20.12. *A process X_t is called simple if there are a strictly increasing sequence of real numbers t_n, $n \geq 0$, such that $t_0 = 0$, $\lim_{n\to\infty} t_n = \infty$, and a sequence of bounded random variables ξ_n, $n \geq 0$, such that ξ_n is $\mathcal{F}_{t_n}$-measurable for every n and*

$$X_t(\omega) = \xi_0(\omega)\chi_{\{0\}}(t) + \sum_{n=0}^{\infty} \xi_n(\omega)\chi_{(t_n, t_{n+1}]}(t) \quad for \ \omega \in \Omega, t \geq 0. \quad (20.2)$$

The class of all simple processes will be denoted by $\mathcal{L}_0$.

It is clear that $\mathcal{L}_0 \subseteq \mathcal{L}^*$. We shall first define the stochastic integral for simple processes. Then we shall extend the definition to all the integrands from $\mathcal{L}^*$ with the help of the following lemma.

Lemma 20.13. *The space $\mathcal{L}_0$ is dense in $\mathcal{L}^*$ in the metric $d_{\mathcal{H}}$ of the space $\mathcal{H}$.*

The lemma states that, given a process $X_t \in \mathcal{L}^*$, we can find a sequence of simple processes X_t^n such that $\lim_{n\to\infty} d_{\mathcal{H}}(X_t^n, X_t) = 0$. We shall only prove this for X_t continuous for almost all ω, the general case being somewhat more complicated.

Proof. It is sufficient to show that for each integer m there is a sequence of simple processes X_t^n such that

$$\lim_{n\to\infty} ||X_t^n - X_t||_{\mathcal{H}_m} = 0. \quad (20.3)$$

Indeed, if this is the case, then for each m we can find a simple process $X_t^{(m)}$ such that $||X_t^{(m)} - X_t||_{\mathcal{H}_m} \leq 1/m$. Then $\lim_{m\to\infty} d_{\mathcal{H}}(X_t^{(m)}, X_t) = 0$ as required. Let m be fixed, and

$$X_t^n(\omega) = X_0(\omega)\chi_{\{0\}}(t) + \sum_{k=0}^{n-1} X_{km/n}(\omega)\chi_{(km/n,(k+1)m/n]}(t).$$

This sequence converges to X_t almost surely uniformly in $t \in [0, m]$, since X_t is continuous almost surely. If X_t is bounded on the interval $[0, m]$ (that is, $|X_t(\omega)| \le c$ for all $\omega \in \Omega$, $t \in [0, m]$), then, by the Lebesgue Dominated Convergence Theorem, $\lim_{n\to\infty} ||X_t^n - X_t||_{\mathcal{H}_m} = 0$. If X_t is not necessarily bounded, it can be first approximated by bounded processes as follows. Let

$$Y_t^n(\omega) = \begin{cases} -n & \text{if } X_t(\omega) < -n, \\ X_t(\omega) & \text{if } -n \le X_t(\omega) \le n, \\ n & \text{if } X_t(\omega) > n. \end{cases}$$

Note that Y_t^n are continuous progressively measurable processes, which are bounded on $[0, m]$. Moreover, $\lim_{n\to\infty} ||Y_t^n - X_t||_{\mathcal{H}_m} = 0$. Each of the processes Y_t^n can, in turn, be approximated by a sequence of simple processes. Therefore, (20.3) holds for some sequence of simple processes. Thus, we have shown that for an almost surely continuous progressively measurable process X_t, there is a sequence of simple processes X_t^n such that $\lim_{n\to\infty} d_{\mathcal{H}}(X_t^n, X_t) = 0$. □

20.4 Definition and Basic Properties of the Stochastic Integral

We first define the stochastic (Ito) integral for a simple process,

$$X_t(\omega) = \xi_0(\omega)\chi_{\{0\}}(t) + \sum_{n=0}^{\infty} \xi_n(\omega)\chi_{(t_n,t_{n+1}]}(t) \quad \text{for } \omega \in \Omega, t \ge 0. \quad (20.4)$$

Definition 20.14. *The stochastic integral $I_t(X)$ of the process X_t is defined as*

$$I_t(X) = \sum_{n=0}^{m(t)-1} \xi_n(M_{t_{n+1}} - M_{t_n}) + \xi_{m(t)}(M_t - M_{t_{m(t)}}),$$

where $m(t)$ is the unique integer such that $t_{m(t)} \le t < t_{m(t)+1}$.

When it is important to stress the dependence of the integral on the martingale, we shall denote it by $I_t^M(X)$. While the same process can be represented in the form (20.4) with different ξ_n and t_n, the definition of the integral does not depend on the particular representation.

Let us study some properties stochastic integral. First, note that $I_0(X) = 0$ almost surely. It is clear that the integral is linear in the integrand, that is,

$$I_t(aX + bY) = aI_t(X) + bI_t(Y) \quad (20.5)$$

for any $X, Y \in \mathcal{L}_0$ and $a, b \in \mathbb{R}$. Also, $I_t(X)$ is continuous almost surely since M_t is continuous. Let us show that $I_t(X)$ is a martingale. If $0 \le s < t$, then

$$E((I_t(X) - I_s(X))|\mathcal{F}_s)$$

$$= E(\xi_{m(s)-1}(M_{t_{m(s)}} - M_s) + \sum_{n=m(s)}^{m(t)-1} \xi_n(M_{t_{n+1}} - M_{t_n}) + \xi_n(M_t - M_{t_{m(t)}})|\mathcal{F}_s).$$

Since ξ_n is $\mathcal{F}_{t_n}$-measurable and M_t is a martingale, the conditional expectation with respect to $\mathcal{F}_s$ of each of the terms on the right-hand side is equal to zero. Therefore, $E(I_t(X) - I_s(X)|\mathcal{F}_s) = 0$, which proves that I_t is a martingale.

The process $I_t(X)$ is square-integrable since M_t is square-integrable and the random variables ξ_n are bounded. Let us find its quadratic variation. Let $0 \le s < t$. Assume that $t_{m(t)} > s$ (the case when $t_{m(t)} \le s$ can be treated similarly). Then,

$$E(I_t^2(X) - I_s^2(X)|\mathcal{F}_s)$$
$$= E((I_t(X) - I_s(X))^2|\mathcal{F}_s)$$
$$= E((\xi_{m(s)}(M_{t_{m(s)+1}} - M_s) + \sum_{n=m(s)+1}^{m(t)-1} \xi_n(M_{t_{n+1}} - M_{t_n})$$
$$+ \xi_{m(t)}(M_t - M_{t_{m(t)}}))^2|\mathcal{F}_s)$$
$$= E(\xi_{m(s)}^2(M_{t_{m(s)+1}} - M_s)^2 + \sum_{n=m(s)+1}^{m(t)-1} \xi_n^2(M_{t_{n+1}} - M_{t_n})^2$$
$$+ \xi_{m(t)}^2(M_t - M_{t_{m(t)}})^2|\mathcal{F}_s)$$
$$= E(\xi_{m(s)}^2(\langle M\rangle_{t_{m(s)+1}} - \langle M\rangle_s) + \sum_{n=m(s)+1}^{m(t)-1} \xi_n^2(\langle M\rangle_{t_{n+1}} - \langle M\rangle_{t_n})$$
$$+ \xi_{m(t)}^2(\langle M\rangle_t - \langle M\rangle_{t_{m(t)}})|\mathcal{F}_s) = E(\int_s^t X_u^2 d\langle M\rangle_u|\mathcal{F}_s).$$

This implies that the process $I_t^2(X) - \int_0^t X_u^2 d\langle M\rangle_u$ is a martingale. Since the process $\int_0^t X_u^2 d\langle M\rangle_u$ is $\mathcal{F}_t$-adapted (as follows from the definition of a simple process), we conclude from the uniqueness of the Doob-Meyer Decomposition that $\langle I(X)\rangle_t = \int_0^t X_u^2 d\langle M\rangle_u$. Also, by setting $s = 0$ in the calculation above and taking expectation on both sides,

$$EI_t^2(X) = E\int_0^t X_u^2 d\langle M\rangle_u. \tag{20.6}$$

Recall that we have the metric $d_\mathcal{M}$ given by the family of norms $||\cdot||_n$ on the space $\mathcal{M}_2^c$ of martingales, and the metric $d_\mathcal{H}$ given by the family of norms

$|| \cdot ||_{\mathcal{H}_n}$ on the space $\mathcal{L}^*$ of integrands. So far, we have defined the stochastic integral as a mapping from the subspace $\mathcal{L}_0$ into $\mathcal{M}_2^c$,

$$\mathcal{I} : \mathcal{L}_0 \to \mathcal{M}_2^c.$$

Equation (20.6) implies that $\mathcal{I}$ is an isometry between $\mathcal{L}_0$ and its image $\mathcal{I}(\mathcal{L}_0) \subseteq \mathcal{M}_2^c$, with the norms $|| \cdot ||_{\mathcal{H}_n}$ and $|| \cdot ||_n$ respectively. Therefore, it is an isometry with respect to the metrics $d_{\mathcal{H}}$ and $d_{\mathcal{M}}$, that is

$$d_{\mathcal{M}}(I_t(X), I_t(Y)) = d_{\mathcal{H}}(X, Y)$$

for any $X, Y \in \mathcal{L}_0$. Since $\mathcal{L}_0$ is dense in $\mathcal{L}^*$ in the metric $d_{\mathcal{H}}$ (Lemma 20.13), and the space $\mathcal{M}_2^c$ is complete (Lemma 20.6), we can now extend the mapping $\mathcal{I}$ to an isometry between $\mathcal{L}^*$ (with the metric $d_{\mathcal{H}}$) and a subset of $\mathcal{M}_2^c$ (with the metric $d_{\mathcal{M}}$),

$$\mathcal{I} : \mathcal{L}^* \to \mathcal{M}_2^c.$$

Definition 20.15. *The stochastic integral of a process $X_t \in \mathcal{L}^*$ is the unique (up to indistinguishability) martingale $I_t(X) \in \mathcal{M}_2^c$ such that*

$$\lim_{Y \to X, Y \in \mathcal{L}_0} d_{\mathcal{M}}(I_t(X), I_t(Y)) = 0.$$

Given a pair of processes $X_t, Y_t \in \mathcal{L}^*$, we can find two sequences $X_t^n, Y_t^n \in \mathcal{L}_0$ such that $X_t^n \to X_t$ and $Y_t^n \to Y_t$ in $\mathcal{L}^*$. Then $aX_t^n + bY_t^n \to aX_t + bY_t$ in $\mathcal{L}^*$, which justifies (20.5) for any $X, Y \in \mathcal{L}^*$.

For $X_t \in \mathcal{L}_0$, we proved that

$$\mathrm{E}(I_t^2(X) - I_s^2(X)|\mathcal{F}_s) = \mathrm{E}(\int_s^t X_u^2 d\langle M \rangle_u |\mathcal{F}_s). \tag{20.7}$$

If $X_t \in \mathcal{L}^*$, we can find a sequence X_t^n such that $X_t^n \to X_t$ in $\mathcal{L}^*$. For any $A \in \mathcal{F}_s$,

$$\int_A (I_t^2(X) - I_s^2(X))dP = \lim_{n \to \infty} \int_A (I_t^2(X^n) - I_s^2(X^n))dP \tag{20.8}$$

$$= \lim_{n \to \infty} \int_A \int_s^t (X_u^n)^2 d\langle M \rangle_u = \int_A \int_s^t X_u^2 d\langle M \rangle_u.$$

This proves that (20.7) holds for all $X_t \in \mathcal{L}^*$. By Lemma 20.11, the process $\int_0^t X_u^2 d\langle M \rangle_u$ is $\mathcal{F}_t$-adapted. Thus, due to the uniqueness in the Doob-Meyer Decomposition, for all $X \in \mathcal{L}^*$,

$$\langle I(X) \rangle_t = \int_0^t X_u^2 d\langle M \rangle_u. \tag{20.9}$$

Remark 20.16. We shall also deal with stochastic integrals over a segment $[s, t]$, where $0 \leq s \leq t$. Namely, let a process X_u be defined for $u \in [s, t]$. We can consider the process $\widetilde{X}_u$ which is equal to X_u for $s \le u \le t$ and to zero for $u < s$ and $u > t$. If $\widetilde{X}_u \in \mathcal{L}^*$, we can define

$$\int_s^t X_u dM_u = \int_0^t \widetilde{X}_u dM_u.$$

Clearly, for $X_u \in \mathcal{L}^*$, $\int_s^t X_u dM_u = I_t(X) - I_s(X)$.

20.5 Further Properties of the Stochastic Integral

We start this section with a formula similar to (20.9), but which applies to the cross-variation of two stochastic integrals.

Lemma 20.17. *Let* $M_t^1, M_t^2 \in \mathcal{M}_2^c$, $X_t^1 \in \mathcal{L}^*(M^1)$, *and* $X_t^2 \in \mathcal{L}^*(M^2)$. *Then*

$$\langle I^{M^1}(X^1), I^{M^2}(X^2) \rangle_t = \int_0^t X_s^1 X_s^2 d\langle M^1, M^2 \rangle_s, \quad t \ge 0, \quad almost\ surely.$$
(20.10)

We only sketch the proof of this lemma, referring the reader to "Brownian Motion and Stochastic Calculus" by I. Karatzas and S. Shreve for a more detailed exposition. We need the Kunita-Watanabe Inequality, which states that under the assumptions of Lemma 20.17,

$$\int_0^t |X_s^1 X_s^2| dV_{[0,s]}^1(\langle M^1, M^2 \rangle)$$

$$\le \left(\int_0^t (X_s^1)^2 d\langle M^1 \rangle_s\right)^{1/2} \left(\int_0^t (X_s^2)^2 d\langle M^2 \rangle_s\right)^{1/2}, \quad t \ge 0, \quad almost\ surely,$$

where $V_{[0,s]}^1(\langle M^1, M^2 \rangle)$ is the first total variation of the process $\langle M^1, M^2 \rangle_t$ over the interval $[0, s]$. In particular, the Kunita-Watanabe Inequality justifies the existence of the integral on the right-hand side of (20.10).

As we did with (20.7), we can show that for $0 \le s \le t < \infty$,

$$E((I_t^{M^1}(X^1) - I_s^{M^1}(X^1))(I_t^{M^2}(X^2) - I_s^{M^2}(X^2))|\mathcal{F}_s)$$

$$= E\left(\int_s^t X_u^1 X_u^2 d\langle M^1 M^2 \rangle_u |\mathcal{F}_s\right)$$

for simple processes $X_t^1, X_t^2 \in \mathcal{L}_0$. This implies that (20.10) holds for simple processes X_t^1 and X_t^2. If $X_t^1 \in \mathcal{L}^*(M^1)$, $X_t^2 \in \mathcal{L}^*(M^2)$, then they can be approximated by simple processes as in the proof of (20.9). The transition from the statement for simple processes to (20.10) can be justified using the Kunita-Watanabe Inequality.

The following lemma will be used in the next section to define the stochastic integral with respect to a local martingale.

Lemma 20.18. *Let $M_t^1, M_t^2 \in \mathcal{M}_2^c$ (with the same filtration), $X_t^1 \in \mathcal{L}^*(M^1)$, and $X_t^2 \in \mathcal{L}^*(M^2)$. Let τ be a stopping time such that*

$$M_{t\wedge\tau}^1 = M_{t\wedge\tau}^2, \quad X_{t\wedge\tau}^1 = X_{t\wedge\tau}^2 \quad \text{for } 0 \le t < \infty \quad \text{almost surely.}$$

Then $I_{t\wedge\tau}^{M^1}(X^1) = I_{t\wedge\tau}^{M^2}(X^2)$ for $0 \le t < \infty$ almost surely.

Proof. Let $Y_t = X_{t\wedge\tau}^1 = X_{t\wedge\tau}^2$ and $N_t = M_{t\wedge\tau}^1 = M_{t\wedge\tau}^2$. Take an arbitrary $t \ge 0$. By the formula for cross-variation of two integrals,

$$\langle I^{M^i}(X^i), I^{M^j}(X^j)\rangle_{t\wedge\tau} = \int_0^{t\wedge\tau} X_s^i X_s^j d\langle M^i, M^j\rangle_s = \int_0^t Y_s^2 d\langle N\rangle_s,$$

where $1 \le i, j \le 2$. Therefore,

$$\langle I^{M^1}(X^1) - I^{M^2}(X^2)\rangle_{t\wedge\tau}$$

$$= \langle I^{M^1}(X^1)\rangle_{t\wedge\tau} + \langle I^{M^2}(X^2)\rangle_{t\wedge\tau} - 2\langle I^{M^1}(X^1), I^{M^2}(X^2)\rangle_{t\wedge\tau} = 0.$$

Lemma 20.3 now implies that $I_s^{M^1}(X^1) = I_s^{M^2}(X^2)$ for all $0 \le s \le t \wedge \tau$ almost surely. Since t was arbitrary, $I_s^{M^1}(X^1) = I_s^{M^2}(X^2)$ for $0 \le s < \tau$ almost surely, which is equivalent to the desired result. $\qquad\square$

The next lemma will be useful when applying the Ito formula (to be defined later in this chapter) to stochastic integrals.

Lemma 20.19. *Let $M_t \in \mathcal{M}_2^c$, $Y_t \in \mathcal{L}^*(M)$, and $X_t \in \mathcal{L}^*(I^M(Y))$. Then $X_t Y_t \in \mathcal{L}^*(M)$ and*

$$\int_0^t X_s d\left(\int_0^s Y_u dM_u\right) = \int_0^t X_s Y_s dM_s. \tag{20.11}$$

Proof. Since $\langle I^M(Y)\rangle_t = \int_0^t Y_s^2 d\langle M\rangle_s$, we have

$$\mathrm{E}\int_0^t X_s^2 Y_s^2 d\langle M\rangle_s = \mathrm{E}\int_0^t X_s^2 d\langle I^M(Y)\rangle_s < \infty,$$

which shows that $X_t Y_t \in \mathcal{L}^*(M)$. Let us examine the quadratic variation of the difference between the two sides of (20.11). By the formula for cross-variation of two integrals,

$$\langle I^{I^M(Y)}(X) - I^M(XY)\rangle_t$$

$$= \langle I^{I^M(Y)}(X)\rangle_t + \langle I^M(XY)\rangle_t - 2\langle I^{I^M(Y)}(X), I^M(XY)\rangle_t$$

$$= \int_0^t X_s^2 d\langle I^M(Y)\rangle_s + \int_0^t X_s^2 Y_s^2 d\langle M\rangle_s - 2\int_0^t X_s^2 Y_s d\langle I^M(Y), M\rangle_s$$

$$= \int_0^t X_s^2 Y_s^2 d\langle M\rangle_s + \int_0^t X_s^2 Y_s^2 d\langle M\rangle_s - 2\int_0^t X_s^2 Y_s^2 d\langle M\rangle_s = 0.$$

By Lemma 20.3, (20.11) holds. $\qquad\square$

20.6 Local Martingales

In this section we define the stochastic integral with respect to continuous local martingales.

Definition 20.20. *Let X_t, $t \in \mathbb{R}^+$, be a process adapted to a filtration $\mathcal{F}_t$. Then $(X_t, \mathcal{F}_t)$ is called a local martingale if there is a non-decreasing sequence of stopping times $\tau_n : \Omega \to [0, \infty]$ such that $\lim_{n \to \infty} \tau_n = \infty$ almost surely, and the process $(X_{t \wedge \tau_n}, \mathcal{F}_t)$ is a martingale for each n.*

This method of introducing a non-decreasing sequence of stopping times, which convert a local martingale into a martingale, is called localization.

The space of equivalence classes of local martingales whose sample paths are continuous almost surely and which satisfy $X_0 = 0$ almost surely will be denoted by $\mathcal{M}^{c,\text{loc}}$. It is easy to see that $\mathcal{M}^{c,\text{loc}}$ is a vector space (see Problem 3). It is also important to note that a local martingale may be integrable and yet fail to be a martingale (see Problem 4).

Now let us define the quadratic variation of a continuous local martingale $(X_t, \mathcal{F}_t) \in \mathcal{M}^{c,\text{loc}}$. We introduce the notation $X_t^{(n)} = X_{t \wedge \tau_n}$. Then, for $m \leq n$, as in the example before Lemma 20.3,

$$\langle X^{(m)} \rangle_t = \langle X^{(n)} \rangle_{t \wedge \tau_m}.$$

This shows that $\langle X^{(m)} \rangle_t$ and $\langle X^{(n)} \rangle_t$ agree on the interval $0 \leq t \leq \tau_m(\omega)$ for almost all ω. Since $\tau_m \to \infty$ almost surely, we can define the limit $\langle X \rangle_t = \lim_{m \to \infty} \langle X^{(m)} \rangle_t$, which is a non-decreasing adapted process whose sample paths are continuous almost surely. The process $\langle X \rangle_t$ is called the quadratic variation of the local martingale X_t. This is justified by the fact that

$$(X^2 - \langle X \rangle)_{t \wedge \tau_n} = (X_t^{(n)})^2 - \langle X^{(n)} \rangle_t \in \mathcal{M}_2^c.$$

That is, $X_t^2 - \langle X \rangle_t$ is a local martingale. Let us show that the process $\langle X \rangle_t$ does not depend on the choice of the sequence of stopping times τ_n.

Lemma 20.21. *Let $X_t \in \mathcal{M}^{c,\text{loc}}$. There exists a unique (up to indistinguishability) non-decreasing adapted continuous process Y_t such that $Y_0 = 0$ almost surely and $X_t^2 - Y_t \in \mathcal{M}^{c,\text{loc}}$.*

Proof. The existence part was demonstrated above. Let us suppose that there are two processes Y_t^1 and Y_t^2 with the desired properties. Then $M_t = Y_t^1 - Y_t^2$ belongs to $\mathcal{M}^{c,\text{loc}}$ (since $\mathcal{M}^{c,\text{loc}}$ is a vector space) and is a process of bounded variation. Let τ_n be a non-decreasing sequence of stopping times which tend to infinity, such that $M_t^{(n)} = M_{t \wedge \tau_n}$ is a martingale for each n. Then $M_t^{(n)}$ is also a process of bounded variation. By Corollary 20.8, $M_t^{(n)} = 0$ for all t almost surely. Since $\tau_n \to \infty$, this implies that Y_t^1 and Y_t^2 are indistinguishable. $\square$

The cross-variation of two local martingales can be defined by the same formula (20.1) as in the square-integrable case. It is also not difficult to see that $\langle X, Y \rangle_t$ is the unique (up to indistinguishability) adapted continuous process of bounded variation, such that $\langle X, Y \rangle_0 = 0$ almost surely, and $X_t Y_t - \langle X, Y \rangle_t \in \mathcal{M}^{c,\mathrm{loc}}$.

Let us now define the stochastic integral with respect to a continuous local martingale $M_t \in \mathcal{M}^{c,\mathrm{loc}}$. We can also extend the class of integrands. Namely, we shall say that $X_t \in \mathcal{P}^*$ if X_t is a progressively measurable process such that for every $0 \le t < \infty$,

$$\int_0^t X_s^2(\omega) d\langle M \rangle_s(\omega) < \infty \quad \text{almost surely.}$$

More precisely, we can view $\mathcal{P}^*$ as the set of equivalence classes of such processes, with two elements X_t^1 and X_t^2 representing the same class if and only if $\int_0^t (X_t^1 - X_t^2)^2 d\langle M \rangle_s = 0$ almost surely for every t.

Let us consider a sequence of stopping times $\tau_n : \Omega \to [0, \infty]$ with the following properties:

1. The sequence τ_n is non-decreasing and $\lim_{n \to \infty} \tau_n = \infty$ almost surely.
2. For each n, the process $M_t^{(n)} = M_{t \wedge \tau_n}$ is in $\mathcal{M}_2^c$.
3. For each n, the process $X_t^{(n)} = X_{t \wedge \tau_n}$ is in $\mathcal{L}^*(M^{(n)})$.

For example, such a sequence can be constructed as follows. Let τ_n^1 be a non-decreasing sequence such that $\lim_{n \to \infty} \tau_n^1 = \infty$ almost surely and the process $(X_{t \wedge \tau_n^1}, \mathcal{F}_t)$ is a martingale for each n. Define

$$\tau_n^2(\omega) = \inf\{t : \int_0^t X_s^2(\omega) d\langle M \rangle_s(\omega) = n\},$$

where the infimum of an empty set is equal to $+\infty$. It is clear that the sequence of stopping times $\tau_n = \tau_n^1 \wedge \tau_n^2$ has the properties (1)–(3).

Given a sequence τ_n with the above properties, a continuous local martingale $M_t \in \mathcal{M}^{c,\mathrm{loc}}$, and a process $X_t \in \mathcal{P}^*$, we can define

$$I_t^M(X) = \lim_{n \to \infty} I_t^{M^{(n)}}(X^{(n)}).$$

For almost all ω, the limit exists for all t. Indeed, by Lemma 20.18, almost surely,

$$I_t^{M^{(m)}}(X^{(m)}) = I_t^{M^{(n)}}(X^{(n)}), \quad 0 \le t \le \tau_m \wedge \tau_n.$$

Let us show that the limit does not depend on the choice of the sequence of stopping times, thus providing a correct definition of the integral with respect to a local martingale. If $\widetilde{\tau}_n$ and $\overline{\tau}_n$ are two sequences of stopping times with properties (1)–(3), and $\widetilde{M}_t^{(n)}, \widetilde{X}_t^{(n)}, \overline{M}_t^{(n)}$, and $\overline{X}_t^{(n)}$ are the corresponding processes, then

$$I_t^{\widetilde{M}^{(n)}}(\widetilde{X}^{(n)}) = I_t^{\overline{M}^{(n)}}(\overline{X}^{(n)}), \quad 0 \le t \le \widetilde{\tau}_n \wedge \overline{\tau}_n,$$

again by Lemma 20.18. Therefore, the limit in the definition of the integral $I_t^M(X)$ does not depend on the choice of the sequence of stopping times.

It is clear from the definition of the integral that $I_t^M(X) \in \mathcal{M}^{c,\mathrm{loc}}$, and that it is linear in the argument, that is, it satisfies (20.5) for any $X, Y \in \mathcal{P}^*$ and $a, b \in \mathbb{R}$. The formula for cross-variation of two integrals with respect to local martingales is the same as in the square-integrable case, as can be seen using localization. Namely, if $M_t, N_t \in \mathcal{M}^{c,\mathrm{loc}}$, $X_t \in \mathcal{P}^*(M)$, and $Y_t \in \mathcal{P}^*(N)$, then for almost all ω,

$$\langle I^M(X), I^N(Y) \rangle_t = \int_0^t X_s Y_s d\langle M, N \rangle_s, \quad 0 \le t < \infty.$$

Similarly, by using localization, it is easy to see that (20.11) remains true if $M_t \in \mathcal{M}^{c,\mathrm{loc}}$, $Y_t \in \mathcal{P}^*(M)$, and $X_t \in \mathcal{P}^*(I^M(Y))$.

Remark 20.22. Let X_u, $s \le u \le t$, be such that the process $\widetilde{X}_u \in \mathcal{P}^*$, where $\widetilde{X}_u$ is equal to X_u for $s \le u \le t$, and to zero otherwise. In this case we can define $\int_s^t X_u dM_u = \int_0^t \widetilde{X}_u dM_u$ as in the case of square-integrable martingales.

20.7 Ito Formula

In this section we shall prove a formula which may be viewed as the analogue of the Fundamental Theorem of Calculus, but is now applied to martingale-type processes with unbounded first variation.

Definition 20.23. *Let X_t, $t \in \mathbb{R}^+$, be a process adapted to a filtration $\mathcal{F}_t$. Then $(X_t, \mathcal{F}_t)$ is a continuous semimartingale if X_t can be represented as*

$$X_t = X_0 + M_t + A_t, \tag{20.12}$$

where $M_t \in \mathcal{M}^{c,\mathrm{loc}}$, A_t is a continuous process adapted to the same filtration such that the total variation of A_t on each finite interval is bounded almost surely, and $A_0 = 0$ almost surely.

Theorem 20.24 (Ito Formula). *Let $f \in C^2(\mathbb{R})$ and let $(X_t, \mathcal{F}_t)$ be a continuous semimartingale as in (20.12). Then, for any $t \ge 0$, the equality*

$$f(X_t) = f(X_0) + \int_0^t f'(X_s) dM_s + \int_0^t f'(X_s) dA_s + \frac{1}{2} \int_0^t f''(X_s) d\langle M \rangle_s \tag{20.13}$$

holds almost surely.

Remark 20.25. The first integral on the right-hand side is a stochastic integral, while the other two integrals must be understood in the Lebesgue-Stieltjes sense. Since both sides are continuous functions of t for almost all ω, the processes on the left- and right-hand sides are indistinguishable.

Proof of Theorem 20.24. We shall prove the result under stronger assumptions. Namely, we shall assume that $M_t = W_t$ and that f is bounded together with its first and second derivatives. The proof in the general case is similar, but somewhat more technical. In particular, it requires the use of localization. Thus we assume that

$$X_t = X_0 + W_t + A_t,$$

and wish to prove that

$$f(X_t) = f(X_0) + \int_0^t f'(X_s)dW_s + \int_0^t f'(X_s)dA_s + \frac{1}{2}\int_0^t f''(X_s)ds. \quad (20.14)$$

Let $\sigma = \{t_0, t_1, \ldots, t_n\}$, $0 = t_0 \le t_1 \le \cdots \le t_n = t$, be a partition of the interval $[0, t]$ into n subintervals. By the Taylor formula,

$$f(X_t) = f(X_0) + \sum_{i=1}^{n}(f(X_{t_i}) - f(X_{t_{i-1}}))$$

$$= f(X_0) + \sum_{i=1}^{n} f'(X_{t_{i-1}})(X_{t_i} - X_{t_{i-1}}) + \frac{1}{2}\sum_{i=1}^{n} f''(\xi_i)(X_{t_i} - X_{t_{i-1}})^2, \quad (20.15)$$

where $\min(X_{t_{i-1}}, X_{t_i}) \le \xi_i \le \max(X_{t_{i-1}}, X_{t_i})$ is such that

$$f(X_{t_i}) - f(X_{t_{i-1}}) = f'(X_{t_{i-1}})(X_{t_i} - X_{t_{i-1}}) + \frac{1}{2}f''(\xi_i)(X_{t_i} - X_{t_{i-1}})^2.$$

Note that we can take $\xi_i = X_{t_{i-1}}$ if $X_{t_{i-1}} = X_{t_i}$. If $X_{t_{i-1}} \ne X_{t_i}$, we can solve the above equation for $f''(\xi_i)$, and therefore we may assume that $f''(\xi_i)$ is measurable.

Let $Y_s = f'(X_s)$, $0 \le s \le t$, and define the simple process Y_s^σ by

$$Y_s^\sigma = f'(X_0)\chi_{\{0\}}(s) + \sum_{i=1}^{n} f'(X_{t_{i-1}})\chi_{(t_{i-1}, t_i]}(s) \quad \text{for } 0 \le s \le t.$$

Note that $\lim_{\delta(\sigma) \to 0} Y_s^\sigma(\omega) = Y_s(\omega)$, where the convergence is uniform on $[0, t]$ for almost all ω since the process Y_s is continuous almost surely.

Let us examine the first sum on the right-hand side of (20.15),

$$\sum_{i=1}^{n} f'(X_{t_{i-1}})(X_{t_i} - X_{t_{i-1}})$$

$$= \sum_{i=1}^{n} f'(X_{t_{i-1}})(W_{t_i} - W_{t_{i-1}}) + \sum_{i=1}^{n} f'(X_{t_{i-1}})(A_{t_i} - A_{t_{i-1}})$$

$$= \int_0^t Y_s^\sigma dW_s + \int_0^t Y_s^\sigma dA_s = S_1^\sigma + S_2^\sigma,$$

where S_1^σ and S_2^σ denote the stochastic and the ordinary integral, respectively. Since

$$E(\int_0^t (Y_s^\sigma - Y_s)dW_s)^2 = E\int_0^t (Y_s^\sigma - Y_s)^2 ds \to 0,$$

we obtain

$$\lim_{\delta(\sigma)\to 0} S_1^\sigma = \lim_{\delta(\sigma)\to 0} \int_0^t Y_s^\sigma dW_s = \int_0^t Y_s dW_s \text{ in } L^2(\Omega, \mathcal{F}, P).$$

We can apply the Lebesgue Dominated Convergence Theorem to the Lebesgue-Stieltjes integral (which is just a difference of two Lebesgue integrals) to obtain

$$\lim_{\delta(\sigma)\to 0} S_2^\sigma = \lim_{\delta(\sigma)\to 0} \int_0^t Y_s^\sigma dA_s = \int_0^t Y_s dA_s \text{ almost surely.}$$

Now let us examine the second sum on the right-hand side of (20.15):

$$\frac{1}{2}\sum_{i=1}^n f''(\xi_i)(X_{t_i} - X_{t_{i-1}})^2 = \frac{1}{2}\sum_{i=1}^n f''(\xi_i)(W_{t_i} - W_{t_{i-1}})^2$$

$$+ \sum_{i=1}^n f''(\xi_i)(W_{t_i} - W_{t_{i-1}})(A_{t_i} - A_{t_{i-1}}) \quad (20.16)$$

$$+ \frac{1}{2}\sum_{i=1}^n f''(\xi_i)(A_{t_i} - A_{t_{i-1}})^2 = S_3^\sigma + S_4^\sigma + S_5^\sigma.$$

The last two sums on the right-hand side of this formula tend to zero almost surely as $\delta(\sigma) \to 0$. Indeed,

$$|\sum_{i=1}^n f''(\xi_i)(W_{t_i} - W_{t_{i-1}})(A_{t_i} - A_{t_{i-1}}) + \frac{1}{2}\sum_{i=1}^n f''(\xi_i)(A_{t_i} - A_{t_{i-1}})^2|$$

$$\leq \sup_{x\in\mathbb{R}} f''(x)(\max_{1\leq i\leq n}(|W_{t_i} - W_{t_{i-1}}|) + \frac{1}{2}\max_{1\leq i\leq n}(|A_{t_i} - A_{t_{i-1}}|))\sum_{i=1}^n |A_{t_i} - A_{t_{i-1}}|,$$

which tends to zero almost surely since W_t and A_t are continuous and A_t has bounded variation.

It remains to deal with the first sum on the right-hand side of (20.16). Let us compare it with the sum

$$\tilde{S}_3^\sigma = \frac{1}{2}\sum_{i=1}^n f''(X_{t_{i-1}})(W_{t_i} - W_{t_{i-1}})^2,$$

and show that the difference converges to zero in L^1. Indeed,

$$E|\sum_{i=1}^n f''(\xi_i)(W_{t_i} - W_{t_{i-1}})^2 - \sum_{i=1}^n f''(X_{t_{i-1}})(W_{t_i} - W_{t_{i-1}})^2|$$

$$\leq (E(\max_{1\leq i\leq n}(f''(\xi_i) - f''(X_{t_{i-1}}))^2)^{1/2}(E(\sum_{i=1}^{n}(W_{t_i} - W_{t_{i-1}})^2)^2)^{1/2}.$$

The first factor here tends to zero since f'' is continuous and bounded. The second factor is bounded since

$$E(\sum_{i=1}^{n}(W_{t_i} - W_{t_{i-1}})^2)^2 = 3\sum_{i=1}^{n}(t_i - t_{i-1})^2 + \sum_{i\neq j}(t_i - t_{i-1})(t_j - t_{j-1})$$

$$\leq 3(\sum_{i=1}^{n}(t_i - t_{i-1}))(\sum_{j=1}^{n}(t_j - t_{j-1})) = 3t^2,$$

which shows that $(S_3^\sigma - \widetilde{S}_3^\sigma) \to 0$ in $L^1(\Omega, \mathcal{F}, P)$ as $\delta(\sigma) \to 0$. Let us compare $\widetilde{S}_3^\sigma$ with the sum

$$\overline{S}_3^\sigma = \frac{1}{2}\sum_{i=1}^{n}f''(X_{t_{i-1}})(t_i - t_{i-1}),$$

and show that the difference converges to zero in L^2. Indeed, similarly to the proof of Lemma 18.24,

$$E[\sum_{i=1}^{n}f''(X_{t_{i-1}})(W_{t_i} - W_{t_{i-1}})^2 - \sum_{i=1}^{n}f''(X_{t_{i-1}})(t_i - t_{i-1})]^2$$

$$= \sum_{i=1}^{n}E([f''(X_{t_{i-1}})^2][(W_{t_i} - W_{t_{i-1}})^2 - (t_i - t_{i-1})]^2)$$

$$\leq \sup_{x\in\mathbb{R}}|f''(x)|^2(\sum_{i=1}^{n}E(W_{t_i} - W_{t_{i-1}})^4 + \sum_{i=1}^{n}(t_i - t_{i-1})^2)$$

$$= 4\sup_{x\in\mathbb{R}}|f''(x)|^2\sum_{i=1}^{n}(t_i - t_{i-1})^2 \leq 4\sup_{x\in\mathbb{R}}|f''(x)|^2\max_{1\leq i\leq n}(t_i - t_{i-1})\sum_{i=1}^{n}(t_i - t_{i-1})$$

$$= 4\sup_{x\in\mathbb{R}}|f''(x)|^2 t\delta(\sigma),$$

where the first equality is justified by

$$E[f''(X_{t_{i-1}})((W_{t_i} - W_{t_{i-1}})^2 - (t_i - t_{i-1}))$$

$$f''(X_{t_{j-1}})((W_{t_j} - W_{t_{j-1}})^2 - (t_j - t_{j-1}))]$$

$$= E[f''(X_{t_{i-1}})((W_{t_i} - W_{t_{i-1}})^2 - (t_i - t_{i-1}))$$

$$E(f''(X_{t_{j-1}})((W_{t_j} - W_{t_{j-1}})^2 - (t_j - t_{j-1}))|\mathcal{F}_{j-1})] = 0 \quad \text{if} \quad i < j.$$

Thus, we see that $(\widetilde{S}_3^\sigma - \overline{S}_3^\sigma) \to 0$ in $L^2(\Omega, \mathcal{F}, P)$ as $\delta(\sigma) \to 0$. It is also clear that

$$\lim_{\delta(\sigma)\to 0}\overline{S}_3^\sigma = \frac{1}{2}\int_0^t f''(X_s)ds \quad \text{almost surely.}$$

Let us return to formula (20.15), which we can now write as

$$f(X_t) = f(X_0) + S_1^\sigma + S_2^\sigma + (S_3^\sigma - \widetilde{S}_3^\sigma) + (\widetilde{S}_3^\sigma - \overline{S}_3^\sigma) + \overline{S}_3^\sigma + S_4^\sigma + S_5^\sigma.$$

Take a sequence $\sigma(n)$ with $\lim_{n\to\infty} \delta(\sigma(n)) = 0$. We saw that

$$\lim_{n\to\infty} S_1^{\sigma(n)} = \int_0^t f'(X_s)dW_s \quad \text{in } L^2(\Omega, \mathcal{F}, \mathrm{P}), \tag{20.17}$$

$$\lim_{n\to\infty} S_2^{\sigma(n)} = \int_0^t f'(X_s)dA_s \quad \text{almost surely}, \tag{20.18}$$

$$\lim_{n\to\infty} (S_3^{\sigma(n)} - \widetilde{S}_3^{\sigma(n)}) = 0 \quad \text{in } L^1(\Omega, \mathcal{F}, \mathrm{P}), \tag{20.19}$$

$$\lim_{n\to\infty} (\widetilde{S}_3^{\sigma(n)} - \overline{S}_3^{\sigma(n)}) = 0 \quad \text{in } L^2(\Omega, \mathcal{F}, \mathrm{P}), \tag{20.20}$$

$$\lim_{n\to\infty} \overline{S}_3^{\sigma(n)} = \frac{1}{2} \int_0^t f''(X_s)ds \quad \text{almost surely}, \tag{20.21}$$

$$\lim_{n\to\infty} S_4^{\sigma(n)} = \lim_{n\to\infty} S_5^{\sigma(n)} = 0 \quad \text{almost surely}. \tag{20.22}$$

We can replace the sequence $\sigma(n)$ by a subsequence for which all the equalities (20.17)–(20.22) hold almost surely. This justifies (20.14). $\quad\square$

Remark 20.26. The stochastic integral on the right-hand side of (20.13) belongs to $\mathcal{M}^{c,\mathrm{loc}}$, while the Lebesgue-Stieltjes integrals are continuous adapted processes with bounded variation. Therefore, the class of semimartingales is invariant under the composition with twice continuously differentiable functions.

Example. Let $f \in C^2(\mathbb{R})$, A_t and B_t be progressively measurable processes such that $\int_0^t A_s^2 ds < \infty$ and $\int_0^t |B_s|ds < \infty$ for all t almost surely, and X_t a semimartingale of the form

$$X_t = X_0 + \int_0^t A_s dW_s + \int_0^t B_s ds.$$

Applying the Ito formula, we obtain

$$f(X_t) = f(X_0) + \int_0^t f'(X_s)A_s dW_s + \int_0^t f'(X_s)B_s ds + \frac{1}{2}\int_0^t f''(X_s)A_s^2 ds,$$

where the relation $\int_0^t f'(X_s)d(\int_0^s A_u dW_u) = \int_0^t f'(X_s)A_s dW_s$ is justified by formula (20.11) applied to local martingales.

This is one of the most common applications of the Ito formula, particularly when the processes A_t and B_t can be represented as $A_t = \sigma(t, X_t)$ and $B_t = v(t, X_t)$ for some smooth functions σ and v, in which case X_t is called a diffusion process with time-dependent coefficients.

We state the following multi-dimensional version of the Ito formula, whose proof is very similar to that of Theorem 20.24.

Theorem 20.27. *Let $M_t = (M_t^1, \ldots, M_t^d)$ be a vector of continuous local martingales, that is $(M_t^i, \mathcal{F}_t)_{t \in \mathbb{R}^+}$ is a local martingale for each $1 \leq i \leq d$. Let $A_t = (A_t^1, \ldots, A_t^d)$ be a vector of continuous processes adapted to the same filtration such that the total variation of A_t^i on each finite interval is bounded almost surely, and $A_0^i = 0$ almost surely. Let $X_t = (X_t^1, \ldots, X_t^d)$ be a vector of adapted processes such that $X_t = X_0 + M_t + A_t$, and let $f \in C^{1,2}(\mathbb{R}^+ \times \mathbb{R}^d)$. Then, for any $t \geq 0$, the equality*

$$f(t, X_t) = f(0, X_0) + \sum_{i=1}^{d} \int_0^t \frac{\partial}{\partial x_i} f(s, X_s) dM_s^i + \sum_{i=1}^{d} \int_0^t \frac{\partial}{\partial x_i} f(s, X_s) dA_s^i$$

$$+ \frac{1}{2} \sum_{i,j=1}^{d} \int_0^t \frac{\partial^2}{\partial x_i \partial x_j} f(s, X_s) d\langle M^i, M^j \rangle_s + \int_0^t \frac{\partial}{\partial s} f(s, X_s) ds$$

holds almost surely.

Let us apply this theorem to a pair of processes $X_t^i = X_0^i + M_t^i + A_t^i$, $i = 1, 2$, and the function $f(x_1, x_2) = x_1 x_2$.

Corollary 20.28. *If $(X_t^1, \mathcal{F}_t)$ and $(X_t^2, \mathcal{F}_t)$ are continuous semimartingales, then*

$$X_t^1 X_t^2 = X_0^1 X_0^2 + \int_0^t X_s^1 dM_s^2 + \int_0^t X_s^1 dA_s^2$$

$$+ \int_0^t X_s^2 dM_s^1 + \int_0^t X_s^2 dA_s^1 + \langle M^1, M^2 \rangle_t.$$

Using the shorthand notation $\int_0^t Y_s dX_s = \int_0^t Y_s dM_s + \int_0^t Y_s dA_s$ for a process Y_s and a semimartingale X_s, we can rewrite the above formula as

$$\int_0^t X_s^1 dX_s^2 = X_t^1 X_t^2 - X_0^1 X_0^2 - \int_0^t X_s^2 dX_s^1 - \langle M^1, M^2 \rangle_s. \qquad (20.23)$$

This is the integration by parts formula for the Ito integral.

20.8 Problems

1. Prove that if X_t is a continuous non-random function, then the stochastic integral $I_t(X) = \int_0^t X_s dW_s$ is a Gaussian process.

2. Let W_t be a one-dimensional Brownian motion defined on a probability space $(\Omega, \mathcal{F}, \mathrm{P})$. Prove that there is a unique orthogonal random measure Z with values in $L^2(\Omega, \mathcal{F}, \mathrm{P})$ defined on a $([0, 1], \mathcal{B}([0, 1])$ such that $Z([s, t]) = W_t - W_s$ for $0 \leq s \leq t \leq 1$. Prove that

$$\int_0^1 \varphi(t)dZ(t) = \int_0^1 \varphi(t)dW_t$$

for any function φ that is continuous on $[0,1]$.

3. Prove that if $X_t, Y_t \in \mathcal{M}^{c,\text{loc}}$, then $aX_t + bY_t \in \mathcal{M}^{c,\text{loc}}$ for any constants a and b.

4. Give an example of a local martingale which is integrable, yet fails to be a martingale.

5. Let W_t be a one-dimensional Brownian motion relative to a filtration $\mathcal{F}_t$. Let τ be a stopping time of $\mathcal{F}_t$ with $\mathrm{E}\tau < \infty$. Prove the Wald Identities

$$\mathrm{E}W_\tau = 0, \quad \mathrm{E}W_\tau^2 = \mathrm{E}\tau.$$

(Note that the Optional Sampling Theorem can not be applied directly since τ may be unbounded.)

6. Find the distribution function of the random variable $\int_0^1 W_t^n dW_t$, where W_t is a one-dimensional Brownian motion.

Stochastic Differential Equations

21.1 Existence of Strong Solutions to Stochastic Differential Equations

Stochastic differential equations arise when modeling prices of financial instruments, a variety of physical systems, and in many other branches of science. As we shall see in the next section, there is a deep relationship between stochastic differential equations and linear elliptic and parabolic partial differential equations.

As an example, let us try to model the motion of a small particle suspended in a liquid. Let us denote the position of the particle at time t by X_t. The liquid need not be at rest, and the velocity field will be denoted by $v(t, x)$, where t is time and x is a point in space. If we neglect the diffusion, the equation of motion is simply $dX_t/dt = v(t, X_t)$, or, formally, $dX_t = v(t, X_t)dt$.

On the other hand, if we assume that macroscopically the liquid is at rest, then the position of the particle can change only due to the molecules of liquid hitting the particle, and X_t would be modeled by the 3-dimensional Brownian motion, $X_t = W_t$, or, formally, $dX_t = dW_t$. More generally, we could write $dX_t = \sigma(t, X_t)dW_t$, where we allow σ to be non-constant, since the rate at which the molecules hit the particle may depend on the temperature and density of the liquid, which, in turn, are functions of space and time.

If both the effects of advection and diffusion are present, we can formally write the stochastic differential equation

$$dX_t = v(t, X_t)dt + \sigma(t, X_t)dW_t. \qquad (21.1)$$

The vector v is called the drift vector, and σ, which may be a scalar or a matrix, is called the dispersion coefficient (matrix).

Now we shall specify the assumptions on v and σ, in greater generality than necessary for modeling the motion of a particle, and assign a strict meaning to the stochastic differential equation above. The main idea is to write the

L. Koralov and Y.G. Sinai, *Theory of Probability and Random Processes,*   311
Universitext, DOI 10.1007/978-3-540-68829-7_21,
© Springer-Verlag Berlin Heidelberg 2012

formal expression (21.1) in the integral form, in which case the right-hand side becomes a sum of an ordinary and a stochastic integral.

We assume that X_t takes values in the d-dimensional space, while W_t is an r-dimensional Brownian motion relative to a filtration $\mathcal{F}_t$. Let v be a Borel-measurable function from $\mathbb{R}^+ \times \mathbb{R}^d$ to $\mathbb{R}^d$, and σ a Borel-measurable function from $\mathbb{R}^+ \times \mathbb{R}^d$ to the space of $d \times r$ matrices. Thus, Eq. (21.1) can be re-written as

$$dX_t^i = v_i(t, X_t)dt + \sum_{j=1}^{r} \sigma_{ij}(t, X_t)dW_t^j, \quad 1 \le i \le d. \tag{21.2}$$

Let us further assume that the underlying filtration $\mathcal{F}_t$ satisfies the usual conditions and that we have a random d-dimensional vector ξ which is $\mathcal{F}_0$-measurable (and consequently independent of the Brownian motion W_t). This random vector is the initial condition for the stochastic differential equation (21.1).

Definition 21.1. *Suppose that the functions v and σ, the filtration, the Brownian motion, and the random variable ξ satisfy the assumptions above. A process X_t with continuous sample paths defined on the probability space $(\Omega, \mathcal{F}, \mathrm{P})$ is called a strong solution to the stochastic differential equation (21.1) with the initial condition ξ if:*

(1) X_t is adapted to the filtration $\mathcal{F}_t$.
(2) $X_0 = \xi$ almost surely.
(3) For every $0 \le t < \infty$, $1 \le i \le d$, and $1 \le j \le r$,

$$\int_0^t (|v_i(s, X_s)| + |\sigma_{ij}(s, X_s)|^2)ds < \infty \quad almost\ surely$$

(which implies that $\sigma_{ij}(t, X_t) \in \mathcal{P}^*(W_t^j)$).
(4) For every $0 \le t < \infty$, the integral version of (21.2) holds almost surely:

$$X_t^i = X_0^i + \int_0^t v_i(s, X_s)ds + \sum_{j=1}^{r} \int_0^t \sigma_{ij}(s, X_s)dW_s^j, \quad 1 \le i \le d.$$

(Since the processes on both sides are continuous, they are indistinguishable.)

We shall refer to the solutions of stochastic differential equations as diffusion processes. Customarily the term "diffusion" refers to a strong Markov family of processes with continuous paths, with the generator being a second order partial differential operator (see Sect. 21.4). As will be discussed later in this chapter, under certain conditions on the coefficients, the solutions to stochastic differential equations form strong Markov families.

As with ordinary differential equations (ODE's), the first natural question which arises is that of the existence and uniqueness of strong solutions. As with ODE's, we shall require the Lipschitz continuity of the coefficients in the space variable, and certain growth estimates at infinity. The local Lipschitz continuity is sufficient to guarantee the local uniqueness of the solutions (as in the case of ODE's), while the uniform Lipschitz continuity and the growth conditions are needed for the global existence of solutions.

Theorem 21.2. *Suppose that the coefficients v and σ in Eq. (21.1) are Borel-measurable functions on $\mathbb{R}^+ \times \mathbb{R}^d$ and are uniformly Lipschitz continuous in the space variable. That is, for some constant c_1 and all $t \in \mathbb{R}^+$, $x, y \in \mathbb{R}^d$,*

$$|v_i(t,x) - v_i(t,y)| \leq c_1||x - y||, \quad 1 \leq i \leq d, \tag{21.3}$$

$$|\sigma_{ij}(t,x) - \sigma_{ij}(t,y)| \leq c_1||x - y||, \quad 1 \leq i \leq d,\ 1 \leq j \leq r. \tag{21.4}$$

Assume also that the coefficients do not grow faster than linearly, that is,

$$|v_i(t,x)| \leq c_2(1 + ||x||), \quad |\sigma_{ij}(t,x)| \leq c_2(1 + ||x||), \quad 1 \leq i \leq d,\ 1 \leq j \leq r, \tag{21.5}$$

for some constant c_2 and all $t \in \mathbb{R}^+$, $x \in \mathbb{R}^d$. Let W_t be a Brownian motion relative to a filtration $\mathcal{F}_t$, and ξ an $\mathcal{F}_0$-measurable $\mathbb{R}^d$-valued random vector that satisfies

$$E||\xi||^2 < \infty.$$

Then there exists a strong solution to Eq. (21.1) with the initial condition ξ. The solution is unique in the sense that any two strong solutions are indistinguishable processes.

Remark 21.3. If we assume that (21.3) and (21.4) hold, then (21.5) is equivalent to the boundedness of

$$|v_i(t,0)|, \quad |\sigma_{ij}(t,0)|, \quad 1 \leq i \leq d,\ 1 \leq j \leq r,$$

as functions of t.

We shall prove the uniqueness part of Theorem 21.2 and indicate the main idea for the proof of the existence part. To prove uniqueness we need the Gronwall Inequality, which we formulate as a separate lemma (see Problem 1).

Lemma 21.4. *If a function $f(t)$ is continuous and non-negative on $[0, t_0]$, and*

$$f(t) \leq K + L \int_0^t f(s)ds$$

holds for $0 \leq t \leq t_0$, with K and L positive constants, then

$$f(t) \leq Ke^{Lt}$$

for $0 \leq t \leq t_0$.

Proof of Theorem 21.2 (uniqueness part). Assume that both X_t and Y_t are strong solutions relative to the same Brownian motion, and with the same initial condition. We define the sequence of stopping times as follows:

$$\tau_n = \inf\{t \geq 0 : \max(||X_t||, ||Y_t||) \geq n\}.$$

For any t and t_0 such that $0 \leq t \leq t_0$,

$$E||X_{t \wedge \tau_n} - Y_{t \wedge \tau_n}||^2$$

$$= E||\int_0^{t \wedge \tau_n} (v(s, X_s) - v(s, Y_s))ds$$

$$+ \int_0^{t \wedge \tau_n} (\sigma(s, X_s) - \sigma(s, Y_s))dW_s||^2 \leq 2E[\int_0^{t \wedge \tau_n} ||v(s, X_s) - v(s, Y_s)||ds]^2$$

$$+2E\sum_{i=1}^{d}[\sum_{j=1}^{r}\int_0^{t \wedge \tau_n} (\sigma_{ij}(s, X_s) - \sigma_{ij}(s, Y_s))dW_s^j]^2$$

$$\leq 2tE\int_0^{t \wedge \tau_n} ||v(s, X_s) - v(s, Y_s)||^2 ds$$

$$+2E\sum_{i=1}^{d}\sum_{j=1}^{r}\int_0^{t \wedge \tau_n} |\sigma_{ij}(s, X_s) - \sigma_{ij}(s, Y_s)|^2 ds$$

$$\leq (2dt + 2rd)c_1^2 \int_0^{t \wedge \tau_n} E||X_{s \wedge \tau_n} - Y_{s \wedge \tau_n}||^2 ds$$

$$\leq (2dt_0 + 2rd)c_1^2 \int_0^{t} E||X_{s \wedge \tau_n} - Y_{s \wedge \tau_n}||^2 ds.$$

By Lemma 21.4 with $K = 0$ and $L = (2dt_0 + 2rd)c_1^2$,

$$E||X_{t \wedge \tau_n} - Y_{t \wedge \tau_n}||^2 = 0 \quad \text{for } 0 \leq t \leq t_0,$$

and, since t_0 can be taken to be arbitrarily large, this equality holds for all $t \geq 0$. Thus, the processes $X_{t \wedge \tau_n}$ and $Y_{t \wedge \tau_n}$ are modifications of one another, and, since they are continuous almost surely, they are indistinguishable. Now let $n \to \infty$, and notice that $\lim_{n \to \infty} \tau_n = \infty$ almost surely. Therefore, X_t and Y_t are indistinguishable. $\square$

The existence of strong solutions can be proved using the Method of Picard Iterations. Namely, we define a sequence of processes $X_t^{(n)}$ by setting $X_t^{(0)} \equiv \xi$ and

$$X_t^{(n+1)} = \xi + \int_0^t v(s, X_s^{(n)})ds + \int_0^t \sigma(s, X_s^{(n)})dW_s, \quad t \geq 0$$

for $n \geq 0$. It is then possible to show that the integrals on the right-hand side are correctly defined for all n, and that the sequence of processes $X_t^{(n)}$

converges to a process X_t for almost all ω uniformly on any interval $[0, t_0]$. The process X_t is then shown to be the strong solution of Eq. (21.1) with the initial condition ξ.

Example (Black and Scholes formula). In this example we consider a model for the behavior of the price of a financial instrument (a share of stock, for example) and derive a formula for the price of an option. Let X_t be the price of a stock at time t. We assume that the current price (at time $t = 0$) is equal to P. One can distinguish two phenomena responsible for the change of the price over time. One is that the stock prices grow on average at a certain rate r, which, if we were to assume that r was constant, would lead to the equation $dX_t = rX_t dt$, since the rate of change is proportional to the price of the stock.

Let us, for a moment, assume that $r = 0$, and focus on the other phenomenon affecting the price change. One can argue that the randomness in X_t is due to the fact that every time someone buys the stock, the price increases by a small amount, and every time someone sells the stock, the price decreases by a small amount. The intervals of time between one buyer or seller and the next are also small, and whether the next person will be a buyer or a seller is a random event. It is also reasonable to assume that the typical size of a price move is proportional to the current price of the stock. We described intuitively the model for the evolution of the price X_t as a random walk, which will tend to the process defined by the equation $dX_t = \sigma X_t dW_t$ if we make the time step tend to zero. (This is a result similar to the Donsker Theorem, which states that the measure induced by a properly scaled simple symmetric random walk tends to the Wiener measure.) Here, σ is the volatility which we assumed to be a constant.

When we superpose the above two effects, we obtain the equation

$$dX_t = rX_t dt + \sigma X_t dW_t \qquad (21.6)$$

with the initial condition $X_0 = P$. Let us emphasize that this is just a particular model for the stock price behavior, which may or may not be reasonable, depending on the situation. For example, when we modeled X_t as a random walk, we did not take into account that the presence of informed investors may cause it to be non-symmetric, or that the transition from the random walk to the diffusion process may be not justified if, with small probability, there are exceptionally large price moves.

Using the Ito formula (Theorem 20.27), with the martingale W_t and the function $f(t, x) = P \exp(\sigma x + (r - \frac{1}{2}\sigma^2)t)$, we obtain

$$f(t, W_t) = P \exp(\sigma W_t + (r - \frac{1}{2}\sigma^2)t) =$$

$$P + \int_0^t rP \exp(\sigma W_s + (r - \frac{1}{2}\sigma^2)s)ds + \int_0^t \sigma P \exp(\sigma W_s + (r - \frac{1}{2}\sigma^2)s)dW_s.$$

This means that
$$X_t = P\exp(\sigma W_t + (r - \tfrac{1}{2}\sigma^2)t)$$
is the solution of (21.6).

A *European call option* is the right to buy a share of the stock at an agreed price S (strike price) at an agreed time $t > 0$ (expiration time). The value of the option at time t is therefore equal to $(X_t - S)^+ = (X_t - S)\chi_{\{X_t \geq S\}}$ (if $X_t \leq S$, then the option becomes worthless). Assume that the behavior of the stock price is governed by (21.6), where r and σ were determined empirically based on previous observations. Then the expected value of the option at time t will be
$$V_t = \mathrm{E}(P\exp(\sigma W_t + (r - \tfrac{1}{2}\sigma^2)t) - S)^+$$
$$= \frac{1}{\sqrt{2\pi t}} \int_{-\infty}^{\infty} e^{-\frac{x^2}{2t}}(Pe^{\sigma x + (r-\frac{1}{2}\sigma^2)t} - S)^+ dx.$$

The integral on the right-hand side of this formula can be simplified somewhat, but we leave this as an exercise for the reader.

Finally, the current value of the option may be less than the expected value at time t. This is due to the fact that the money spent on the option at the present time could instead be invested in a no-risk security with an interest rate γ, resulting in a larger buying power at time t. Therefore the expected value V_t should be discounted by the factor $e^{-\gamma t}$ to obtain the current value of the option. We obtain the Black and Scholes formula for the value of the option
$$V_0 = \frac{e^{-\gamma t}}{\sqrt{2\pi t}} \int_{-\infty}^{\infty} e^{-\frac{x^2}{2t}}(Pe^{\sigma x + (r-\frac{1}{2}\sigma^2)t} - S)^+ dx.$$

Example (A Linear Equation). Let W_t be a Brownian motion on a probability space $(\Omega, \mathcal{F}, \mathrm{P})$ relative to a filtration $\mathcal{F}_t$. Let ξ be a square-integrable random variable measurable with respect to $\mathcal{F}_0$.

Consider the following one-dimensional stochastic differential equation with time-dependent coefficients
$$dX_t = (a(t)X_t + b(t))dt + \sigma(t)dW_t. \tag{21.7}$$

The initial data is $X_0 = \xi$. If $a(t), b(t)$, and $\sigma(t)$ are bounded measurable functions, by Theorem 21.2 this equation has a unique strong solution. In order to find an explicit formula for the solution, let us first solve the homogeneous ordinary differential equation
$$y'(t) = a(t)y(t)$$
with the initial data $y(0) = 1$. The solution to this equation is $y(t) = \exp(\int_0^t a(s)ds)$, as can be verified by substitution. We claim that the solution of (21.7) is

$$X_t = y(t)(\xi + \int_0^t \frac{b(s)}{y(s)}ds + \int_0^t \frac{\sigma(s)}{y(s)}dW_s). \tag{21.8}$$

Note that if $\sigma \equiv 0$, we recover the formula for the solution of a linear ODE, which can be obtained by the method of variation of constants. If we formally differentiate the right-hand side of (21.8), we obtain the expression on the right-hand side of (21.7). In order to justify this formal differentiation, let us apply Corollary 20.28 to the pair of semimartingales

$$X_t^1 = y(t) \quad \text{and} \quad X_t^2 = \xi + \int_0^t \frac{b(s)}{y(s)}ds + \int_0^t \frac{\sigma(s)}{y(s)}dW_s.$$

Thus,

$$X_t = y(t)(\xi + \int_0^t \frac{b(s)}{y(s)}ds + \int_0^t \frac{\sigma(s)}{y(s)}dW_s)$$

$$= \xi + \int_0^t y(s)d(\int_0^s \frac{b(u)}{y(u)}du) + \int_0^t y(s)d(\int_0^s \frac{\sigma(u)}{y(u)}dW_u) + \int_0^t X_s^2 dy(s)$$

$$= \xi + \int_0^t b(s)ds + \int_0^t \sigma(s)dW_s + \int_0^t a(s)X_s ds,$$

where we used (20.11) to justify the last equality. We have thus demonstrated that X_t is the solution to (21.7) with initial data $X_0 = \xi$.

Example (the Ornstein-Uhlenbeck Process). Consider the stochastic differential equation

$$dX_t = -aX_t dt + \sigma dW_t, \quad X_0 = \xi. \tag{21.9}$$

This is a particular case of (21.7) with $a(t) \equiv -a$, $b(t) \equiv 0$, and $\sigma(t) \equiv \sigma$. By (21.8), the solution is

$$X_t = e^{-at}\xi + \sigma \int_0^t e^{-a(t-s)}dW_s.$$

This process is called the Ornstein-Uhlenbeck Process with parameters (a, σ) and initial condition ξ. Since the integrand $e^{-a(t-s)}$ is a deterministic function, the integral is a Gaussian random process independent of ξ (see Problem 1, Chap. 20). If ξ is Gaussian, then X_t is a Gaussian process. Its expectation and covariance can be easily calculated:

$$m(t) = EX_t = e^{-at}E\xi,$$

$$b(s,t) = E(X_s X_t) = e^{-as}e^{-at}E\xi^2 + \sigma^2 \int_0^{s \wedge t} e^{-a(s-u)-a(t-u)}du$$

$$= e^{-a(s+t)}(E\xi^2 + \sigma^2 \frac{e^{2as \wedge t} - 1}{2a}).$$

In particular, if ξ is Gaussian with $E\xi = 0$ and $E\xi^2 = \frac{\sigma^2}{2a}$, then

$$b(s,t) = \frac{\sigma^2 e^{-a|s-t|}}{2a}.$$

Since the covariance function of the process depends on the difference of the arguments, the process is wide-sense stationary, and since it is Gaussian, it is also strictly stationary.

21.2 Dirichlet Problem for the Laplace Equation

In this section we show that solutions to the Dirichlet problem for the Laplace equation can be expressed as functionals of the Wiener process.

Let D be an open bounded domain in $\mathbb{R}^d$, and let $f : \partial D \to \mathbb{R}$ be a continuous function defined on the boundary. We shall consider the following partial differential equation

$$\Delta u(x) = 0 \quad \text{for } x \in D \tag{21.10}$$

with the boundary condition

$$u(x) = f(x) \quad \text{for } x \in \partial D. \tag{21.11}$$

This pair, Eq. (21.10) and the boundary condition (21.11), is referred to as the Dirichlet problem for the Laplace equation with the boundary condition $f(x)$.

By a solution of the Dirichlet problem we mean a function u which satisfies (21.10), (21.11) and belongs to $C^2(D) \cap C(\overline{D})$.

Let W_t be a d-dimensional Brownian motion relative to a filtration $\mathcal{F}_t$. Without loss of generality we may assume that $\mathcal{F}_t$ is the augmented filtration constructed in Sect. 19.4. Let $W_t^x = x + W_t$. For a point $x \in \overline{D}$, let τ^x be the first time the process W_t^x reaches the boundary of D, that is

$$\tau^x(\omega) = \inf\{t \geq 0 : W_t^x(\omega) \notin D\}.$$

In Sect. 19.6 we showed that the function

$$u(x) = Ef(W_{\tau^x}^x) \tag{21.12}$$

defined in $\overline{D}$ is harmonic inside D, that is, it belongs to $C^2(D)$ and satisfies (21.10). From the definition of $u(x)$ it is clear that it satisfies (21.11). It remains to study the question of continuity of $u(x)$ at the points of the boundary of D.

Let

$$\sigma^x(\omega) = \inf\{t > 0 : W_t^x(\omega) \notin D\}.$$

Note that here $t > 0$ on the right-hand side, in contrast to the definition of τ^x. Let us verify that σ^x is a stopping time. Define an auxiliary family of stopping times

$$\tau^{x,s}(\omega) = \inf\{t \geq s : W_t^x(\omega) \notin D\}$$

(see Lemma 13.15). Then, for $t > 0$,

$$\{\sigma^x \leq t\} = \bigcup_{n=1}^{\infty}\{\tau^{x,\frac{1}{n}} \leq t\} \in \mathcal{F}_t.$$

In addition,

$$\{\sigma^x = 0\} = \bigcap_{m=1}^{\infty}\bigcup_{n=1}^{\infty}\{\tau^{0,\frac{1}{n}} \leq 1/m\} \in \bigcap_{m=1}^{\infty}\mathcal{F}_{1/m} = \mathcal{F}_{0+} = \mathcal{F}_0,$$

where we have used the right-continuity of the augmented filtration. This demonstrates that σ^x is a stopping time. Also note that since $\{\sigma^x = 0\} \in \mathcal{F}_0$, the Blumenthal Zero-One Law implies that $P(\sigma^x = 0)$ is either equal to one or to zero.

Definition 21.5. *A point $x \in \partial D$ is called regular if $P(\sigma^x = 0) = 1$, and irregular if $P(\sigma^x = 0) = 0$.*

Regularity means that a typical Brownian path which starts at $x \in \partial D$ does not immediately enter D and stay there for an interval of time.

Example. Let $D = \{x \in \mathbb{R}^d, 0 < ||x|| < 1\}$, where $d \geq 2$, that is, D is a punctured unit ball. The boundary of D consists of the unit sphere and the origin. Since Brownian motion does not return to zero for $d \geq 2$, the origin is an irregular point for Brownian motion in D.

Similarly, let $D = B^d\backslash\{x \in \mathbb{R}^d : x_2 = \ldots = x_d = 0\}$. ($D$ is the set of points in the unit ball that do not belong to the x_1-axis.) The boundary of D consists of the unit sphere and the segment $\{x \in \mathbb{R}^d : -1 < x_1 < 1, x_2 = \ldots = x_d = 0\}$. If $d \geq 3$, the segment consists of irregular points.

Example. Let $x \in \partial D$, $y \in \mathbb{R}^d$, $||y|| = 1$, $0 < \theta \leq \pi$, and $r > 0$. The cone with vertex at x, direction y, opening θ, and radius r is the set

$$C_x(y, \theta, r) = \{z \in \mathbb{R}^d : ||z - x|| \leq r, (z - x, y) \geq ||z - x|| \cos\theta\}.$$

We shall say that a point $x \in \partial D$ satisfies the exterior cone condition if there is a cone $C_x(y, \theta, r)$ with y, θ, and r as above such that $C_x(y, \theta, r) \subseteq \mathbb{R}^d \setminus D$. It is not difficult to show (see Problem 8) that if x satisfies the exterior cone condition, then it is regular. In particular, if D is a domain with a smooth boundary, then all the points of ∂D are regular.

The question of regularity of a point $x \in \partial D$ is closely related to the continuity of the function u given by (21.12) at x.

Theorem 21.6. *Let D be a bounded open domain in $\mathbb{R}^d$, $d \geq 2$, and $x \in \partial D$. Then x is regular if and only if for any continuous function $f : \partial D \to \mathbb{R}$, the function u defined by (21.12) is continuous at x, that is*

$$\lim_{y \to x, y \in \overline{D}} \mathrm{E} f(W^y_{\tau^y}) = f(x). \tag{21.13}$$

Proof. Assume that x is regular. First, let us show that, with high probability, a Brownian trajectory which starts near x exits D fast. Take ε and δ such that $0 < \varepsilon < \delta$, and define an auxiliary function

$$g^\delta_\varepsilon(y) = \mathrm{P}(W^y_t \in D \text{ for } \varepsilon \leq t \leq \delta).$$

This is a continuous function of $y \in \overline{D}$, since the indicator function of the set $\{\omega : W^y_t(\omega) \in D \text{ for } \varepsilon \leq t \leq \delta\}$ tends to the indicator function of the set $\{\omega : W^{y_0}_t(\omega) \in D \text{ for } \varepsilon \leq t \leq \delta\}$ almost surely as $y \to y_0$. Note that

$$\lim_{\varepsilon \downarrow 0} g^\delta_\varepsilon(y) = \mathrm{P}(W^y_t \in D \text{ for } 0 < t \leq \delta) = \mathrm{P}(\sigma^y > \delta),$$

which implies that the right-hand side is an upper semicontinuous function of y, since it is a limit of a decreasing sequence of continuous functions. Therefore,

$$\limsup_{y \to x, y \in \overline{D}} \mathrm{P}(\tau^y > \delta) \leq \limsup_{y \to x, y \in \overline{D}} \mathrm{P}(\sigma^y > \delta) \leq \mathrm{P}(\sigma^x > \delta) = 0,$$

since x is a regular point. We have thus demonstrated that

$$\lim_{y \to x, y \in \overline{D}} \mathrm{P}(\tau^y > \delta) = 0$$

for any $\delta > 0$.

Next we show that, with high probability, a Brownian trajectory which starts near x exits D through a point on the boundary which is also near x. Namely, we wish to show that for $r > 0$,

$$\lim_{y \to x, y \in \overline{D}} \mathrm{P}(\|x - W^y_{\tau^y}\| > r) = 0. \tag{21.14}$$

Take an arbitrary $\varepsilon > 0$. We can then find $\delta > 0$ such that

$$\mathrm{P}(\max_{0 \leq t \leq \delta} \|W_t\| > r/2) < \varepsilon/2.$$

We can also find a neighborhood U of x such that $\|y - x\| < r/2$ for $y \in U$, and

$$\sup_{y \in \overline{D} \cap U} \mathrm{P}(\tau^y > \delta) < \varepsilon/2.$$

Combining the last two estimates, we obtain

$$\sup_{y \in \overline{D} \cap U} P(\|x - W^y_{\tau_y}\| > r) < \varepsilon,$$

which justifies (21.14).

Now let f be a continuous function defined on the boundary, and let $\varepsilon > 0$. Take $r > 0$ such that $\sup_{z \in \partial D, \|z-x\| \le r} |f(x) - f(z)| < \varepsilon$. Then

$$|Ef(W^y_{\tau_y}) - f(x)| \le$$

$$\sup_{z \in \partial D, \|z-x\| \le r} |f(x) - f(z)| + 2P(\|x - W^y_{\tau_y}\| > r) \sup_{z \in \partial D} |f(y)|.$$

The first term on the right-hand side here is less than ε, while the second one tends to zero as $y \to x$ by (21.14). We have thus demonstrated that (21.13) holds.

Now let us prove that x is regular if (21.13) holds for every continuous f. Suppose that x is not regular. Since $\sigma^x > 0$ almost surely, and a Brownian trajectory does not return to the origin almost surely for $d \ge 2$, we conclude that $\|W^x_{\sigma^x} - x\| > 0$ almost surely. We can then find $r > 0$ such that

$$P(\|W^x_{\sigma^x} - x\| \ge r) > 1/2.$$

Let S_n be the sphere centered at x with radius $r_n = 1/n$. We claim that if $r_n < r$, there is a point $y_n \in S_n \cap D$ such that

$$P(\|W^{y_n}_{\tau_{y_n}} - x\| \ge r) > 1/2. \qquad (21.15)$$

Indeed, let τ^x_n be the first time the process W^x_t reaches S_n. Let μ_n be the measure on $S_n \cap D$ defined by $\mu_n(A) = P(\tau^x_n < \sigma^x; W^x_{\tau^x_n} \in A)$, where A is a Borel subset of $S_n \cap D$. Then, due to the Strong Markov Property of Brownian motion,

$$1/2 < P(\|W^x_{\sigma^x} - x\| \ge r) = \int_{S_n \cap D} P(\|W^y_{\tau_y} - x\| \ge r) d\mu_n(y)$$

$$\le \sup_{y \in S_n \cap D} P(\|W^y_{\tau_y} - x\| \ge r),$$

which justifies (21.15). Now we can take a continuous function f such that $0 \le f(y) \le 1$ for $y \in \partial D$, $f(x) = 1$, and $f(y) = 0$ if $\|y - x\| \ge r$. By (21.15),

$$\limsup_{n \to \infty} Ef(W^{y_n}_{\tau_{y_n}}) \le 1/2 < f(x),$$

which contradicts (21.13). $\qquad \square$

Now we can state the existence and uniqueness result.

Theorem 21.7. *Let D be a bounded open domain in $\mathbb{R}^d$, $d \ge 2$, and f a continuous function on ∂D. Assume that all the points of ∂D are regular. Then the Dirichlet problem for the Laplace equation (21.10)–(21.11) has a unique solution. The solution is given by (21.12).*

Proof. The existence follows from Theorem 21.6. If u_1 and u_2 are two so-lutions, then $u = u_1 - u_2$ is a solution to the Dirichlet problem with zero boundary condition. A harmonic function which belongs to $C^2(D) \cap C(\overline{D})$ takes the maximal and the minimal values on the boundary of the domain. This implies that u is identically zero, that is $u_1 = u_2$.   □

Probabilistic techniques can also be used to justify the existence and uniqueness of solutions to more general elliptic and parabolic partial differen-tial equations. However, we shall now assume that the boundary of the domain is smooth, and thus we can bypass the existence and uniqueness questions, instead referring to the general theory of PDE's. In the next section we shall demonstrate that the solutions to PDE's can be expressed as functionals of the corresponding diffusion processes.

21.3 Stochastic Differential Equations and PDE's

First we consider the case in which the drift and the dispersion matrix do not depend on time. Let X_t be the strong solution of the stochastic differential equation

$$dX_t^i = v_i(X_t)dt + \sum_{j=1}^r \sigma_{ij}(X_t)dW_t^j, \quad 1 \le i \le d, \tag{21.16}$$

with the initial condition $X_0 = x \in \mathbb{R}^d$, where the coefficients v and σ satisfy the assumptions of Theorem 21.2. In fact, Eq. (21.16) defines a family of pro-cesses X_t which depend on the initial point x and are defined on a common probability space. When the dependence of the process on the initial point needs to be emphasized, we shall denote the process by X_t^x. (The superscript x is not to be confused with the superscript i used to denote the i-th component of the process.)

Let $a_{ij}(x) = \sum_{k=1}^r \sigma_{ik}(x)\sigma_{jk}(x) = (\sigma\sigma^*)_{ij}(x)$. This is a square non-negative definite symmetric matrix which will be called the diffusion matrix corresponding to the family of processes X_t^x. Let us consider the differential operator L which acts on functions $f \in C^2(\mathbb{R}^d)$ according to the formula

$$Lf(x) = \frac{1}{2}\sum_{i=1}^d\sum_{j=1}^d a_{ij}(x)\frac{\partial^2 f(x)}{\partial x_i \partial x_j} + \sum_{i=1}^d v_i(x)\frac{\partial f(x)}{\partial x_i}. \tag{21.17}$$

This operator is called the infinitesimal generator of the family of diffusion processes X_t^x. Let us show that for $f \in C^2(\mathbb{R}^d)$ which is bounded together with its first and second partial derivatives,

$$Lf(x) = \lim_{t\downarrow 0} \frac{E[f(X_t^x) - f(x)]}{t}. \tag{21.18}$$

In fact, the term "infinitesimal generator" of a Markov family of processes X_t^x usually refers to the right-hand side of this formula. (The Markov property of the solutions to SDE's will be discussed below.) By the Ito Formula, the expectation on the right-hand side of (21.18) is equal to

$$\mathrm{E}[\int_0^t Lf(X_s^x)ds + \int_0^t \sum_{i=1}^d \sum_{j=1}^r \frac{\partial f(X_s^x)}{\partial x_i}\sigma_{ij}(X_s^x)dW_s^j] = \mathrm{E}[\int_0^t Lf(X_s^x)ds],$$

since the expectation of the stochastic integral is equal to zero. Since Lf is bounded, the Dominated Convergence Theorem implies that

$$\lim_{t\downarrow 0} \frac{\mathrm{E}[\int_0^t Lf(X_s^x)ds]}{t} = Lf(x).$$

The coefficients of the operator L can be obtained directly from the law of the process X_t instead of the representation of the process as a solution of the stochastic differential equation. Namely,

$$v_i(x) = \lim_{t\downarrow 0} \frac{\mathrm{E}[(X_t^x)^i - x_i]}{t}, \quad a_{ij}(x) = \lim_{t\downarrow 0} \frac{\mathrm{E}[((X_t^x)^i - x_i)((X_t^x)^j - x_j)]}{t}.$$

We leave the proof of this statement to the reader.

Now let L be any differential operator given by (21.17). Let D be a bounded open domain in $\mathbb{R}^d$ with a smooth boundary ∂D. We shall consider the following partial differential equation

$$Lu(x) + q(x)u(x) = g(x) \quad \text{for } x \in D, \tag{21.19}$$

with the boundary condition

$$u(x) = f(x) \quad \text{for } x \in \partial D. \tag{21.20}$$

This pair, Eq. (21.19) and the boundary condition (21.20), is referred to as the Dirichlet problem for the operator L with the potential $q(x)$, the right-hand side $g(x)$, and the boundary condition $f(x)$. We assume that the coefficients $a_{ij}(x)$, $v_i(x)$ of the operator L, and the functions $q(x)$ and $g(x)$ are continuous on the closure of D (denoted by $\overline{D}$), while $f(x)$ is assumed to be continuous on ∂D.

Definition 21.8. *An operator L of the form (21.17) is called uniformly elliptic on D if there is a positive constant k such that*

$$\sum_{i=1}^d \sum_{j=1}^d a_{ij}(x)y_iy_j \geq k||y||^2 \tag{21.21}$$

for all $x \in D$ and all vectors $y \in \mathbb{R}^d$.

We shall use the following fact from the theory of partial differential equations (see "Partial Differential Equations" by A. Friedman, for example).

Theorem 21.9. *If a_{ij}, v_i, q, and g are Lipschitz continuous on $\overline{D}$, f is continuous on ∂D, the operator L is uniformly elliptic on D, and $q(x) \leq 0$ for $x \in \overline{D}$, then there is a unique solution $u(x)$ to (21.19)–(21.20) in the class of functions which belong to $C^2(D) \cap C(\overline{D})$.*

Let $\sigma_{ij}(x)$ and $v_i(x)$, $1 \leq i \leq d$, $1 \leq j \leq r$, be Lipschitz continuous on $\overline{D}$. It is not difficult to see that we can then extend them to bounded Lipschitz continuous functions on the entire space $\mathbb{R}^d$ and define the family of processes X_t^x according to (21.16). Let τ_D^x be the stopping time equal to the time of the first exit of the process X_t^x from the domain D, that is

$$\tau_D^x = \inf\{t \geq 0 : X_t^x \notin D\}.$$

By using Lemma 20.18, we can see that the stopped process $X_{t \wedge \tau_D^x}^x$ and the stopping time τ_D^x do not depend on the values of $\sigma_{ij}(x)$ and $v_i(x)$ outside of $\overline{D}$.

When L is the generator of the family of diffusion processes, we shall express the solution $u(x)$ to (21.19)–(21.20) as a functional of the process X_t^x. First, we need a technical lemma.

Lemma 21.10. *Suppose that $\sigma_{ij}(x)$ and $v_i(x)$, $1 \leq i \leq d$, $1 \leq j \leq r$, are Lipschitz continuous on $\overline{D}$, and the generator of the family of processes X_t^x is uniformly elliptic in D. Then*

$$\sup_{x \in \overline{D}} \mathrm{E}\tau_D^x < \infty.$$

Proof. Let B be an open ball so large that $\overline{D} \subset B$. Since the boundary of D is smooth and the coefficients $\sigma_{ij}(x)$ and $v_i(x)$ are Lipschitz continuous in $\overline{D}$, we can extend them to Lipschitz continuous functions on $\overline{B}$ in such a way that L becomes uniformly elliptic on B. Let $\varphi \in C^2(B) \cap C(\overline{B})$ be the solution of the equation $L\varphi(x) = 1$ for $x \in B$ with the boundary condition $\varphi(x) = 0$ for $x \in \partial B$. The existence of the solution is guaranteed by Theorem 21.9. By the Ito Formula,

$$\mathrm{E}\varphi(X_{t \wedge \tau_D^x}^x) - \varphi(x) = \mathrm{E} \int_0^{t \wedge \tau_D^x} L\varphi(X_s^x)ds = \mathrm{E}t \wedge \tau_D^x.$$

(The use of the Ito Formula is justified by the fact that φ is twice continuously differentiable in a neighborhood of $\overline{D}$, and thus there is a function $\psi \in C_0^2(\mathbb{R}^2)$, which coincides with φ in a neighborhood of $\overline{D}$. Theorem 20.27 can now be applied to the function ψ.)

Thus,

$$\sup_{x \in \overline{D}} \mathrm{E}\left(t \wedge \tau_D^x\right) \leq 2 \sup_{x \in \overline{D}} |\varphi(x)|,$$

which implies the lemma if we let $t \to \infty$. $\qquad\square$

Theorem 21.11. *Suppose that $\sigma_{ij}(x)$ and $v_i(x)$, $1 \le i \le d$, $1 \le j \le r$, are Lipschitz continuous on $\overline{D}$, and the generator L of the family of processes X_t^x is uniformly elliptic on D. Assume that the potential $q(x)$, the right-hand side $g(x)$, and the boundary condition $f(x)$ of the Dirichlet problem (21.19)– (21.20) satisfy the assumptions of Theorem 21.9. Then the solution to the Dirichlet problem can be written as follows:*

$$u(x) = \mathrm{E}[f(X_{\tau_D^x}^x) \exp(\int_0^{\tau_D^x} q(X_s^x)ds) - \int_0^{\tau_D^x} g(X_s^x) \exp(\int_0^s q(X_u^x)du)ds].$$

Proof. As before, we can extend $\sigma_{ij}(x)$ and $v_i(x)$ to Lipschitz continuous bounded functions on $\mathbb{R}^d$, and the potential $q(x)$ to a continuous function on $\mathbb{R}^d$, satisfying $q(x) \le 0$ for all x. Assume at first that $u(x)$ can be extended as a C^2 function to a neighborhood of $\overline{D}$. Then it can be extended as a C^2 function with compact support to the entire space $\mathbb{R}^d$. We can apply the integration by parts (20.23) to the pair of semimartingales $u(X_t^x)$ and $\exp(\int_0^t q(X_s^x)ds)$. In conjunction with (20.11) and the Ito formula,

$$u(X_t^x) \exp(\int_0^t q(X_s^x)ds) = u(x) + \int_0^t u(X_s^x) \exp(\int_0^s q(X_u^x)du) q(X_s^x)ds$$

$$+ \int_0^t \exp(\int_0^s q(X_u^x)du) Lu(X_s^x)ds$$

$$+ \sum_{i=1}^d \sum_{j=1}^r \int_0^t \exp(\int_0^s q(X_u^x)du) \frac{\partial u}{\partial x_i}(X_s^x) \sigma_{ij}(X_s^x)dW_s^j.$$

Notice that, by (21.19), $Lu(X_s^x) = g(X_s^x) - q(X_s^x)u(X_s^x)$ for $s \le \tau_D^x$. Therefore, after replacing t by $t \wedge \tau_D^x$ and taking the expectation on both sides, we obtain

$$\mathrm{E}(u(X_{t \wedge \tau_D^x}^x) \exp(\int_0^{t \wedge \tau_D^x} q(X_s^x)ds)) = u(x)$$

$$+ \mathrm{E} \int_0^{t \wedge \tau_D^x} g(X_s^x) \exp(\int_0^s q(X_u^x)du)ds.$$

By letting $t \to \infty$, which is justified by the Dominated Convergence Theorem, since $\mathrm{E}\tau_D^x$ is finite, we obtain

$$u(x) = \mathrm{E}[u(X_{\tau_D^x}^x) \exp(\int_0^{\tau_D^x} q(X_s^x)ds) - \mathrm{E} \int_0^{\tau_D^x} g(X_s^x) \exp(\int_0^s q(X_u^x)du)ds].$$

$$(21.22)$$

Since $X_{\tau_D^x}^x \in \partial D$ and $u(x) = f(x)$ for $x \in \partial D$, this is exactly the desired expression for $u(x)$.

At the beginning of the proof, we assumed that $u(x)$ can be extended as a C^2 function to a neighborhood of $\overline{D}$. In order to remove this assumption, we consider a sequence of domains $D_1 \subseteq D_2 \subseteq \dots$ with smooth boundaries, such that $\overline{D}_n \subset D$ and $\bigcup_{n=1}^\infty D_n = D$. Let $\tau_{D_n}^x$ be the stopping times corresponding to the domains D_n. Then $\lim_{n\to\infty} \tau_{D_n}^x = \tau_D^x$ almost surely for all $x \in D$. Since u is twice differentiable in D, which is an open neighborhood of $\overline{D}_n$, we have

$$u(x) = \mathrm{E}[u(X_{\tau_{D_n}^x})\exp(\int_0^{\tau_{D_n}^x} q(X_s^x)ds) - \mathrm{E}\int_0^{\tau_{D_n}^x} g(X_s^x)\exp(\int_0^s q(X_u^x)du)ds]$$

for $x \in D_n$. By taking the limit as $n \to \infty$ and using the dominated convergence theorem, we obtain (21.22). $\qquad\square$

Example. Let us consider the partial differential equation

$$Lu(x) = -1 \quad \text{for } x \in D$$

with the boundary condition

$$u(x) = 0 \quad \text{for } x \in \partial D.$$

By Theorem 21.11, the solution to this equation is simply the expectation of the time it takes for the process to exit the domain, that is $u(x) = \mathrm{E}\tau_D^x$.

Example. Let us consider the partial differential equation

$$Lu(x) = 0 \quad \text{for } x \in D$$

with the boundary condition

$$u(x) = f \quad \text{for } x \in \partial D.$$

By Theorem 21.11, the solution of this equation is

$$u(x) = \mathrm{E}f(X_{\tau_D^x}) = \int_{\partial D} f(y)d\mu_x(y),$$

where $\mu_x(A) = \mathrm{P}(X_{\tau_D^x} \in A)$, $A \in \mathcal{B}(\partial D)$, is the measure on ∂D induced by the random variable $X_{\tau_D^x}$.

Now let us explore the relationship between diffusion processes and parabolic partial differential equations. Let L be a differential operator with time-dependent coefficients, which acts on functions $f \in C^2(\mathbb{R}^d)$ according to the formula

$$Lf(x) = \frac{1}{2}\sum_{i=1}^{d}\sum_{j=1}^{d} a_{ij}(t,x)\frac{\partial^2 f(x)}{\partial x_i \partial x_j} + \sum_{i=1}^{d} v_i(t,x)\frac{\partial f(x)}{\partial x_i}.$$

We shall say that L is uniformly elliptic on $D \subseteq \mathbb{R}^{1+d}$ (with t considered as a parameter) if

$$\sum_{i=1}^{d}\sum_{j=1}^{d} a_{ij}(t,x)y_iy_j \geq k||y||^2$$

for some positive constant k, all $(t,x) \in D$, and all vectors $y \in \mathbb{R}^d$. Without loss of generality, we may assume that a_{ij} form a symmetric matrix, in which case $a_{ij}(t,x) = (\sigma\sigma^*)_{ij}(t,x)$ for some matrix $\sigma(t,x)$.

Let $T_1 < T_2$ be two moments of time. We shall be interested in the solutions to the backward parabolic equation

$$\frac{\partial u(t,x)}{\partial t} + Lu(t,x) + q(t,x)u(t,x) = g(t,x) \quad \text{for } (t,x) \in (T_1,T_2) \times \mathbb{R}^d \quad (21.23)$$

with the terminal condition

$$u(T_2,x) = f(x) \quad \text{for } x \in \mathbb{R}^d. \tag{21.24}$$

The function $u(t,x)$ is called the solution to the Cauchy problem (21.23)–(21.24). Let us formulate an existence and uniqueness theorem for the solutions to the Cauchy problem (see "Partial Differential Equations" by A. Friedman, for example).

Theorem 21.12. *Assume that $q(t,x)$ and $g(t,x)$ are bounded, continuous, and uniformly Lipschitz continuous in the space variables on $(T_1,T_2] \times \mathbb{R}^d$, and that $\sigma_{ij}(t,x)$ and $v_i(t,x)$ are continuous and uniformly Lipschitz continuous in the space variables on $(T_1,T_2] \times \mathbb{R}^d$. Assume that they do not grow faster than linearly, that is (21.5) holds, and that $f(x)$ is bounded and continuous on $\mathbb{R}^d$. Also assume that the operator L is uniformly elliptic on $(T_1,T_2] \times \mathbb{R}^d$.*

Then there is a unique solution $u(t,x)$ to the problem (21.23)–(21.24) in the class of functions which belong to $C^{1,2}((T_1,T_2) \times \mathbb{R}^d) \cap C_b((T_1,T_2] \times \mathbb{R}^d)$. (These are the functions which are bounded and continuous in $(T_1,T_2] \times \mathbb{R}^d$, and whose partial derivative in t and all second order partial derivatives in x are continuous in $(T_1,T_2) \times \mathbb{R}^d$.)

Remark 21.13. In textbooks on PDE's, this theorem is usually stated under the assumption that σ_{ij} and v_i are bounded. As will be explained below, by using the relationship between PDE's and diffusion processes, it is sufficient to assume that σ_{ij} and v_i do not grow faster than linearly.

Let us now express the solution to the Cauchy problem as a functional of the corresponding diffusion process. For $t \in (T_1,T_2]$, define $X_s^{t,x}$ to be the solution to the stochastic differential equation

$$dX_s^i = v_i(t+s,X_s)ds + \sum_{j=1}^{r}\sigma_{ij}(t+s,X_s)dW_s^j, \quad 1 \leq i \leq d, \quad s \leq T_2-t, \quad (21.25)$$

with the initial condition $X_0^{t,x} = x$. Let

$$a_{ij}(t,x) = \sum_{k=1}^{r} \sigma_{ik}(t,x)\sigma_{jk}(t,x) = (\sigma\sigma^*)_{ij}(t,x).$$

Theorem 21.14. *Suppose that the assumptions regarding the operator L and the functions $q(t,x)$, $g(t,x)$, and $f(x)$, formulated in Theorem 21.12, are satisfied. Then the solution to the Cauchy problem can be written as follows:*

$$u(t,x) = \mathrm{E}[f(X_{T_2-t}^{t,x})\exp(\int_0^{T_2-t} q(t+s,X_s^{t,x})ds)$$
$$- \int_0^{T_2-t} g(t+s,X_s^{t,x})\exp(\int_0^s q(t+u,X_u^{t,x})du)ds].$$

This expression for $u(t,x)$ is called the Feynman-Kac formula.

The proof of Theorem 21.14 is the same as that of Theorem 21.11, and therefore is left to the reader.

Remark 21.15. Let us assume that we have Theorems 21.12 and 21.14 only for the case in which the coefficients are bounded.

Given $\sigma_{ij}(t,x)$ and $v_i(t,x)$, which are continuous, uniformly Lipschitz continuous in the space variables, and do not grow faster than linearly, we can find continuous functions $\sigma_{ij}^n(t,x)$ and $v_i^n(t,x)$ which are uniformly Lipschitz continuous in the space variables and bounded on $(T_1,T_2] \times \mathbb{R}^d$, and which coincide with $\sigma_{ij}(t,x)$ and $v_i(t,x)$, respectively, for $||x|| \le n$.

Let $u^n(t,x)$ be the solution to the corresponding Cauchy problem. By Theorem 21.14 for the case of bounded coefficients, it is possible to show that u^n converge point-wise to some function u, which is a solution to the Cauchy equation with the coefficients which do not grow faster than linearly, and that this solution is unique. The details of this argument are left to the reader.

In order to emphasize the similarity between the elliptic and the parabolic problems, consider the processes $Y_t^{x,t_0} = (t+t_0,X_t^x)$ with values in $\mathbb{R}^{1+d}$ and initial conditions (t_0,x). Then the operator $\partial/\partial t + L$, which acts on functions defined on $\mathbb{R}^{1+d}$, is the infinitesimal generator for this family of processes.

Let us now discuss fundamental solutions to parabolic PDE'a and their relation to the transition probability densities of the corresponding diffusion processes.

Definition 21.16. *A non-negative function $G(t,r,x,y)$ defined for $t < r$ and $x,y \in \mathbb{R}^d$ is called a fundamental solution to the backward parabolic equation*

$$\frac{\partial u(t,x)}{\partial t} + Lu(t,x) = 0, \tag{21.26}$$

if for fixed t,r, and x, the function $G(t,r,x,y)$ belongs to $L^1(\mathbb{R}^d, \mathcal{B}(\mathbb{R}^d), \lambda)$, where λ is the Lebesgue measure, and for any $f \in C_b(\mathbb{R}^d)$, the function

$$u(t, x) = \int_{\mathbb{R}^d} G(t, r, x, y) f(y) dy$$

belongs to $C^{1,2}((-\infty, r) \times \mathbb{R}^d) \cap C_b((-\infty, r] \times \mathbb{R}^d)$ and is a solution to (21.26) with the terminal condition $u(r, x) = f(x)$.

Suppose that $\sigma_{ij}(t, x)$ and $v_i(t, x)$, $1 \leq i \leq d$, $1 \leq j \leq r$, $(t, x) \in \mathbb{R}^{1+d}$, are continuous, uniformly Lipschitz continuous in the space variables, and do not grow faster than linearly. It is well-known that in this case the fundamental solution to (21.26) exists and is unique (see "Partial Differential Equations of Parabolic Type" by A. Friedman). Moreover, for fixed r and y, the function $G(t, r, x, y)$ belongs to $C^{1,2}((-\infty, r) \times \mathbb{R}^d)$ and satisfies (21.26). Let us also consider the following equation, which is formally adjoint to (21.26)

$$-\frac{\partial \widetilde{u}(r, y)}{\partial r} + \frac{1}{2} \sum_{i=1}^{d} \sum_{j=1}^{d} \frac{\partial^2}{\partial y_i \partial y_j} [a_{ij}(r, y) \widetilde{u}(r, y)] - \sum_{i=1}^{d} \frac{\partial}{\partial y_i} [v_i(r, y) \widetilde{u}(r, y)] = 0,$$

(21.27)

where $\widetilde{u}(r, y)$ is the unknown function. If the partial derivatives

$$\frac{\partial a_{ij}(r, y)}{\partial y_i}, \frac{\partial^2 a_{ij}(r, y)}{\partial y_i \partial y_j}, \frac{\partial v_i(r, y)}{\partial y_i}, \quad 1 \leq i, j \leq d, \tag{21.28}$$

are continuous, uniformly Lipschitz continuous in the space variables, and do not grow faster than linearly, then for fixed t and x the function $G(t, r, x, y)$ belongs to $C^{1,2}((t, \infty) \times \mathbb{R}^d)$ and satisfies (21.27).

Let $X_s^{t,x}$ be the solution to Eq. (21.25), and let $\mu(t, r, x, dy)$ be the distribution of the process at time $r > t$. Let us show that under the above conditions on σ and v, the measure $\mu(t, r, x, dy)$ has a density, that is

$$\mu(t, r, x, dy) = \rho(t, r, x, y) dy, \tag{21.29}$$

where $\rho(t, r, x, y) = G(t, r, x, y)$. It is called the transition probability density for the process $X_s^{t,x}$. (It is exactly the density of the Markov transition function, which is defined in the next section for the time-homogeneous case.) In order to prove (21.29), take any $f \in C_b(\mathbb{R}^d)$ and observe that

$$\int_{\mathbb{R}^d} f(y) \mu(t, r, x, dy) = \int_{\mathbb{R}^d} f(y) G(t, r, x, y) dy,$$

since both sides are equal to the solution to the same backward parabolic PDE evaluated at the point (t, x) due to Theorem 21.14 and Definition 21.16. Therefore, the measures $\mu(t, r, x, dy)$ and $G(t, r, x, y) dy$ coincide (see Problem 4, Chap. 8). We formalize the above discussion in the following lemma.

Lemma 21.17. *Suppose that $\sigma_{ij}(t, x)$ and $v_i(t, x)$, $1 \leq i \leq d$, $1 \leq j \leq r$, $(t, x) \in \mathbb{R}^{1+d}$, are continuous, uniformly Lipschitz continuous in the space variables, and do not grow faster than linearly.*

Then, the family of processes $X_s^{t,x}$ defined by (21.25) has transition probability density $\rho(t,r,x,y)$, which for fixed r and y satisfies equation (21.26) (backward Kolmogorov equation). If, in addition, the partial derivatives in (21.28) are continuous, uniformly Lipschitz continuous in the space variables, and do not grow faster than linearly, then, for fixed t and x, the function $\rho(t,r,x,y)$ satisfies equation (21.27) (forward Kolmogorov equation).

Now consider a process whose initial distribution is not necessarily concentrated in a single point.

Lemma 21.18. *Assume that the distribution of a square-integrable $\mathbb{R}^d$-valued random variable ξ is equal to μ, where μ is a measure with continuous density p_0. Assume that the coefficients v_i and σ_{ij} and their partial derivatives in (21.28) are continuous, uniformly Lipschitz continuous in the space variables, and do not grow faster than linearly. Let X_t^μ be the solution to (21.16) with initial condition $X_0^\mu = \xi$.*

Then the distribution of X_t^μ, for fixed t, has a density $p(t,x)$ which belongs to $C^{1,2}((0,\infty) \times \mathbb{R}^d) \cap C_b([0,\infty) \times \mathbb{R}^d)$ and is the solution of the forward Kolmogorov equation

$$(-\frac{\partial}{\partial t} + L^*)p(t,x) = 0$$

with initial condition $p(0,x) = p_0(x)$.

Sketch of the Proof. Let $\tilde{\mu}_t$ be the measure induced by the process at time t, that is, $\tilde{\mu}_t(A) = \mathrm{P}(X_t^\mu \in A)$ for $A \in \mathcal{B}(\mathbb{R}^d)$. We can view $\tilde{\mu}$ as a generalized function (element of $\mathcal{S}'(\mathbb{R}^{1+d})$), which acts on functions $f \in \mathcal{S}(\mathbb{R}^{1+d})$ according to the formula

$$(\tilde{\mu}, f) = \int_0^\infty \int_{\mathbb{R}^d} f(t,x)d\tilde{\mu}_t(x)dt.$$

Now let $f \in \mathcal{S}(\mathbb{R}^{1+d})$, and apply Ito's formula to $f(t, X_t^\mu)$. After taking expectation on both sides,

$$\mathrm{E}f(t, X_t^\mu) = \mathrm{E}f(0, X_0^\mu) + \int_0^t \mathrm{E}(\frac{\partial f}{\partial s} + Lf)(s, X_s^\mu)ds.$$

If f is equal to zero for all sufficiently large t, we obtain

$$0 = \int_{\mathbb{R}^d} f(0,x)d\mu(x) + (\tilde{\mu}, \frac{\partial f}{\partial t} + Lf),$$

or, equivalently,

$$((-\frac{\partial}{\partial t} + L^*)\tilde{\mu}, f) + \int_{\mathbb{R}^d} f(0,x)d\mu(x) = 0. \qquad (21.30)$$

A generalized function $\tilde{\mu}$, such that (21.30) is valid for any infinitely smooth function with compact support, is called a generalized solution to the equation

$$(-\frac{\partial}{\partial t} + L^*)\tilde{\mu} = 0$$

with initial data μ. Since the partial derivatives in (21.28) are continuous, uniformly Lipschitz continuous in the space variables, and do not grow faster than linearly, and μ has a continuous density $p_0(x)$, the equation

$$(-\frac{\partial}{\partial t} + L^*)p(t, x) = 0$$

with initial condition $p(0, x) = p_0(x)$ has a unique solution in $C^{1,2}((0, \infty) \times \mathbb{R}^d) \cap C_b([0, \infty) \times \mathbb{R}^d)$. Since $\tilde{\mu}_t$ is a finite measure for each t, it can be shown that the generalized solution $\tilde{\mu}$ coincides with the classical solution $p(t, x)$. Then it can be shown that for t fixed, $p(t, x)$ is the density of the distribution of X_t^μ. $\qquad\square$

21.4 Markov Property of Solutions to SDE's

In this section we prove that solutions to stochastic differential equations form Markov families.

Theorem 21.19. *Let X_t^x be the family of strong solutions to the stochastic differential equation (21.16) with the initial conditions $X_0^x = x$. Let L be the infinitesimal generator for this family of processes. If the coefficients v_i and σ_{ij} are Lipschitz continuous and do not grow faster than linearly, and L is uniformly elliptic in $\mathbb{R}^d$, then X_t^x is a Markov family.*

Proof. Let us show that $p(t, x, \Gamma) = P(X_t^x \in \Gamma)$ is Borel-measurable as a function of $x \in \mathbb{R}^d$ for any $t \geq 0$ and any Borel set $\Gamma \subseteq \mathbb{R}^d$. When $t = 0$, $P(X_0^x \in \Gamma) = \chi_\Gamma(x)$, so it is sufficient to consider the case $t > 0$. First assume that Γ is closed. In this case, we can find a sequence of bounded continuous functions $f_n \in C_b(\mathbb{R}^d)$ such that $f_n(y)$ converge to $\chi_\Gamma(y)$ monotonically from above. By the Lebesgue Dominated Convergence Theorem,

$$\lim_{n \to \infty} \int_{\mathbb{R}^d} f_n(y)p(t, x, dy) = \int_{\mathbb{R}^d} \chi_\Gamma(y)p(t, x, dy) = p(t, x, \Gamma).$$

By Theorem 21.14, the integral $\int_{\mathbb{R}^d} f_n(y)p(t, x, dy)$ is equal to $u(0, x)$, where u is the solution of the equation

$$(\frac{\partial}{\partial t} + L)u = 0 \qquad\qquad (21.31)$$

with the terminal condition $u(t, x) = f(x)$. Since the solution is a smooth (and therefore measurable) function of x, $p(t, x, \Gamma)$ is a limit of measurable functions, and therefore measurable. Closed sets form a π-system, while the collection of sets Γ for which $p(t, x, \Gamma)$ is measurable is a Dynkin system. Therefore, $p(t, x, \Gamma)$ is measurable for all Borel sets Γ by Lemma 4.13. The second condition of Definition 19.2 is clear.

To verify the third condition of Definition 19.2, it suffices to show that

$$\mathrm{E}(f(X^x_{s+t})|\mathcal{F}_s) = \int_{\mathbb{R}^d} f(y)p(t, X^x_s, dy) \qquad (21.32)$$

for any $f \in C_b(\mathbb{R}^d)$. Indeed, we can approximate χ_Γ by a monotonically non-increasing sequence of functions from $C_b(\mathbb{R}^d)$, and, if (21.32) is true, by the Conditional Dominated Convergence Theorem,

$$\mathrm{P}(X^x_{s+t} \in \Gamma|\mathcal{F}_s) = p(t, X^x_s, \Gamma) \ \text{ almost surely.}$$

In order to prove (21.32), we can assume that $s, t > 0$, since otherwise the statement is obviously true. Let u be the solution to (21.31) with the terminal condition $u(s+t, x) = f(x)$. By Theorem 21.14, the right-hand side of (21.32) is equal to $u(s, X^x_s)$ almost surely. By the Ito formula,

$$u(s + t, X^x_{s+t}) = u(0, x) + \sum_{i=1}^d \sum_{j=1}^r \int_0^{s+t} \frac{\partial u}{\partial x_i}(X^x_u)\sigma_{ij}(X^x_u)dW^j_u.$$

After taking conditional expectation on both sides,

$$\mathrm{E}(f(X^x_{s+t})|\mathcal{F}_s) = u(0, x) + \sum_{i=1}^d \sum_{j=1}^r \int_0^s \frac{\partial u}{\partial x_i}(X^x_u)\sigma_{ij}(X^x_u)dW^j_u$$

$$= u(s, X^x_s) = \int_{\mathbb{R}^d} f(y)p(t, X^x_s, dy).$$

$\square$

Remark 21.20. Since $p(t, X^x_s, \Gamma)$ is $\sigma(X^x_s)$-measurable, it follows from the third property of Definition 19.2 that

$$\mathrm{P}(X^x_{s+t} \in \Gamma|\mathcal{F}_s) = \mathrm{P}(X^x_{s+t} \in \Gamma|X^x_s).$$

Thus, Theorem 21.19 implies that X^x_t is a Markov process for each fixed x.

We state the following theorem without a proof.

Theorem 21.21. Under the conditions of Theorem 21.19, the family of processes X^x_t is a strong Markov family.

Given a Markov family of processes X^x_t, we can define two families of Markov transition operators. The first family, denoted by P_t, acts on bounded measurable functions. It is defined by

$$(P_t f)(x) = \mathbb{E} f(X_t^x) = \int_{\mathbb{R}^d} f(y) p(t, x, dy),$$

where p is the Markov transition function. From the definition of the Markov property, we see that $P_t f$ is again a bounded measurable function.

The second family of operators, denoted by P_t^*, acts on probability measures. It is defined by

$$(P_t^* \mu)(C) = \int_{\mathbb{R}^d} \mathbb{P}(X_t^x \in C) d\mu(x) = \int_{\mathbb{R}^d} p(t, x, C) d\mu(x).$$

It is clear that the image of a probability measure μ under P_t^* is again a probability measure. The operators P_t and P_t^* are adjoint. Namely, if f is a bounded measurable function and μ is a probability measure, then

$$\int_{\mathbb{R}^d} (P_t f)(x) d\mu(x) = \int_{\mathbb{R}^d} f(x) d(P_t^* \mu)(x). \tag{21.33}$$

Indeed, by the definitions of P_t and P_t^*, this formula is true if f is an indicator function of a measurable set. Therefore, it is true for finite linear combinations of indicator functions. An arbitrary bounded measurable function can, in turn, be uniformly approximated by finite linear combinations of indicator functions, which justifies (21.33).

Definition 21.22. *A measure μ is said to be invariant for a Markov family X_t^x if $P_t^* \mu = \mu$ for all $t \geq 0$.*

Let us answer the following question: when is a measure μ invariant for the family of diffusion processes X_t^x that solve (21.16) with initial conditions $X_0^x = x$? Let the coefficients of the generator L satisfy the conditions stated in Lemma 21.19. Assume that μ is an invariant measure. Then the right-hand side of (21.33) does not depend on t, and therefore neither does the left hand side. In particular,

$$\int_{\mathbb{R}^d} (P_t f - f)(x) d\mu(x) = 0.$$

Let f belong to the Schwartz space $\mathcal{S}(\mathbb{R}^d)$. In this case,

$$\int_{\mathbb{R}^d} L f(x) d\mu(x) = \int_{\mathbb{R}^d} \lim_{t \downarrow 0} \frac{(P_t f - f)(x)}{t} d\mu(x)$$

$$= \lim_{t \downarrow 0} \int_{\mathbb{R}^d} \frac{(P_t f - f)(x)}{t} d\mu(x) = 0,$$

where the first equality is due to (21.18) and the second one to the Dominated Convergence Theorem. Note that we can apply the Dominated Convergence Theorem, since $(P_t f - f)/t$ is uniformly bounded for $t > 0$ if $f \in \mathcal{S}(\mathbb{R}^d)$, as is clear from the discussion following (21.18).

We can rewrite the equality $\int_{\mathbb{R}^d} Lf(x)d\mu(x) = 0$ as $(L^*\mu, f) = 0$, where $L^*\mu$ is the following generalized function

$$L^*\mu = \frac{1}{2}\sum_{i=1}^{d}\sum_{j=1}^{d}\frac{\partial^2}{\partial x_i \partial x_j}[a_{ij}(x)\mu(x)] - \sum_{i=1}^{d}\frac{\partial}{\partial x_i}[v_i(x)\mu(x)].$$

Here, $a_{ij}(x)\mu(x)$ and $v_i(x)\mu(x)$ are the generalized functions corresponding to the signed measures whose densities with respect to μ are equal to $a_{ij}(x)$ and $v_i(x)$, respectively. The partial derivatives here are understood in the sense of generalized functions. Since $f \in \mathcal{S}(\mathbb{R}^d)$ was arbitrary, we conclude that $L^*\mu = 0$.

The converse is also true: if $L^*\mu = 0$, then μ is an invariant measure for the family of diffusion processes X_t^x. We leave this statement as an exercise for the reader.

Example. Let X_t^x be the family of solutions to the stochastic differential equation

$$dX_t^x = dW_t - X_t^x dt$$

with the initial data $X_t^x = x$. (See Sect. 21.1, in which we discussed the Ornstein-Uhlenbeck process.) The generator for this family of processes and the adjoint operator are given by

$$Lf(x) = \frac{1}{2}f''(x) - xf'(x) \quad \text{and} \quad L^*\mu(x) = \frac{1}{2}\mu''(x) + (x\mu(x))'.$$

It is not difficult to see that the only probability measure that satisfies $L^*\mu = 0$ is that whose density with respect to the Lebesgue measure is equal to $p(x) = \frac{1}{\sqrt{\pi}}\exp(-x^2)$. Thus, the invariant measure for the family of Ornstein-Uhlenbeck processes is $\mu(dx) = \frac{1}{\sqrt{\pi}}\exp(-x^2)\lambda(dx)$, where λ is the Lebesgue measure.

21.5 A Problem in Homogenization

Given a parabolic partial differential equation with variable (e.g. periodic) coefficients, it is often possible to describe asymptotic properties of its solutions (as $t \to \infty$) in terms of solutions to a simpler equation with constant coefficients. Similarly, for large t, solutions to a stochastic differential equation with variable coefficients may exhibit similar properties to those for an SDE with constant coefficients.

In order to state one such homogenization result, let us consider the $\mathbb{R}^d$-valued process X_t which satisfies the following stochastic differential equation

$$dX_t = v(X_t)dt + dW_t \tag{21.34}$$

with initial condition $X_0 = \xi$, where ξ is a bounded random variable, $v(x) = (v_1(x),\ldots,v_d(x))$ is a vector field on $\mathbb{R}^d$, and $W_t = (W_t^1,\ldots,W_t^d)$ is a d-dimensional Brownian motion. We assume that the vector field v is smooth, periodic ($v(x + z) = v(x)$ for $z \in \mathbb{Z}^d$) and incompressible (div$v = 0$). Let T^d be the unit cube in $\mathbb{R}^d$,

$$T^d = \{x \in \mathbb{R}^d : 0 \leq x_i < 1, \ i = 1\ldots, d\}$$

(we may glue the opposite sides to make it into a torus). Let us assume that $\int_{T^d} v_i(x)dx = 0$, $1 \leq i \leq d$, that is the "net drift" of the vector field is equal to zero. Notice that we can consider X_t as a process with values on the torus.

Although the solution to (21.34) cannot be written out explicitly, we can describe the asymptotic behavior of X_t for large t. Namely, consider the $\mathbb{R}^d$-valued process Y_t defined by

$$Y_t^i = \sum_{1 \leq j \leq d} \sigma_{ij} W_t^j, \quad 1 \leq i, j \leq d,$$

with some coefficients σ_{ij}. Due to the scaling property of Brownian motion, for any positive ε, the distribution of the process $Y_t^\varepsilon = \sqrt{\varepsilon} Y_{t/\varepsilon}$ is the same as that of the original process Y_t. Let us now apply the same scaling transformation to the process X_t. Thus we define

$$X_t^\varepsilon = \sqrt{\varepsilon} X_{t/\varepsilon}.$$

Let P_X^ε be the measure on $C([0, \infty))$ induced by the process X_t^ε, and P_Y the measure induced by the process Y_t. It turns out that for an appropriate choice of the coefficients σ_{ij}, the measures P_X^ε converge weakly to P_Y when $\varepsilon \to 0$. In particular, for t fixed, X_t^ε converges in distribution to a Gaussian random variable with covariance matrix $a_{ij} = (\sigma\sigma^*)_{ij}$.

We shall not prove this statement in full generality, but instead study only the behavior of the covariance matrix of the process X_t^ε as $\varepsilon \to 0$ (or, equivalently, of the process X_t as $t \to \infty$). We shall show that $\mathrm{E}(X_t^i X_t^j)$ grows linearly, and identify the limit of $\mathrm{E}(X_t^i X_t^j)/t$ as $t \to \infty$. An additional simplifying assumption will concern the distribution of ξ.

Let L be the generator of the process X_t which acts on functions $u \in C^2$ (T^d) (the class of smooth periodic functions) according to the formula

$$Lu(x) = \frac{1}{2}\Delta u(x) + (v, \nabla u)(x).$$

If u is periodic, then so is Lu, and therefore we can consider L as an operator on $C^2(T^d)$ with values in $C(T^d)$. Consider the following partial differential equations for unknown periodic functions u_i, $1 \leq i \leq d$,

$$L(u_i(x) + x_i) = 0, \tag{21.35}$$

where x_i is the i-th coordinate of the vector x. These equations can be rewritten as

$$Lu_i(x) = -v_i(x).$$

Note that the right-hand side is a periodic function. It is well-known in the general theory of elliptic PDE's that this equation has a solution in $C^2(T^d)$ (which is then unique up to an additive constant) if and only if the right-hand side is orthogonal to the kernel of the adjoint operator (see "Partial Differential Equations" by A. Friedman). In other words, to establish the existence of a solution we need to check that

$$\int_{T^d} -v_i(x)g(x)dx = 0 \text{ if } g \in C^2(T^d) \text{ and } L^*g(x) = \frac{1}{2}\Delta g(x) - \text{div}(gv)(x) = 0.$$

It is easy to see that the only $C^2(T^d)$ solutions to the equation $L^*g = 0$ are constants, and thus the existence of solutions to (21.35) follows from $\int_{T^d} v_i(x)dx = 0$. Since we can add an arbitrary constant to the solution, we can define $u_i(x)$ to be the solution to (21.35) for which $\int_{T^d} u_i(x)dx = 0$.

Now let us apply Ito's formula to the function $u_i(x) + x_i$ of the process X_t:

$$u_i(X_t) + X_t^i - u_i(X_0) - X_0^i = \int_0^t \sum_{k=1}^d \frac{\partial(u_i + x_i)}{\partial x_k}(X_s)dW_s^k$$

$$+ \int_0^t L(u_i + x_i)(X_s)ds = \int_0^t \sum_{k=1}^d \frac{\partial(u_i + x_i)}{\partial x_k}(X_s)dW_s^k,$$

since the ordinary integral vanishes due to (21.35). Let $g_t^i = u_i(X_t) - u_i(X_0) - X_0^i$. Thus,

$$X_t^i + g_t^i = \int_0^t \sum_{k=1}^d \frac{\partial(u_i + x_i)}{\partial x_k}(X_s)dW_s^k.$$

Similarly, using the index j instead of i, we can write

$$X_t^j + g_t^j = \int_0^t \sum_{k=1}^d \frac{\partial(u_j + x_j)}{\partial x_k}(X_s)dW_s^k.$$

Let us multiply the right-hand sides of these equalities, and take expectations. With the help of Lemma 20.17 we obtain

$$\mathrm{E}\left(\int_0^t \sum_{k=1}^d \frac{\partial(u_i + x_i)}{\partial x_k}(X_s)dW_s^k \int_0^t \sum_{k=1}^d \frac{\partial(u_j + x_j)}{\partial x_k}(X_s)dW_s^k\right)$$

$$= \int_0^t \mathrm{E}((\nabla u_i, \nabla u_j)(X_s) + \delta_{ij})ds,$$

where $\delta_{ij} = 1$ if $i = j$, and $\delta_{ij} = 0$ if $i \neq j$.

Notice that, since v is periodic, we can consider (21.34) as an equation for a process on the torus T^d. Let us assume that $X_0 = \xi$ is uniformly distributed on the unit cube (and, consequently, when we consider X_t as a process on the torus, X_0 is uniformly distributed on the unit torus). Let $p_0(x) \equiv 1$ be the density of this distribution. Since $L^* p_0(x) = 0$, the density of X_s on the torus is also equal to p_0. (Here we used Lemma 21.18, modified to allow for processes to take values on the torus.) Consequently,

$$\int_0^t E((\nabla u_i, \nabla u_j)(X_s) + \delta_{ij}) ds = \int_0^t \int_{T^d} ((\nabla u_i, \nabla u_j)(x) + \delta_{ij}) dx ds$$

$$= t \int_{T^d} ((\nabla u_i, \nabla u_j)(x) + \delta_{ij}) dx.$$

Thus,

$$E((X_t^i + g_t^i)(X_t^j + g_t^j))/t = \int_{T^d} ((\nabla u_i, \nabla u_j)(x) + \delta_{ij}) dx. \tag{21.36}$$

Lemma 21.23. *Under the above assumptions,*

$$E(X_t^i X_t^j)/t \to \int_{T^d} ((\nabla u_i, \nabla u_j)(x) + \delta_{ij}) dx \quad \text{as} \quad t \to \infty. \tag{21.37}$$

Proof. The difference between (21.37) and (21.36) is the presence of the bounded processes g_t^i and g_t^j in expectation on the left-hand side of (21.36). The desired result follows from the following simple lemma.

Lemma 21.24. *Let f_t^i and h_t^i, $1 \le i \le d$, be two families of random processes. Suppose*

$$E\left((f_t^i + h_t^i)(f_t^j + h_t^j)\right) = \phi^{ij}. \tag{21.38}$$

Also suppose there is a constant c such that

$$t E(h_t^i)^2 \le c. \tag{21.39}$$

Then,

$$\lim_{t \to \infty} E(f_t^i f_t^j) = \phi^{ij}.$$

Proof. By (21.38) with $i = j$,

$$E(f_t^i)^2 = \phi^{ii} - E(h_t^i)^2 - 2E(f_t^i h_t^i). \tag{21.40}$$

By (21.40) and (21.39), we conclude that there exists a constant c' such that

$$E(f_t^i)^2 < c' \quad \text{for all} \quad t > 1. \tag{21.41}$$

By (21.38),

$$E(f_t^i f_t^j) - \phi^{ij} = -E(h_t^i h_t^j) - E(f_t^i h_t^j) - E(f_t^j h_t^i). \tag{21.42}$$

By the Schwartz Inequality, (21.39) and (21.41), the right-hand side of (21.42) tends to zero as $t \to \infty$. □

To complete the proof of Lemma 21.23 it suffices to take $f_t^i = X_t^i/\sqrt{t}$, $h_t^i = g_t^i/\sqrt{t}$, and apply Lemma 21.24. □

21.6 Problems

1. Prove the Gronwall Inequality (Lemma 21.4).
2. Let W_t be a one-dimensional Brownian motion. Prove that the process

$$X_t = (1-t) \int_0^t \frac{dW_s}{1-s}, \quad 0 \le t < 1,$$

is the solution of the stochastic differential equation

$$dX_t = \frac{X_t}{t-1} dt + dW_t, \quad 0 \le t < 1, \quad X_0 = 0.$$

3. For the process X_t defined in Problem 2, prove that there is the almost sure limit

$$\lim_{t \to 1-} X_t = 0.$$

Define $X_1 = 0$. Prove that the process X_t, $0 \le t \le 1$ is Gaussian, and find its correlation function. Prove that X_t is a Brownian Bridge (see Problem 13, Chap. 18).

4. Consider two European call options with the same strike price for the same stock (i.e., r, σ, P and S are the same for the two options). Assume that the risk-free interest rate γ is equal to zero. Is it true that the option with longer time till expiration is more valuable?

5. Let W_t be a one-dimensional Brownian motion, and $Y_t = e^{-t/2} W(e^t)$. Find a, σ and ξ such that Y_t has the same finite-dimensional distributions as the solution of (21.9).

6. Let W_t be a two-dimensional Brownian motion, and τ the first time when W_t hits the unit circle, $\tau = \inf(t : ||W_t|| = 1)$. Find $E\tau$.

7. Prove that if a point satisfies the exterior cone condition, then it is regular.

8. Prove that regularity is a local condition. Namely, let D_1 and D_1 be two domains, and let $x \in \partial D_1 \cap \partial D_2$. Suppose that there is an open neighborhood U of x such that $U \cap \partial D_1 = U \cap \partial D_1$. Then x is a regular boundary point for D_1 if and only if it is a regular boundary point for D_2.

9. Let W_t be a two-dimensional Brownian motion. Prove that for any $x \in \mathbb{R}^2$, $||x|| > 0$, we have

$$P(\text{there is } t \ge 0 \text{ such that } W_t = x) = 0.$$

Prove that for any $\delta > 0$

$$P(\text{there is } t \ge 0 \text{ such that } ||W_t - x|| \le \delta) = 1.$$

10. Let W_t be a d-dimensional Brownian motion, where $d \ge 3$. Prove that

$$\lim_{t \to \infty} ||W_t|| = \infty$$

almost surely.

11. Let $W_t = (W_t^1, W_t^2)$ be a two-dimensional Brownian motion, and τ the first time when W_t hits the unit square centered at the origin, $\tau = \inf(t : \min(|W_t^1|, |W_t^2|) = 1/2)$. Find $E\tau$.

12. Let D be the open unit disk in $\mathbb{R}^2$ and $u^\varepsilon \in C^2(D) \cap C(\overline{D})$ the solution of the following Dirichlet problem

$$\varepsilon \Delta u^\varepsilon + \frac{\partial u^\varepsilon}{\partial x_1} = 0,$$

$$u(x) = f(x) \quad \text{for } x \in \partial D,$$

where f is a continuous function on ∂D. Find the limit $\lim_{\varepsilon \downarrow 0} u^\varepsilon(x_1, x_2)$ for $(x_1, x_2) \in D$.

13. Let X_t be the strong solution to the stochastic differential equation

$$dX_t = v(X_t)dt + \sigma(X_t)dW_t$$

with the initial condition $X_0 = 1$, where v and σ are Lipschitz continuous functions on $\mathbb{R}$. Assume that $\sigma(x) \geq c > 0$ for some constant c and all $x \in \mathbb{R}$. Find a non-constant function f such that $f(X_t)$ is a local martingale.

14. Let X_t, v and σ be the same as in the previous problem. For which functions v and σ do we have

$$P(\text{there is } t \in [0, \infty) \text{ such that } X_t = 0) = 1?$$

Gibbs Random Fields

22.1 Definition of a Gibbs Random Field

The notion of Gibbs random fields was formalized by mathematicians relatively recently. Before that, these fields were known in physics, particularly in statistical physics and quantum field theory. Later, it was understood that Gibbs fields play an important role in many applications of probability theory. In this section we define the Gibbs fields and discuss some of their properties.

We shall deal with random fields with a finite state space X defined over $\mathbb{Z}^d$. The realizations of the field will be denoted by $\omega = (\omega_k, k \in \mathbb{Z}^d)$, where ω_k is the value of the field at the site k.

Let V and W be two finite subsets of $\mathbb{Z}^d$ such that $V \subset W$ and $\mathrm{dist}(V, \mathbb{Z}^d \backslash W) > R$ for a given positive constant R. We can consider the following conditional probabilities:

$$P(\omega_k = i_k, k \in V | \omega_k = i_k, k \in W \backslash V), \quad \text{where } i_k \in X \text{ for } k \in W.$$

Definition 22.1. *A random field is called a Gibbs field with memory R if, for any finite sets V and W as above, these conditional probabilities (whenever they are defined) depend only on those of the values i_k for which $\mathrm{dist}(k, V) \le R$.*

Note that the Gibbs fields can be viewed as generalizations of Markov chains. Indeed, consider a Markov chain on a finite state space. The realizations of the Markov chain will be denoted by $\omega = (\omega_k, k \in \mathbb{Z})$. Let k_1, k_2, l_1 and l_2 be integers such that $k_1 < l_1 \le l_2 < k_2$. Consider the conditional probabilities

$$f(i_{k_1}, \dots, i_{l_1-1}, i_{l_2+1}, \dots, i_{k_2}) =$$

$$P(\omega_{l_1} = i_{l_1}, \dots, \omega_{l_2} = i_{l_2} | \omega_{k_1} = i_{k_1}, \dots, \omega_{l_1-1} = i_{l_1-1}, \omega_{l_2+1} = i_{l_2+1}, \dots, \omega_{k_2} = i_{k_2})$$

with $i_{l_1}, \dots, i_{l_2}$ fixed. It is easy to check that whenever f is defined, it depends only on i_{l_1-1} and i_{l_2+1} (see Problem 12, Chap. 5). Thus, a Markov chain is a Gibbs field with $d = 1$ and $R = 1$.

L. Koralov and Y.G. Sinai, *Theory of Probability and Random Processes*, 341
Universitext, DOI 10.1007/978-3-540-68829-7_22,
© Springer-Verlag Berlin Heidelberg 2012

Let us introduce the notion of the interaction energy. Let $N_{d,R}$ be the number of points of $\mathbb{Z}^d$ that belong to the closed ball of radius R centered at the origin. Let U be a real-valued function defined on $X^{N_{d,R}}$. As arguments of U we shall always take the values of the field in a ball of radius R centered at one of the points of $\mathbb{Z}^d$. We shall use the notation $U(\omega_k; \omega_{k'}, 0 < |k' - k| \le R)$ for the value of U on a realization ω in the ball centered at k and call U the interaction energy with radius R.

For a finite set $V \subset \mathbb{Z}^d$, its R-neighborhood will be denoted by V^R,

$$V^R = \{k : \operatorname{dist}(V, k) \le R\}.$$

Definition 22.2. *A Gibbs field with memory $2R$ is said to correspond to the interaction energy U if*

$$P(\omega_k = i_k, k \in V | \omega_k = i_k, k \in V^{2R} \setminus V)$$
$$= \frac{\exp(-\sum_{k \in V^R} U(i_k; i_{k'}, 0 < |k' - k| \le R))}{Z(i_k, k \in V^{2R} \setminus V)}, \quad (22.1)$$

where $Z = Z(i_k, k \in V^{2R} \setminus V)$ is the normalization constant, which is called the partition function,

$$Z(i_k, k \in V^{2R} \setminus V) = \sum_{\{i_k, k \in V\}} \exp(-\sum_{k \in V^R} U(i_k; i_{k'}, 0 < |k' - k| \le R)).$$

The equality (22.1) for the conditional probabilities is sometimes called the Dobrushin-Lanford-Ruelle or, simply, DLR equation after three mathematicians who introduced the general notion of a Gibbs random field. The minus sign is adopted from statistical physics.

Let $\Omega(V)$ be the set of configurations $(\omega_k, k \in V)$. The sum

$$\sum_{k \in V^R} U(\omega_k; \omega_{k'}, 0 < |k' - k| \le R)$$

is called the energy of the configuration $\omega \in \Omega(V)$. It is defined as soon as we have the boundary conditions ω_k, $k \in V^{2R} \setminus V$.

Theorem 22.3. *For any interaction energy U with finite radius, there exists at least one Gibbs field corresponding to U.*

Proof. Take a sequence of cubes V_i centered at the origin with sides of length $2i$. Fix arbitrary boundary conditions $\{\omega_k, k \in V_i^{2R} \setminus V_i\}$, for example $\omega_k = x$ for all $k \in V_i^{2R} \setminus V_i$, where x is a fixed element of X, and consider the probability distribution $P_{V_i}(\cdot | \omega_k, k \in V_i^{2R} \setminus V_i)$ on the finite set $\Omega(V_i)$ given by (22.1) (with V_i instead of V).

Fix V_j. For $i > j$, the probability distribution $P_{V_i}(\cdot | \omega_k, k \in V_i^{2R} \setminus V_i)$ induces a probability distribution on the set $\Omega(V_j)$. The space of such probability distributions is tight. (The set $\Omega(V_j)$ is finite, and we can consider an arbitrary metric on it. The property of tightness does not depend on the particular metric.)

Take a subsequence $\{j_s^{(1)}\}$ such that the induced probability distributions on $\Omega(V_1)$ converge to a limit $Q^{(1)}$. Then find a subsequence $\{j_s^{(2)}\} \subseteq \{j_s^{(1)}\}$ such that the induced probability distributions on the space $\Omega(V_2)$ converge to a limit $Q^{(2)}$. Since $\{j_s^{(2)}\} \subset \{j_s^{(1)}\}$, the probability distribution induced by $Q^{(2)}$ on the space $\Omega(V_1)$ coincides with $Q^{(1)}$. Arguing in the same way, we can find a subsequence $\{j_s^{(m)}\} \subseteq \{j_s^{(m-1)}\}$, for any $m \geq 1$, such that the probability distributions induced by $P_{V_{j(m)}}(\cdot|\omega_k, k \notin V_{j(m)})$ on $\Omega(V_m)$ converge to a limit, which we denote by $Q^{(m)}$. Since $\{j_s^{(m)}\} \subseteq \{j_s^{(m-1)}\}$, the probability distribution on $\Omega(V_{m-1})$ induced by $Q^{(m)}$ coincides with $Q^{(m-1)}$.

Then, for the sequence of probability distributions

$$P_{V_{j_m^{(m)}}}(\cdot|\omega_k, k \in V_{j_m^{(m)}}^{2R} \setminus V_{j_m^{(m)}})$$

corresponding to the diagonal subsequence $\{j_m^{(m)}\}$, we have the following property:

For each m, the restrictions of the probability distributions to the set $\Omega(V_m)$ converge to a limit $Q^{(m)}$, and the probability distribution induced by $Q^{(m)}$ on $\Omega(V_{m-1})$ coincides with $Q^{(m-1)}$. The last property is a version of the Consistency Conditions, and by the Kolmogorov Consistency Theorem, there exists a probability distribution Q defined on the natural σ-algebra of subsets of the space Ω of all possible configurations $\{\omega_k, k \in \mathbb{Z}^d\}$ whose restriction to each $\Omega(V_m)$ coincides with $Q^{(m)}$ for any $m \geq 1$.

It remains to prove that Q is generated by a Gibbs random field corresponding to U. Let V be a finite subset of $\mathbb{Z}^d$, W a finite subset of $\mathbb{Z}^d$ such that $V^{2R} \subseteq W$, and let the values $\omega_k = i_k$ be fixed for $k \in W \setminus V$. We need to consider the conditional probabilities

$$q = Q\{\omega_k = i_k, k \in V | \omega_k = i_k, k \in W \setminus V\}.$$

In fact, it is more convenient to deal with the ratio of the conditional probabilities corresponding to two different configurations, $\omega_k = \bar{i}_k$ and $\omega_k = \bar{\bar{i}}_k$, $k \in V$, which is equal to

$$
\begin{aligned}
q_1 &= \frac{Q\{\omega_k = \bar{i}_k, k \in V | \omega_k = i_k, k \in W \setminus V\}}{Q\{\omega_k = \bar{\bar{i}}_k, k \in V | \omega_k = i_k, k \in W \setminus V\}} \\
&= \frac{Q\{\omega_k = \bar{i}_k, k \in V, \omega_k = i_k, k \in W \setminus V\}}{Q\{\omega_k = \bar{\bar{i}}_k, k \in V, \omega_k = i_k, k \in W \setminus V\}}.
\end{aligned}
\tag{22.2}
$$

It follows from our construction that the probabilities Q in this ratio are the limits found with the help of probability distributions $P_{V_{j_m^{(m)}}}(\cdot|\omega_k, k \in V_{j_m^{(m)}}^{2R} \setminus V_{j_m^{(m)}})$. We can express the numerator in (22.2) as follows:

$$Q\{\omega_k = \bar{i}_k, k \in V, \omega_k = i_k, k \in W\backslash V\}$$

$$= \lim_{m \to \infty} P_{V_{j_m^{(m)}}}(\omega_k = \bar{i}_k, k \in V, \omega_k = i_k, k \in W\backslash V | \omega_k = i_k, k \in V_{j_m^{(m)}}^{2R}\backslash V_{j_m^{(m)}})$$

$$= \lim_{m \to \infty} \frac{\displaystyle\sum_{\{i_k, k \in V_{j_m^{(m)}}\backslash W\}} \exp(-\sum_{k \in V_{j_m^{(m)}}^R} U(i_k; i_{k'}, 0 < |k' - k| \leq R))}{Z(i_k, k \in V_{j_m^{(m)}}^{2R}\backslash V_{j_m^{(m)}})}.$$

A similar expression for the denominator in (22.2) is also valid. The difference between the expressions for the numerator and the denominator is that, in the first case, $i_k = \bar{i}_k$ for $k \in V$, while in the second, $i_k = \bar{\bar{i}}_k$ for $k \in V$.

Therefore, the corresponding expressions $U(i_k; i_{k'}, |k' - k| \leq R)$ for k such that dist $(k, V) > R$ coincide in both cases, and

$$q_1 = \frac{r_1}{r_2},$$

where

$$r_1 = \exp(-\sum_{k \in V^R} U(i_k; i_{k'}, 0 < |k' - k| \leq R)), \quad i_k = \bar{i}_k \text{ for } k \in V,$$

while

$$r_2 = \exp(-\sum_{k \in V^R} U(i_k; i_{k'}, 0 < |k' - k| \leq R)), \quad i_k = \bar{\bar{i}}_k \text{ for } k \in V.$$

This is the required expression for q_1, which implies that Q is a Gibbs field. $\square$

22.2 An Example of a Phase Transition

Theorem 22.3 is an analogue of the theorem on the existence of stationary distributions for finite Markov chains. In the ergodic case, this distribution is unique. In the case of multi-dimensional time, however, under very general conditions there can be different random fields corresponding to the same function U. The related theory is connected to the theory of phase transitions in statistical physics.

If $X = \{-1, 1\}$, $R = 1$ and $U(i_0; i_k, |k| = 1) = \pm\beta\sum_{|k|=1}(i_0 - i_k)^2$, the corresponding Gibbs field is called the Ising model with inverse temperature β (and zero magnetic field). The plus sign corresponds to the so-called ferromagnetic model; the minus sign corresponds to the so-called anti-ferromagnetic model. Again, the terminology comes from statistical mechanics. We shall consider here only the case of the ferromagnetic Ising model and prove the following theorem.

Theorem 22.4. *Consider the following interaction energy over* $\mathbb{Z}^2$:

$$U(\omega_0; \omega_k, |k| = 1) = \beta \sum_{|k|=1} (\omega_0 - \omega_k)^2.$$

If β *is sufficiently large, there exist at least two different Gibbs fields corresponding to* U.

Proof. As before, we consider the increasing sequence of squares V_i and plus-minus boundary conditions, i.e., either $\omega_k \equiv +1$, $k \notin V_i$, or $\omega_k \equiv -1$, $k \notin V_i$. The corresponding probability distributions on $\Omega(V_i)$ will be denoted by $P_{V_i}^+$ and $P_{V_i}^-$ respectively. We shall show that $P_{V_i}^+(\omega_0 = +1) \geq 1 - \varepsilon(\beta)$, $P_{V_i}^-(\omega_0 = -1) \geq 1 - \varepsilon(\beta)$, $\varepsilon(\beta) \to 0$ as $\beta \to \infty$. In other words, the Ising model displays strong memory of the boundary conditions for arbitrarily large i. Sometimes this kind of memory is called the long-range order. It is clear that the limiting distributions constructed with the help of the sequences $P_{V_i}^+$ and $P_{V_i}^-$ are different, which constitutes the statement of the theorem.

We shall consider only $P_{V_i}^+$, since the case of $P_{V_i}^-$ is similar. We shall show that a typical configuration with respect to $P_{V_i}^+$ looks like a "sea" of $+1$'s surrounding small "islands" of -1's, and the probability that the origin belongs to this "sea" tends to 1 as $\beta \to \infty$ uniformly in i.

Take an arbitrary configuration $\omega \in \Omega(V_i)$. For each $k \in V_i$ such that $\omega_k = -1$ we construct a unit square centered at k with sides parallel to the coordinate axes, and then we slightly round off the corners of the square.

The union of these squares with rounded corners is denoted by $I(\omega)$. The boundary of $I(\omega)$ is denoted by $B(\omega)$. It consists of those edges of the squares where ω takes different values on different sides of the edge. A connected component of $B(\omega)$ is called a contour.

It is clear that each contour is a closed non-self-intersecting curve. If $B(\omega)$ does not have a contour containing the origin inside the domain it bounds, then $\omega_0 = +1$.

Given a contour Γ, we shall denote the domain it bounds by $\text{int}(\Gamma)$. Let a contour Γ be such that the origin is contained inside $\text{int}(\Gamma)$. The number of such contours of length n does not exceed $n3^{n-1}$.

Indeed, since the origin is inside $\text{int}(\Gamma)$, the contour Γ intersects the semi-axis $\{z_1 = 0\} \cap \{z_2 < 0\}$. Of all the points belonging to $\Gamma \cap \{z_1 = 0\}$, let us select that with the smallest z_2 coordinate and call it the starting point of the contour. Since the contour has length n, there are no more than n choices for its starting point. Once the starting point of the contour is fixed, the edge of Γ containing it is also fixed. It is the horizontal segment centered at the starting point of the contour. Counting from the segment connected to the right end-point of this edge, there are no more than three choices for each next edge, since the contour is not self-intersecting. Therefore, there are no more than $n3^{n-1}$ contours in total.

Lemma 22.5 (Peierls Inequality). *Let* Γ *be a closed curve of length* n. *Then,*

$$P_{V_i}^+(\{\omega \in \Omega(V_i) : \Gamma \subseteq B(\omega)\}) \le e^{-8\beta n}.$$

We shall prove the Peierls Inequality after completing the proof of Theorem 22.4.

Due to the Peierls Inequality, the probability $P_{V_i}^+$ that there is at least one contour Γ with the origin inside $\text{int}(\Gamma)$, is estimated from above by

$$\sum_{n=4}^{\infty} n 3^{n-1} e^{-8\beta n},$$

which tends to zero as $\beta \to \infty$. Therefore, the probability of $\omega_0 = -1$ tends to zero as $\beta \to \infty$. Note that this convergence is uniform in i.   □

Proof of the Peierls Inequality. For each configuration $\omega \in \Omega(V_i)$, we can construct a new configuration $\omega' \in \Omega(V_i)$, where

$$\omega_k' = -\omega_k \quad \text{if } k \in \text{int}(\Gamma),$$

$$\omega_k' = \omega_k \quad \text{if } k \notin \text{int}(\Gamma).$$

For a given Γ, the correspondence $\omega \leftrightarrow \omega'$ is one-to-one.

Let $\omega \in \Omega(V_i)$ be such that $\Gamma \subseteq B(\omega)$. Consider the ratio

$$\frac{P_{V_i}^+(\omega)}{P_{V_i}^+(\omega')} = \frac{\exp(-\beta \sum_{k:\text{dist}(k,V_i)\le 1} \sum_{k':|k'-k|=1}(\omega_k - \omega_{k'})^2)}{\exp(-\beta \sum_{k:\text{dist}(k,V_i)\le 1} \sum_{k':|k'-k|=1}(\omega_k' - \omega_{k'}')^2)}.$$

Note that all the terms in the above ratio cancel out, except those in which k and k' are adjacent and lie on the opposite sides of the contour Γ. For those terms, $|\omega_k - \omega_{k'}| = 2$, while $|\omega_k' - \omega_{k'}'| = 0$. The number of such terms is equal to $2n$ (one term for each side of each of the edges of Γ). Therefore,

$$P_{V_i}^+(\omega) = e^{-8\beta n} P_{V_i}^+(\omega').$$

By taking the sum over all $\omega \in \Omega(V_i)$ such that $\Gamma \subseteq B(\omega)$, we obtain the statement of the lemma.   □

One can show that the Gibbs field is unique if β is sufficiently small. The proof of this statement will not be discussed here.

The most difficult problem is to analyze Gibbs fields in neighborhoods of those values β_{cr} where the number of Gibbs fields changes. This problem is related to the so-called critical percolation problem and to conformal quantum field theory.

Index

L. Koralov and Y.G. Sinai, *Theory of Probability and Random Processes*,
Universitext, DOI 10.1007/978-3-540-68829-7,
© Springer-Verlag Berlin Heidelberg 2012